ADDRESSING THE INTENTIONAL DESTRUCTION OF THE ENVIRONMENT DURING WARFARE UNDER THE ROME STATUTE OF THE INTERNATIONAL CRIMINAL COURT

Series Supranational Criminal Law: Capita Selecta, Volume 18

Published in the series:

1. *Supranational Criminal Law: a System Sui Generis,* Roelof Haveman, Olga Kavran and Julian Nichols (eds.)
2. *Double Jeopardy Without Parameters: Re-characterization in International Criminal Law,* Olaoluwa Olusanya
3. *Harmonization of Criminal Law in Europe,* Erling Johannes Husabø and Asbjørn Strandbakken (eds.)
4. *Sentencing and Sanctioning in Supranational Criminal Law,* Roelof Haveman and Olaoluwa Olusanya (eds.)
5. *Application of International Humanitarian and Human Rights Law to the Armed Conflicts of the Sudan: Complementary or Mutually Exclusive Regimes?,* Mohamed Abdelsalam Babiker
6. *Supranational Criminology: Towards a Criminology of International Crimes,* Alette Smeulers and Roelof Haveman (eds.)
7. *Individual Criminal Liability for the International Crime of Aggression,* Gerhard Kemp
8. *Collective Violence and International Criminal Justice – An Interdisciplinary Approach,* Alette Smeulers (ed.)
9. *International Criminal Law from a Swedish Perspective,* Iain Cameron, Malin Thunberg Schunke, Karin Påle Bartes, Christoffer Wong and Petter Asp
10. *The Implementation of the European Arrest Warrant in the European Union: Law, Policy and Practice,* Massimo Fichera
11. *The Principle of Mutual Recognition in Cooperation in Criminal Matters,* Annika Suominen
12. *The European Public Prosecutor's Office,* Martijn Willem Zwiers
13. *Victimological Approaches to International Crimes: Africa,* Rianne Letschert, Roelof Haveman, Anne-Marie de Brouwer, Antony Pemberton (eds.)
14. *The United States and International Criminal Tribunals,* Harry M. Rhea
15. *EU Sanctions: Law and Policy Issues Concerning Restrictive Measures,* Iain Cameron (ed.)
16. *Whose Responsibility?* Malin Thunberg Schunke
17. *Genocidal Gender and Sexual Violence,* Usta Kaitesi

ADDRESSING THE INTENTIONAL DESTRUCTION OF THE ENVIRONMENT DURING WARFARE UNDER THE ROME STATUTE OF THE INTERNATIONAL CRIMINAL COURT

Steven FREELAND

intersentia
Cambridge – Antwerp – Portland

Intersentia Ltd
Sheraton House | Castle Park
Cambridge | CB3 0AX | United Kingdom
Tel.: +44 1223 370 170 | Email: mail@intersentia.co.uk

Distribution for the UK:
NBN International
Airport Business Centre, 10 Thornbury Road
Plymouth, PL6 7 PP
United Kingdom
Tel.: +44 1752 202 301 | Fax: +44 1752 202 331
Email: orders@nbninternational.com

Distribution for the USA and Canada:
International Specialized Book Services
920 NE 58th Ave. Suite 300
Portland, OR 97213
USA
Tel.: +1 800 944 6190 (toll free)
Email: info@isbs.com

Distribution for Austria:
Neuer Wissenschaftlicher Verlag
Argentinierstraße 42/6
1040 Wien
Austria
Tel.: +43 1 535 61 03 24
Email: office@nwv.at

Distribution for other countries:
Intersentia Publishing nv
Groenstraat 31
2640 Mortsel
Belgium
Tel.: +32 3 680 15 50
Email: mail@intersentia.be

Addressing the Intentional Destruction of the Environment during Warfare under the Rome Statute of the International Criminal Court
Steven Freeland

Cover illustration: Francisco de Goya (1746-1828), Etching, Plate 41 from 'Los desastres de la guerra'

ISBN 978-1-78068-314-0
D/2015/7849/54
NUR 824

British Library Cataloguing in Publication Data. A catalogue record for this book is available from the British Library.

CONTENTS

INTRODUCTION

Acts perpetrated during the course of warfare have, through the ages, led to significant environmental destruction. These have included situations where the natural environment has intentionally been targeted as a 'victim', or has somehow been manipulated to serve as a 'weapon' of warfare. Until recently, such acts were generally regarded as an unfortunate but unavoidable element of armed conflict, despite their potentially disastrous impacts. The existing legislative rules that purport to regulate such acts have largely been ineffective and inappropriate, and have in practical terms done little to deter deliberate environmental destruction, particularly when measured against perceived military advantages. However, as the significance of the environment has come to be more widely understood and recognised, such intentional destruction of the environment is no longer acceptable, particularly given the ongoing development of weapons capable of widespread and significant damage.

This book, which is a revised and updated version of my PhD Dissertation submitted to the Faculty of Law at the University of Maastricht, represents the culmination of many years of contemplation of, and increasing concern for the state of the global environment. In 2003, I was invited to present a key note lecture on 'Armed Conflict and the Environment' at the University of Stetson, Florida, when I was there as a Faculty Advisor of my own University's moot court team participating in the world finals of the International Environmental Law Moot Court Competition. This talk sparked my desire to delve further into this topic. Given my research interests also in the areas of international criminal law and international humanitarian law, it seemed to me that the issue of intentional destruction of the environment during warfare was an area on which I should focus, particularly since it represents a point of intersection of all of these fields.

Several years later I approached my close friend and colleague, Professor Dr Andre Klip, who graciously agreed to take on the role of Principal Supervisor for my PhD dissertation on this topic. I enrolled as a part-time PhD candidate at the University of Maastricht, and thus began a lengthy period of research and writing, unfortunately punctuated by several lengthy periods of inactivity as a result of my responsibilities as Professor of International Law at the University of Western Sydney, as well as other research related activities at a number of institutions and international associations. Yet, eventually, this project was completed although, in my view, it only represents a 'work in progress' and a small stepping stone towards the broader goal of enhancing and incorporating more appropriate international regulatory frameworks designed to prevent all forms of deliberate destruction of the natural environment.

This book therefore examines the current international legal regime relevant to the intentional destruction of the environment during warfare, and argues that such acts should, in appropriate circumstances, be recognised as an international crime and be subject to more effective rules giving rise to international criminal responsibility. It also suggests a framework as to how this might be achieved.

In essence, this book focuses on the following issues:

(a) To what extent do treaty and customary principles of international law currently address the intentional destruction of the environment during warfare?
(b) Is the current position sufficient to allow for criminal accountability for such acts in appropriate circumstances?
(c) What function should the mechanisms of justice under international criminal law, in particular the International Criminal Court, play in addressing this issue?
(d) Is it therefore necessary to amend the Rome Statute of the International Criminal Court in order to more appropriately deal with what might be referred to as 'crimes against the environment'?

As one would expect, there are a number of people to whom I owe sincere thanks. Professor Dr Andre Klip and Dr David Roef were the most challenging and stimulating supervisors one could possibly hope for. Whilst there were times when I thought that they were too demanding, I eventually came to understand how incredibly insightful and helpful their comments and criticisms were. I have learnt a lot from them, not only with respect to legal expertise and research, but also in terms of research supervision. These are lessons that I now try to emulate when I myself supervise PhD candidates. Thank you Andre and David for your support, patience and friendship – I am most grateful.

I also acknowledge and thank the members of the Assessment Committee for my PhD Dissertation, and for their valuable and challenging comments and suggestions.

There are two people who, over the years, have guided me and provided me with the inspiration to continue and expand upon my research and teaching activities

– H.E. Judge Errki Kourula at the International Criminal Court and H.E. Judge Abdul Koroma, for many years a judge at the International Court of Justice. Both of these Judges have encouraged me to pursue this research and I am very honoured to regard them as close friends. They continue to provide perfect role models for all persons concerned with making the world a better and more just place.

I am also grateful to the many people and organizations I have met along the way who share a common concern with me as regards the need for greater protection of the environment, and who offered me the opportunity to present my views on this topic publicly. I should mention in particular Werner Wouter and his colleagues at the Institute for Environmental Security in The Hague, the T.M.C. Asser Institute, and a broad range of colleagues involved in the 'End Ecocide' campaign. The feedback I have received from my various discussions with these and many other institutions has been invaluable.

I would also like to thank my colleagues at the School of Law at University of Western Sydney, and my Dean, Professor Michael Adams, for their support in enabling me to fit the time in to complete this project alongside my other activities at that institution.

I also acknowledge the generous financial support I received from the European Commission as a Marie Curie International Fellow (2013–2014), and the stimulating research environments made available to me as a Visiting Professor at both the iCourts Centre of Excellence for International Courts at the University of Copenhagen, and at the University of Vienna.

I owe thanks also to my publishers, Intersentia, and in particular Hans Kluwer and Tom Scheirs, for their support with regard not only to the publication of this book, but also to an ongoing book series, *Annotated Leading Cases of International Criminal Tribunals*, of which I have the privilege of being a Co-Editor along with Andre Klip.

Of course, thanks also go to my family. My dear 93-year old father Lou, my brother Michael and his family, and my in-laws John and Charlotte, have always been so supportive and understanding of the travel-related and other demands I have placed upon myself, and have been there for me whenever possible. This book is also dedicated to my mother Blanka, who passed away in 2010.

Research projects such as this are heavily time-consuming and require the writer to be selfish and utilise whatever time is available for overly-indulgent purposes – time that should instead be spent with loved ones and friends. I am so lucky to be married to the kindest, most loving and wonderful person, Donna, and to share with her the joy and love of Becky and my own daughter Michaela. My greatest thanks and love are reserved for these three amazing people. They represent everything that is good in the world.

Steven Freeland
Sydney, January 2015

CHAPTER ONE

THE IMPERATIVE TO REGULATE THE INTENTIONAL DESTRUCTION OF THE ENVIRONMENT DURING WARFARE UNDER INTERNATIONAL CRIMINAL LAW

1.1. INTRODUCTION – INTENTIONAL ENVIRONMENTAL DESTRUCTION AS A PART OF WARFARE

History has been witness to many deliberate acts aimed at destroying the natural environment in order to achieve strategic goals and safeguard State sovereignty during the course of a military conflict. Herodotus described how, in the fifth century BC, the retreating Scythians scorched the earth and poisoned the water wells in an effort to slow the advancing Persian army led by Darius. In 146 BC, Roman troops razed the city of Carthage and poisoned the surrounding soil with salt to prevent its future fertilisation.

In August 1945, we witnessed the overwhelmingly destructive capability of weapons technology when the United States detonated two atomic bombs over the cities of Hiroshima and Nagasaki, resulting in massive loss of life and environmental destruction.[1] During the Vietnam War, the United States implemented 'Operation Ranch Hand' to devastating effect to destroy vegetation used by the enemy for cover and sustenance, through the use of chemicals such as Agent Orange.[2]

---

1 The bombing of Nagasaki was described as a 'war crime' by one of Japan's most senior politicians, Shoichi Nakagawa, the then policy chief of the ruling Liberal Democratic Party: Chris Hogg, 'Nagasaki bombing labeled a crime' (18 December 2006) *BBC News* <http://news.bbc.co.uk/2/hi/asia-pacific/6189489.stm> at 18 January 2015.

2 It has been estimated that Operation Ranch Hand destroyed 8 per cent of Vietnam's croplands, 14 per cent of its forests, and 50 per cent of its swamp areas: Ensign Florencio J. Yuzon, 'Deliberate Environmental Modification Through the Use of Chemical and Biological Weapons: "Greening" the Environmental Laws of Armed Conflict to Establish an

More recently still, few can forget the haunting images of over 700 burning Kuwaiti oil well heads, which had been deliberately ignited by retreating Iraqi forces towards the end of the Gulf War in 1991 – a scene that was likened to 'Dante's Inferno'.[3] These actions galvanized the international community to address deliberate attacks on the environment during armed conflict,[4] and gave rise to almost universal condemnation.[5] The then German Chancellor, Helmut Kohl, for example, asserted that this constituted a 'crime against the environment'.[6]

Following the conclusion of the Gulf War, the Saddam Hussein regime, in retaliation for what it saw as the support by the Ma'dan, or Marsh Arabs, for an uprising against his government, built barriers and levees to drain the al-Hawizeh and al-Hammar marshes in southern Iraq, an area some believe to be the site of the biblical Garden of Eden.[7] This effectively destroyed the livelihood of the 500,000 Marsh Arabs who had inhabited the area of this unique ecosystem.[8] Saddam Hussein and eleven other members of his former regime were subsequently indicted by the Supreme Iraqi Criminal Tribunal (Iraqi Special Tribunal) to face charges that, in part, related to these acts of environmental destruction.[9] However, they did not include specific reference to environmental 'crimes', even though the Iraqi Special Tribunal Statute includes an express reference to damage to the natural environment.[10]

Environmentally Protective Regime' (1996) 11 *American University Journal of International Law and Policy* 793, 795–6.

3 Phillip Elmer-Dewitt, 'A Man-Made Hell on Earth', *Time* (London), 18 March 1999, 36.

4 Jessica E. Seacor, 'Environmental Terrorism: Lessons from the Oil Fires of Kuwait' (1994) 10 *American University Journal of International Law and Policy* 481, 488.

5 For example, shortly afterwards, the Parliamentary Assembly of the Council of Europe called for the establishment of a war crimes tribunal to prosecute those responsible for 'this disgraceful attack on the environment': Anthony Leibler, 'Deliberate Wartime Environmental Damage: New Challenges for International Law' (1992) 23 *California Western International Law Journal* 67, 68.

6 Helmut Kohl, Statement by the German Chancellor, *Bulletin* (Bonn), 9 April 1991, 255.

7 David Blair, 'Toll of Infamy: Regime Killed 2 Million', *The Sydney Morning Herald* (Sydney), 16 December 2003, 8.

8 Perhaps as few as 20,000 of the Ma'dan were left in what remained of the wetlands on the eve of the March 2003 invasion of Iraq by United States-led forces: Heather Sharp, 'Iran's 'devastated' Marsh Arabs', *BBC News*, 3 March, 2003 <http://news.bbc.co.uk/2/hi/middle_east/2807821.stm> at 18 January 2015. For a discussion of the possible prosecution of those responsible for this act, see Tara Weinstein, 'Prosecuting Attacks that Destroy the Environment: Environmental Crimes or Humanitarian Atrocities?' (2005) 17 *Georgetown International Environmental Law Review* 697, 714–22.

9 On 30 December 2006, Saddam Hussein was executed, following the death sentence imposed by the Iraqi Special Tribunal, for crimes against humanity in the Shia town of Dujail, where 148 Shia villagers were tortured and killed in reprisal for an assassination attempt against Saddam in 1982.

10 The Iraqi Special Tribunal had jurisdiction over 'any Iraqi national or resident' accused of committing the crimes of genocide, war crimes, crimes against humanity or violations of certain Iraqi laws between 17 July 1968 and 1 May 2003: Statute of the Iraqi Special Tribunal

It has been estimated by Human Rights Watch that, during the course of the 2003 invasion of Iraq, United States and British forces used almost 13,000 cluster bombs[11] – containing almost two million munitions – and over 1.9 tonnes of depleted uranium,[12] causing very significant environmental damage, the extent of which may still not be known for a considerable period of time.[13]

Even now the world is still witnessing a continuing humanitarian and environmental catastrophe in the western region of Darfur in Sudan,[14] which has seen the poisoning of vital water wells and drinking water installations,[15] as part of a deliberate government-supported strategy by the Arab Janjaweed Militia to eliminate or displace the ethnic black Africans living in that region. There had been calls for the International Criminal Court (ICC) to charge

2003, article 10 (English version reproduced in Al-Waqai Al-Iraqiya, Official Gazette of Iraq, Volume 44 <www.loc.gov/law/help/hussein/docs/CoalitionProvisionalAuthorityOrder48en.pdf> at 18 January 2015). The respective definitions of the crimes of genocide, war crimes and crimes against humanity in the Statute of the Iraqi Special Tribunal (articles 11–13) are based on those that appear in the Rome Statute of the International Criminal Court (opened for signature 17 July 1998, 2187 UNTS 3; 37 ILM 999 (entered into force 1 July 2002)) (Rome Statute). Article 13(b)(5) of the Statute of the Iraqi Special Tribunal is identical with article 8(2)(b)(iv) of the Rome Statute, which relates to 'widespread, long-term and severe damage to the natural environment', and is discussed in chapter 4.

11 On 30 May 2008, at a conference in Dublin, 107 States adopted the Convention on Cluster Munitions, opened for signature 3 December 2008, 48 ILM 357 (entered into force 1 August 2010), which imposes restrictions on the use, development and stockpiling of certain types of cluster bombs.

12 Pekka Haavisto, in Symposium, 'The International Responses to the Environmental Impacts of War' (2005) 17 *Georgetown International Environmental Law Review* 565, 581.

13 Depleted uranium is a by-product of the process of 'uranium enrichment', which involves the separation of the three different uranium isotopes (uranium-238, uranium-235 and uranium-234) as a preliminary step towards the use of nuclear fission as a source of energy (uranium-235 is the most suitable for nuclear fission): Erik V. Koppe, *The Use of Nuclear Weapons and the Protection of the Environment During International Armed Conflict* (2006), 18.

14 In early 2013, the United Nations High Commission for Refugees estimated that there were approximately 2.3 million internally displaced persons (IDPs), 140,000 refugees, 7,000 asylum-seekers, and hundreds of thousands persons at risk of statelessness in Sudan, much of it due to the ongoing conflict in the Darfur region: United Nations High Commission for Refugees, '2013 UNHCR country operations profile – Sudan' (2013) <www.unhcr.org/pages/49e483b76.html> at 22 June 2013. See also 'U.S. Warns Sudan Death Toll Could Reach 300,000 To 1 Million' (3 June 2004) UN Wire <http://medilinkz.org/news/news2.asp?NewsID=6776> at 27 March 2007.

15 See United Nations Commission on Human Rights, 'Report of the United Nations High Commissioner for Human Rights and Follow-Up to the World Conference on Human Rights: Situation of Human Rights in Darfur Region of the Sudan' (7 May 2004) E/CN.4/2005/3, para 50 and 73.

Sudanese officials with 'environmental crimes',[16] but this has not happened, although other charges have been laid.[17]

1.2. IMPACT OF ENVIRONMENTAL DESTRUCTION ON HUMAN SECURITY AND CONFLICT

The examples referred to above indicate how the intentional destruction of the environment has already been a part of many warfare situations, even though such destruction can have widespread and long lasting impacts. This section discusses a number of these impacts, thus highlighting the significance of having access to appropriate regulatory measures to restrict and prosecute such acts.

The intentional destruction of the environment can have catastrophic effects, not only in human, but also in ecological terms. For example, nuclear, biological and chemical weapons, as well as potentially killing many thousands of people, have effects that may persist for the environment, in some cases indefinitely.[18] Such acts may also represent a breach of the fundamental human rights of the targeted individuals.[19] The relationship between human security and a safe and habitable environment is vital, particularly in relation to access to natural resources.

Another equally significant, but perhaps not yet fully understood, link between the environment and human conflict is that access to natural resources – or the lack of access – can itself be the trigger for conflict. In both the Democratic Republic of Congo and Haiti, the United Nations Environment Programme (UNEP) has reported that environmental damage has been a major cause of

---

16 Katy Glassborow, 'Court urged to consider environmental crimes', *Institute for War and Peace Reporting* (United Kingdom), 30 August 2007 <http://iwpr.net/?p=acr&s=f&o=338199& apc_state=henh#> at 25 March 2013.

17 In March 2005, the United Nations Security Council referred the situation in Darfur to the Prosecutor of the International Criminal Court pursuant to article 13(b) of the Rome Statute: see United Nations Security Council Resolution 1593 (31 March 2005) UN Doc S/RES/1593/2005 (Algeria, Brazil, China, United States abstaining). There are currently five cases at the Court relating to the situation in Darfur. However, the current ICC Prosecutor, Fatou Bensouda, has recently announced that she is 'suspending' the investigation into the matter: David Smith, 'ICC Chief Prosecutor shelves Darfur war crime probe', *The Guardian* (UK), 14 December 2014, <www.theguardian.com/world/2014/dec/14/icc-darfur-war-crimes-fatou-bensouda-sudan> at 18 January 2015.

18 Weapons of Mass Destruction Commission, *Weapons of Terror: Freeing the World of Nuclear, Biological and Chemical Arms* (2006), 22.

19 For example, Principle 1 of the Declaration of the United Nations Conference on the Human Environment (16 June 1972) UN Doc A/CONF.48/14/Rev.1 (Stockholm Declaration) provides that '[m]an has the fundamental right to freedom, equality, and adequate conditions of life, in an environment of a quality that permits a life of dignity and well-being'.

political unrest and conflict.[20] It has been estimated that, during the 1990s, approximately five million people were killed in armed conflicts that related to the exploitation of natural resources,[21] and that one quarter of the 50 active armed conflicts in 2001 were largely 'motivated' by resources.[22]

In 1990, a research team at the University of Toronto concluded that, in countries as diverse as Haiti, Pakistan, the Philippines and South Africa, 'severe environmental stress multiplied the pain caused by such problems as ethnic strife and poverty'.[23] A water expert has predicted that, in regions initially experiencing low-level conflict, the risk of escalation to full-scale civil war approximately *doubled* immediately following a year of abnormally low rainfall.[24] In similar vein, in 2004, the United Nations High-Level Panel on Threats, Challenges, and Change concluded that:[25]

> [p]overty, infectious diseases, environmental degradation and war feed one another in a deadly cycle … Environmental stress … can contribute to civil violence.

In addition, environmental degradation results in increasing numbers of refugees. The United Nations High Commission for Refugees estimated that, at the end of 2011, 42.5 million people worldwide were considered as forcibly displaced as a result of conflict or persecution, including 15.2 million refugees.[26] In an interview following the release of an earlier (2008) report, the High Commissioner, António Guterres, stated that climate change led to the dislocation of people 'by provoking conflicts over increasingly scarce resources,

20 Alister Doyle, 'UN Aims to Study Link Between Environment, Wars', reproduced in United Nations Environment Programme, 'The Environment in the News', 14 January 2004.

21 Rudy S. Salo, 'When the Logs Roll Over: The Need for an International Convention Criminalizing Involvement in the Global Illegal Timber Trade' (2003) 16 *Georgetown International Environmental Law Review* 127, 142.

22 David R. Francis, 'Fuelling the Fire: "Resource Wars" Spurred by Assets of Developing Nations', *Christian Science Monitor* (Boston), 6 December 2002, A3.

23 Thomas Homer-Dixon, 'Terror in the Weather Forecast', *The New York Times* (New York), 24 April 2007, <www.nytimes.com/2007/04/24/opinion/24homer-dixon.html?_r=1&th&emc=th&oref=s login> at 18 January 2015. The author of that article led the research team.

24 Professor Charles Vörösmarty, 'Drought as a Contributor to Civil War: Results from a Global Spatial Analysis', speech delivered at seminar entitled 'Climate-Security Connections: An Empirical Approach to Risk Assessment', Washington DC, USA, 6 March 2007.

25 Report of the High-Level Panel on Threats, Challenges, and Change, 'A More Secure World: Our Shared Responsibility' (1 December 2004), para 22, UN Doc A/59/565 (2004).

26 United Nations High Commission for Refugees, '2011 Global Trends: A Year of Crisis' (2012) <www.unhcr.org/cgibin/texis/vtx/home/opendocPDFViewer.html?docid=4fd6f87f9&query=Global%20Trends> at 25 March 2013.

such as water' and that its impact on the environment 'was a trigger of extreme poverty and conflict'.[27]

The Iran-Iraq War, the events in Kuwait at the end of the 1991 Gulf War, and the conflicts in African countries such as Sierra Leone, the Democratic Republic of Congo and Liberia, all serve to illustrate how armed conflict that is (partially) driven by disputes over natural resources can subsequently result in significant destruction to the natural environment.[28] In the latter cases, not only were the conflicts fought over natural resources, but the resources themselves (for example, timber and diamonds) became sources of funding for the combating parties, to be used to acquire weapons for the conflict. In 2002, a panel of experts appointed by the United Nations Security Council concluded that many of the serious crimes committed in the Democratic Republic of Congo were both in aid of, and financed by, the profits of the illegal appropriation of natural resources in that country.[29]

These examples illustrate a phenomena sometimes referred to as 'asset wars', where the economic value of natural resources serves to fuel or trigger conflict. Exploitation of these resources represents a source of funding for military activities, giving rise to an unvirtuous circle.[30] The misuse of natural resources such as diamonds, or the scarcity of resources like water, may facilitate potential conflict, a situation which then becomes self-perpetuating and protracted.[31] Environmental degradation can thus be both a cause and a consequence of armed conflict. In addition, during the course of conflict there are further 'knock-on' effects flowing from environmental destruction arising from the actions of the combatants.[32]

---

27 Julian Borger, 'Conflicts fuelled by climate change causing new refugee crisis, warns UN', *The Guardian* (London), 17 June 2008, 15.

28 Betsy Baker, 'Legal Protections for the Environment in Times of Armed Conflict' (1993) 33 *Virginia Journal of International Law* 351, 351.

29 United Nations Security Council, 'Final Report of the Panel of Experts on the Illegal Exploitation of Natural Resources and Other Forms of Wealth of the Democratic Republic of the Congo' (16 October 2002) UN Doc: S/2002/1146.

30 Aaron Ezekiel, 'The Application of International Criminal Law to Resource Exploitation: Ituri, Democratic Republic of the Congo' (2007) 47 *Natural Resources Journal* 225, 234.

31 Report of the Special Representative of the Secretary-General for Children and Armed Conflict, UN Doc A/62/228, 13 August 2007, para 12. See also the section entitled 'Diamonds' in the Trial Chamber decision of the Special Court of Sierra Leone, Judgment, *Prosecutor v. Charles Gankay Taylor*, Case No. SCSL-03-01-T, Trial Chamber II, 18 May 2012, para 2046–2173.

32 See Stephanie Nebehay, 'Dirty Water Provokes Hepatitis Outbreak in Darfur', *Reliefweb*, 9 August 2004 <http://reliefweb.int/report/sudan/sudan-dirty-water-provokes-hepatitis-outbreak-darfur> at 18 January 2015, who describes how the refugee camps that were set up in Darfur as a result of the conflict in that region are struggling with additional problems from the lack of safe drinking water.

Moreover, the very nature of armed conflict and its adverse effects on the livelihood of communities and destruction of the natural environment fuel a spiralling vicious cycle of poverty and further violence, leaving desperate individuals, who are often children, with no choice but to themselves become active participants in the conflicts and to offer their services to a 'cause'.[33] This contributes to the tragically high number of 'child soldiers' now engaged in armed conflict.[34]

While there is, of course, much more work to be done to accurately determine the nature and extent of the link between environmental degradation, poverty and political and social conflict, the logic of some form of connection appears undeniable. In 1992, the United Nations Security Council concluded that:[35]

> [t]he absence of war and military conflicts amongst States does not in itself ensure international peace and security. The non-military sources of instability in the economic, social, humanitarian and *ecological* fields have become threats to international peace and security. The United Nations membership as a whole needs to give the highest priority to the solution of these matters.

This also indicates how environmental degradation can lead to social upheaval and tensions, thus representing a threat to national security.[36] This is one reason why combatants in a conflict may seek to intentionally 'target' the environment. Many States now view environmental concerns 'in strategic terms'.[37] It has even been suggested that the world is 'only one international environmental disaster that implicates environmental security away from' the development of customary rules that may permit the legal use of 'environmental armed force' as

33 Steven Freeland, 'Mere Children or Weapons of War – Child Soldiers and International Law' (2008) 29 *University of La Verne Law Review* 19, 27–8.

34 In a 2004 report, the NGO Coalition to Stop the Use of Child Soldiers found that children were 'fighting in almost every major conflict, in both government and opposition forces'. In addition to an estimated 300,000 children who engage in actual military conflict, another 500,000 are 'conscripted' into paramilitary organisations, guerrilla groups and civil militias in over 85 countries: see also Steven Freeland, 'Child Soldiers and International Crimes – How Should International Law be Applied?' (2005) 3 *New Zealand Journal of Public and International Law* 303, 304.

35 United Nations Security Council, 'Note by the President of the Security Council' (31 January 1992) Presidential Statement S/23500, quoted in Catherine Tinker, 'Environmental Security' in The United Nations: Not a Matter for the Security Council' (1992) 59 *Tennessee Law Revue* 787, 787 (emphasis added).

36 Jutta Brunneé, 'Environmental Security in the Twenty-First Century: New Momentum for the Development of International Environmental Law' (1995) 18 *Fordham International Law Journal* 1742, 1742.

37 Rymn James Parsons, 'The Fight to Save the Planet: U.S. Armed Forces, "Greenkeeping," and Enforcement of the Law Pertaining to Environmental Protection During Armed Conflict' (1998) 10 *Georgetown International Environmental Law Review* 441, 444.

a legitimate exercise of the right of self-defence, in order to avert environmental destruction.[38]

The United Nations plays an increasing role in addressing deliberate environmental degradation. On 5 November 2001, the General Assembly declared 6 November of each year as the 'International Day for Preventing the Exploitation of the Environment in War and Armed Conflict'.[39] As indicated, the increasing threats posed by a lack of environmental security adds support to the view that the Security Council might more frequently act under its Chapter VII powers within the United Nations Charter[40] to address environmental issues.[41] These might extend to coercive measures under article 41, and perhaps even the use of military force under article 42, to counter environmental threats, although such action may itself give rise to environmental destruction.[42] In a 2002 report on the 'Prevention of Armed Conflict', the then United Nations Secretary-General Kofi Annan highlighted the role that the General Assembly could also play in addressing the 'conflict prevention dimension' of a number of concerns, including 'environmental degradation'.[43]

In the *Gabcíkovo-Nagymaros Project* Case, the International Court of Justice acknowledged that Hungary's concerns relating to damage to its environment constituted an 'essential interest' of that country, within the context of the international law defence of necessity.[44] Moreover, the effects of environmental destruction can have transboundary consequences, thus undermining the security of neighbouring countries, regions, and even the broader international community.

Of course, it must be noted that, even where man-made environmental damage is not caused within a conflict situation, the impacts may still resonate in terms of future warfare. In April 2007, the Security Council held its first ever debate on the impact of climate change on international peace and security.[45] That same

38 Michael K. Murphy, 'Note: Achieving Economic Security with Swords as Ploughshares: The Modern Use of Force to Combat Environmental Degradation' (1999) 39 *Virginia Journal of International Law Association* 1181, 1214 and 1219.

39 See Steven Freeland, 'Destroying the environment is also a war crime', *The Canberra Times* (Canberra), 6 November 2008, 23.

40 See United Nations Charter, articles 39–51.

41 Murphy, *supra* note 38, 1198.

42 Alexandra Knight, 'Global Environmental Threats: Can the Security Council Protect Our Earth?' (2005) 80:5 *New York University Law Review* 1549, 1553.

43 Report of the Secretary-General, Kofi A. Annan, 'Prevention of Armed Conflict' (2002), para 29.

44 *Gabcíkovo-Nagymaros Project (Hungary v. Slovakia)* (Judgment) [1997] ICJ Rep 7, para 53.

45 See United Nations Security Council, 'Security Council Hold First-Ever Debate on Impact of Climate Change on Peace, Security, Hearing Over 50 Speakers' (17 April 2007) UN Doc SC/9000 <www.un.org/News/Press/docs/2007/sc9000.doc.htm> at 18 January 2015.

month, 11 retired United States generals and admirals released a report,[46] in which they argued that the environmental impacts of climate change constituted a 'threat multiplier' in fragile parts of the world, which would 'exacerbat[e] conditions that lead to failed states – the breeding grounds for extremism and terrorism'.[47] They highlighted the examples of Darfur and Somalia as conflicts arising from scarce natural resources.[48] These sentiments were supported by the Military Advisory Council of the Institute for Environmental Security (comprising a group of serving and retired military officers from Africa, Asia, Europe, Latin America and the United States), which, in October 2009, highlighted the 'conflict and instability implications of climate change'.[49]

United Nations Secretary-General Ban Ki-moon has said that the effects of climate change were partly to blame for the ongoing conflict in Darfur.[50] Shortly after his inauguration as President of the United States in January 2009, Barack Obama spoke of the risk of 'violent conflict' stemming from unchecked global warming.[51] The NGO Christian Aid has predicted that up to one billion people could be forced to leave their homes over the next 50 years as a result of widespread environmental degradation caused by climate change.[52]

The importance of these factors was affirmed as early as 1972 in the Stockholm Declaration – a landmark instrument described as representing 'the birth of international environmental law'[53] – which provided that:[54]

> [t]o defend and improve the human environment for present and future generations has become an imperative goal for mankind – a goal to be pursued together with, and in harmony with, the established and fundamental goals of peace and of world-wide economic and social development

46 The CNA Corporation, 'National Security and the Threat of Climate Change' (April 2007) <www.cna.org/sites/default/files/news/FlipBooks/Climate%20Change%20web/flipviewerxpress.html> at 18 January 2015.

47 'Warming and Global Security', *The New York Times* (Editorial), 20 April 2007 <www.nytimes.com/2007/04/20/opinion/20fri2.html> at 18 January 2015.

48 Ed Pilkington, 'UK to raise climate talks as security council issue', *The Guardian* (London), 16 April 2007, 8.

49 Institute for Environmental Security, Military Advisory Council, 'Climate Change and the Military – First Statement of the Military Advisory Council' (29 October 2009), para 6.

50 'U.N chief: Darfur a victim of climate change' (16 June 2007) *NBC News* <www.nbcnews.com/id/19268452/#.UU8qVFf-jdc> at 18 January 2015.

51 Andrew C. Revkin, 'Bush's environmental legacy in play: Obama reshaping some policies and building on others: Climate Change', *The International Herald Tribune* (Paris), 7–8 February 2009, 5.

52 John Vidal, 'Climate may force 1 billion from their homes', *The Sydney Morning Herald* (Sydney), 15 May 2007 <www.smh.com.au/news/environment/climate-may-force-1-billion-from-their-homes/2007/05/14/1178995077364.html> at 18 January 2015.

53 Yuzon, *supra* note 2, 97.

54 Stockholm Declaration, preamble para 6.

The general discussion above highlights the significance of the state of the natural environment and its potential relationship to armed conflict in many circumstances. In some respects, the intentional destruction of the environment may be a precursor or trigger point for conflict. In other circumstances, however, such intentional environmental damage may be perpetrated as a part of the conduct of warfare. Whilst we must, naturally, remain cognizant of the impacts of the former, it is the latter that this book is primarily concerned with, and which it seeks to address by way of more appropriate international regulation.

1.3. INTENTIONAL TARGETING OF THE ENVIRONMENT DURING WARFARE AND THE DEVELOPMENT OF INTERNATIONAL CRIMINAL JUSTICE

Notwithstanding (or perhaps because of) its significance to human populations, the targeting of the environment has increasingly become a part of the conduct of armed conflict. Such acts represent not only a feature of conflict strategy, but also, as noted, a root cause for the further escalation of the conflict itself.[55] Given the increasingly diverse ways in which armed conflict might be conducted, particularly in light of the development of weapons technology, there is even greater scope for environmental damage in future warfare, perhaps even in ways that we cannot at this time completely fathom.

As a consequence, for the purposes of considering what might represent appropriate regulation of such acts under international criminal law, the discussion that follows in this book assumes an expansive perspective of what is encompassed by 'the environment'. When it comes to drafting the specifics of a crime relating to intentional destruction of the environment during warfare (see chapter 5), a definition of the environment for the purposes of that crime will be proposed.

Actions taken during previous armed conflicts illustrate that the environment has been targeted as a 'victim' of armed conflict – in the sense that it is deliberately damaged. It has also sometimes been utilized as a 'weapon' of warfare to secure a military advantage.

For example, the intentional setting alight of the Kuwaiti oil wells in 1991 highlighted vividly how the environment can become a victim of warfare. This

55 Myron N. Nordquist, 'Panel Discussion on International Environmental Crimes: Problems of Enforceable Norms and Accountability' (1997) 3 *ILSA Journal of International Law* 697.

act deliberately targeted the environment and led to significant damage. On the other hand, the deliberate manipulation or modification of the environment – for example, American military efforts to alter natural weather patterns so as to cause unusual flooding through rainfall during the Vietnam War – illustrates how the environment could be utilized as a weapon in order to seek a military advantage.

Such acts, both in the sense of the environment as a victim but also when utilized as a weapon, have been described as 'environmental warfare.'[56] Both represent not only an unacceptable approach towards the environment, but also may result in irreversible and unpredictable consequences, irrespective of the manner in which the environment is involved. It is submitted that both aspects of environmental damage should be appropriately regulated, and that the existing legal frameworks are inadequate in this regard.

Moreover, the long-term environmental costs of warfare – both in terms of economic[57] as well as health and safety[58] terms – are often impossible to determine for a considerable period of time, and may escalate dramatically beyond the immediate short-term impacts.[59] Following the 1999 bombing of Serbia and Kosovo by NATO forces (Operation Allied Force), the Government of Yugoslavia[60] initiated proceedings in the International Court of Justice against ten NATO member States, seeking an order for provisional measures. In its pleadings, the applicant alleged that the NATO States had violated their obligation 'to protect the environment' and were directly responsible for

56 See, for example, Mansour Jabbari-Gharabagh, 'Type of State Responsibility for Environmental Matters in International Law' (1999) 33 *Revue Juridique Themis* 59, 98. The author (at 98) also quotes Richard Falk, who has defined 'environmental warfare' as 'including all those weapons and tactics which ... intend to destroy the environment *per se* ...'

57 Up until 2007, when it completed its payments to individuals, the United Nations Compensation Commission established to deal with claims, including environmental damage, arising from the actions taken by Iraq during the 1991 Gulf War, had received more than 2.6 million claims from nearly 100 States, on behalf of their nationals, corporations and/or themselves, seeking a total of approximately US$352 billion in compensation: United Nations Compensation Commission, 'The Claims' <www.uncc.ch/theclaims.htm> at 21 June 2013.

58 The State-owned Pancevo petrochemical plant near Belgrade was damaged during Operation Allied Force, resulting in serious pollution and threats to human health, including abnormal levels of bronchitis, as well as cancers and leukemia: see Aaron Schwabach, 'NATO'S War in Kosovo and the Final Report to the Prosecutor of the International Criminal Tribunal for the Former Yugoslavia' (2001) 9 *Tulane Journal of International and Comparative Law* 167, 174 and the corresponding footnotes.

59 The immediate impact of environmental destruction may give rise to lack of food and the outbreak of disease and, in the longer term, whole communities may be destroyed.

60 The name of the Applicant in these proceedings subsequently became 'Serbia and Montenegro'. In 2006, those areas separated to become two independent countries.

significant environmental damage. The applicant alleged, among other things, that:[61]

> [t]he bombing of oil refineries and oil storage tanks as well as chemical plants is bound to produce massive pollution of the environment, posing a threat to human life, plants and animals. The use of weapons containing depleted uranium warheads is having far-reaching consequences for human health

The substance of these claims was not considered by the Court,[62] but they do reflect the potential gravity of assertions that a party to an armed conflict has engaged in intentional environmental destruction.[63] Yet, environmental damage undoubtedly has been overlooked when assessing the toll wrought by warfare.[64] This approach is not sustainable, particularly, as noted, in light of the ongoing development of weapons with unimaginable and ever increasing destructive capability.[65]

It is submitted, therefore, that the deliberate nature of such actions requires stringent legal regulation and accountability. The relevant legal measures must 'stigmatise' such acts, so as to properly reflect the unsustainable risks they entail. Intentional damage to the environment during armed conflict cannot be regarded as immaterial or of secondary importance. The International Court of

---

61 *Legality of Use of Force (Yugoslavia v. France)* (Request for the Indication of Provisional Measures submitted by Applicant, 28 April 1999) para 3 <www.icj-cij.org/docket/files/105/10545.pdf> at 19 January 2015.

62 On 2 June 1999, the International Court of Justice dismissed, on jurisdictional grounds, the cases brought by Yugoslavia against Spain (*Legality of Use of Force (Yugoslavia v. Spain)* (Request for the Indication of Provisional Measures)) [1999] ICJ Rep 761, and against United States of America (*Legality of Use of Force (Yugoslavia v. United States of America)* (Request for Provisional Measures)) [1999] ICJ Rep 916. On 15 December 2004, the Court held that it did not have jurisdiction to entertain the remaining eight claims: see, for example, *Legality of Use of Force (Serbia and Montenegro v. Belgium)* (Preliminary Objections, Judgment) [2004] ICJ Rep 328, para 129.

63 Michael N. Schmitt, 'Humanitarian Law and the Environment' (2000) 28 *Denver Journal of International Law and Policy* 265, 274.

64 A committee established by the Office of the Prosecutor of the International Criminal Tribunal for the former Yugoslavia (ICTY) to consider whether any crimes falling within the jurisdiction of that Tribunal had been committed by NATO forces during Operation Allied Force failed to consider the full environmental consequences of the operation: Steven Freeland, 'The Bombing of Kosovo and the Milosevic Trial: Reflections on Some Legal Issues' [2002] *Australian International Law Journal* 150, 164. The committee's report is discussed in subsequent chapters.

65 For example, it has been asserted that tens of thousands of children born in Vietnam to war veterans and exposed civilians shortly after the Vietnam war had physical and mental deformities that could be directly linked to the widespread spraying of the herbicide Agent Orange by the United States, as well as the continuing presence in the destroyed regions of its poisonous residue, dioxin: Seth Mydans, 'Vietnam Sees War's Legacy in Its Young', *The New York Times* (New York), 16 May 1999, <www.nytimes.com/1999/05/16/world/vietnam-sees-war-s-legacy-in-its-young.html> at 19 January 2015.

Justice has affirmed that, 'the environment is not an abstraction but represents the living space, the quality of life and the very health of human beings, including generations unborn'.[66] This is a truism even during times of warfare.

Of course, warfare by its very nature results in damage to the natural environment, as well as to other 'protected sites'.[67] However, this alone cannot be justification for the *intentional* targeting of the environment. The increasing concern regarding the state of the environment and the need for a system of greater culpability for intentional environmental destruction has previously given rise to proposed crimes such as 'ecocide'[68] and 'geocide',[69] as well as calls for an Ecocide Convention, to be enforced by an International Environmental Court.[70] Whilst these proposals have generally not been limited to situations of armed conflict, they are demonstrative of the need to at least address those circumstances more comprehensively. The appropriate regulation of unjustified damage to the environment during armed conflict has been described as the 'unmet challenge of the 20th century'.[71]

It is argued in this book that the most appropriate method to address the issue is not through environmental protection measures but rather under international criminal law (although it is hoped that this will have positive consequences for

---

66 *Legality of the Threat or Use of Nuclear Weapons* (Advisory Opinion) [1996] ICJ Rep 242, para 29. See also *Gabcíkovo-Nagymaros Project (Hungary v. Slovakia)* (Judgment) [1997] ICJ Rep 7, para 53.

67 For example, similar concerns arise from the deliberate destruction of cultural property during armed conflict. The Hague Convention for the Protection of Cultural Property in the Event of Armed Conflict (opened for signature 14 May 1954, 249 UNTS 240 (entered into force 7 August 1956)) (Cultural Property Convention), obligates States to refrain from any act of hostility directed against cultural property, and to abstain from using such property for military purposes. The Cultural Property Convention does not provide for international criminal responsibility, although article 28 does provide for the possibility of prosecution under national criminal jurisdiction: see also article 15 of the Second Protocol to the Hague Convention of 1954 for the Protection of Cultural Property in the Event of Armed Conflict, opened for signature 26 March 1999, 38 ILM 769 (entered into force 9 March 2004). Some have argued that the deliberate destruction of important cultural property should be made a crime within the jurisdiction of the International Criminal Court: see, for example, Robert Bevan, 'Heritage caught in the crossfire', *The Australian* (Sydney), 24 March 2006, 15.

68 See, for example, Ludwik A. Teclaff, 'Beyond Restoration – the Case of Ecocide' (1994) 34 *Natural Resources Journal* 933; Mark Allan Gray, 'The International Crime of Ecocide' (1996) 26 *California Western International Law Journal* 215.

69 See, for example, Lynn Berat, 'Defending the Right to a Healthy Environment: Toward a Crime of Geocide in International Law' (1993) 11 *Boston University International Law Journal* 327.

70 Mark A. Drumbl, 'Waging War against the World: The Need to Move from War Crimes to Environmental Crimes', in Jay E. Austin and Carl E. Bruch (eds), *The Environmental Consequences of War: Legal, Economic, and Scientific Perspectives* (2000), 620, 641.

71 Parsons, *supra* note 37, 442 and footnote 344.

the environment), so that those who intentionally target the environment during armed conflict can be made criminally liable.

This approach is both timely and appropriate. International criminal law has been developing rapidly, particularly over the past two decades. This has principally been through the operation of the United Nations *ad hoc* international criminal tribunals, as well as a number of 'hybrid' or 'internationalised' courts, established to deal with international crimes perpetrated during specific conflicts.[72] More recently, of course, the world's first *permanent* international criminal institution, the ICC, was established in 2002. The development of this system of international criminal justice has been described as 'one of the few bright spots in the recent history of international law.'[73]

The process of articulating the crimes that exist under international (criminal) law has generally been subject to the traditional principle of legality (*nullum crimen sine lege*), which is reiterated in the Rome Statute.[74] Partially in response to arguments (which were inconclusively dealt with by the judges) by the defendants at the Nuremberg trials, that the *ex post facto* punishment of the crimes in the Charter of that Tribunal[75] offended the *nullum crimen sine*

---

72 The United Nations *ad hoc* Tribunals are the ICTY (established by United Nations Security Council Resolution 827 (25 May 1993) UN Doc S/RES/827/1993 and the International Criminal Tribunal for Rwanda (established by United Nations Security Council Resolution 955 (8 November 1994) UN Doc S/RES/955/1994) (ICTR). In December 2010, the Mechanism for International Criminal Tribunals (MICT) was established to continue the 'obligations and essential functions' of the ICTR and the ICTY: United Nations Security Council Resolution 1966 (22 December 2010) UN Doc S/Res/1966/2010 on the ICTY and ICTR. The 'hybrid' or 'internationalised' criminal tribunals include those that operate or have operated in East Timor (established in 2000 by the United Nations Transitional Administration in East Timor (UNTAET), pursuant to UNTAET Regulations 2000/11 and 2000/15 (6 March 2000 and 6 June 2000 respectively)), Sierra Leone (established by an agreement between the United Nations and the Government of Sierra Leone dated 16 January 2002, pursuant to United Nations Security Council Resolution 1315 (14 August 2000) UN Doc S/RES/1315/2000), and Cambodia (established by an agreement between the United Nations and the Government of Cambodia dated 6 June 2003, pursuant to United Nations General Assembly Resolution 57/228 B (13 May 2003) UN Doc A/RES/57/228 B. There is also The Special Tribunal for Lebanon, which operates in The Netherlands and was established by United Nations Security Council Resolution 1757 (30 May 2007) UN Doc S/RES/1757/2007 on the Situation in the Middle East.

73 Florian Jessberger and Julia Geneuss, 'Down the Drain or Down to Earth? International Criminal Justice under Pressure' (2013) 11 *Journal of International Criminal Justice* 501, 501.

74 See Rome Statute, article 22(1).

75 Charter of the Nuremberg International Military Tribunal, annexed to the 1945 London Agreement for the Establishment of an International Military Tribunal (8 August 1945) 82 UNTS 279, article 6.

lege principle,[76] the ICTY was established to prosecute only crimes that were 'doubtless part of customary international law.'[77]

However, this may not always be the case,[78] since the definition of the crimes specified in (one of) the statutes of the international criminal tribunals may differ from the crime as it existed at the relevant time under customary international law. In addition, the development of these crimes is, in reality, a 'reactive' process. As Theodor Meron has noted that, in the context of discussing the evolving principles of international humanitarian law, '[c]alamitous events and atrocities have repeatedly driven the development of … law'.[79]

This is equally the case with respect to the development of international crimes and the principles of international criminal law. In other words, the acceptance by the international community that certain behaviour is unacceptable and illegal under international criminal law is a responsive act, perhaps also stemming from changing views regarding the conduct of armed conflict. As such, international criminal law is a dynamic area of legal regulation, and continues to evolve over time. It is therefore important to assess the appropriateness of the existing rules of international criminal law from time to time.[80] For a crime to fall within the classification of international crimes – which have been described as *crimen contra omnes*[81] – it is to be regarded as an affront to us all, violating fundamental norms that regulate the peaceful co-existence of the international community.[82] The factors underlying such a determination are, however, dynamic, and will change and evolve over time.

---

76 For example, the concept of 'crimes against humanity' was first codified in the Charter of the Nuremberg International Military Tribunal (article 6(c)), and then applied to actions that took place at an earlier time. It is unclear whether such a crime existed coincidentally under customary international law: see, for example, the judgments of Justice Brennan (587–590) and Justice Toohey (664–677) of the High Court in Australia in *Polyukhovich v Commonwealth and Another* (1991) 172 CLR 501.

77 United Nations Security Council, 'Report of the Secretary-General Pursuant to Paragraph 2 of Security Council Resolution 808' (3 May 1993) UN Doc S/25704, para 33–5. See also Decision on Dragoljub Ojdanić's Motion Challenging Jurisdiction – *Joint Criminal Enterprise, Prosecutor v. Milan Milutinović, Nikola Šainović and Dragoljub Ojdanić*, Case. No. IT-99-37-AR72, Appeals Chamber, 21 May 2003, para 9.

78 With respect to the ICTR, see Kenneth S. Gallant, 'Jurisdiction to Adjudicate and Jurisdiction to Prescribe in International Criminal Courts' (2003) 48 *Villanova Law Review* 764, 783.

79 Theodor Meron, 'The Humanization of Humanitarian Law' (2000) 94 *American Journal of International Law* 239, 243.

80 Tullio Scovazzi, 'State Responsibility for Environmental Harm' (2001) 12 *Yearbook of International Environmental Law* 43, 43.

81 Louis Rene Beres, 'Iraqi Crimes and International Law: The Imperative to Punish' (1993) 21 *Denver Journal of International Law and Policy* 335, 338.

82 Marcos A. Orellana, 'Criminal Punishment for Environmental Damage: Individual and State Responsibility at a Crossroad' (2005) 17 *Georgetown International Environmental Law Review* 673, 689.

The establishment of, and resort to these international(ised) criminal courts indicates a broad acceptance of international criminal law as an appropriate legal mechanism to address grave violations of human rights and crimes that conflict with humanity. A significant element in this evolution was the establishment of the ICC in July 2002. Described by the then United Nations Secretary-General Kofi Annan as 'a giant step forward in the march towards universal human rights and the rule of law',[83] the ICC seeks to address 'the most serious crimes of concern to the international community as a whole'.[84] Such institutions help to further shape and develop important norms that seek to regulate the behaviour of those engaged *inter alia* in armed conflict.

A clear example of the evolving nature of international criminal law (and its codification) is the elaboration of an increasing number of acts involving sexual violence that can constitute the international crime of crimes against humanity.[85] Similarly, we have seen the relatively recent characterization of rape as an act of genocide[86] – the 'crime of crimes'[87] – representing a significant further extension of the scope of international criminal law within the context of an armed conflict.[88]

It is submitted that a similar shift in sentiment is taking place in relation to the intentional destruction of the environment during armed conflict. The universal condemnation of the acts committed by the Iraqi forces in 1991 – involving not only the igniting of the oil wells, but also the dumping of

83 United Nations, 'The International Criminal Court' (Fact Sheet, December 2002) <www.un.org/News/facts/iccfact.htm> at 17 March 2007.

84 Rome Statute, preamble para 4.

85 Compare article 5(g) of the ICTY Statute (Statute of the International Tribunal for the Prosecution of Persons Responsible for Serious Violations of International Humanitarian Law Committed in the Territory of the Former Yugoslavia since 1991, 32 ILM 1159) and article 3(g) of the ICTR Statute (Statute of the International Criminal Tribunal for the Prosecution of Persons Responsible for Genocide and Other Serious Violations of International Humanitarian Law Committed in the Territory of Rwanda and Rwandan Citizens Responsible for Genocide and Other Such Violations Committed in the Territory of Neighbouring States, between 1 January 1994 and 31 December 1994, 33 ILM 1598), both of which specify 'rape' as an act that might constitute a crime against humanity, with article 7(1)(g) of the Rome Statute, which refers not only to rape, but also to 'sexual slavery, enforced prostitution, forced pregnancy, enforced sterilization, or any other form of sexual violence of comparable gravity'.

86 Judgement, *Prosecutor v. Jean-Paul Akayesu*, Case No. ICTR-96-4-T, Trial Chamber, 2 September 1998, para 734.

87 Judgement and Sentence, *Prosecutor v. Jean Kambanda*, Case No. ICTR 97–23-S, Trial Chamber, 4 September 1998, para 16.

88 The crime of genocide requires the existence of a special intent (*dolus specialis*) 'to destroy, in whole or in part, a national, ethnical, racial or religious group, as such': article II of the Convention on the Prevention and Punishment of the Crime of Genocide (opened for signature 9 December 1948, 78 UNTS 277 (entered into force 12 January 1951)) (Genocide Convention); see also Rome Statute, article 6.

up to four million barrels of crude oil into the Persian Gulf[89] – galvanised significant support among the broader community for the establishment of stronger criminal sanctions. Although some commentators regarded such acts as constituting a violation(s) of article 53 of the Fourth 1949 Geneva Convention,[90] no action was initiated under that instrument.[91] Others called for the finalization and implementation of a 'Fifth Geneva Convention' specifically directed towards such acts,[92] arguing that the existing legal regime was shown to be incapable of providing appropriate protection and enforcement in the circumstances.[93]

In the *Legality of the Threat or Use of Nuclear Weapons* Advisory Opinion, one relevant issue was whether, and to what extent, environmental considerations restricted what could be regarded as legitimate military actions. In essence, as articulated by Edith Brown Weiss, the International Court of Justice was faced with three options on this question:[94]

> that environmental considerations apply directly to armed conflicts; that environmental considerations do not apply to armed conflicts unless specifically provided for in an international agreement governing armed conflicts; or, third, in addition to any express provisions in international agreements governing armed conflicts, environmental considerations must be taken into account in applying laws related to armed conflict

The Court affirmed that environmental issues *do* play a role in the overall assessment of military necessity, so that even though:[95]

> existing international law relating to the protection and safeguarding of the environment does not specifically prohibit the use of nuclear weapons ... States must take environmental considerations into account when assessing what is necessary and proportionate in the pursuit of legitimate military objectives. Respect for the

89 Yuzon, *supra* note 2, 794.

90 Geneva Convention Relative to the Protection of Civilian Persons in Time of War, opened for signature 12 August 1949, 75 UNTS 287 (entered into force 21 October 1950) (1949 Geneva Convention IV).

91 Beres, *supra* note 81, footnote 16.

92 This was, for example, suggested at a conference sponsored by the London School of Economics and the British Centre for Defence Studies in June 1991, shortly after the environmental destruction in Kuwait had taken place: see Michael N. Schmitt, *supra* note 63, 269 and the corresponding footnote.

93 Major Walter G. Sharp, Sr., 'The Effective Deterrence of Environmental Damage During Armed Conflict: A Case Analysis of the Persian Gulf War' (1992) 137 *Military Law Review* 1, 3.

94 Edith Brown Weiss in Symposium, 'The International Responses to the Environmental Impacts of War' (2005) 17 *Georgetown International Environmental Law Review* 565, 569.

95 *Legality of the Threat or Use of Nuclear Weapons* (Advisory Opinion) [1996] ICJ Rep 242, para 33 and 30.

> environment is one of the elements that go to assessing whether an action is in conformity with the principles of necessity and proportionality

It has been suggested that this conclusion indicates that the protection of the environment during armed conflict 'is an object and purpose of international law.'[96] However, it remains for examination as to whether the overarching criteria of military necessity – which is itself often an extremely difficult concept to quantify in the circumstances of decision-making – dilutes, within the context of an armed conflict, the importance to be attached to the environment.

This poses some complex questions: do the existing legal rules provide the appropriate level of deterrence, and do they readily create mechanisms that allow for accountability and for sufficient condemnation of the perpetrators? Moreover, do they emphasize sufficiently the gravity and stigma that is associated with acts that intentionally target the environment as a victim, or use the environment as a weapon?

What is clear is that international law increasingly recognizes links between environmental and ecological rights and human rights.[97] It is broadly accepted that customary international law requires that one's actions do not cause significant destruction to the environment so as to threaten fundamental human rights.[98] It logically follows that international legal regulation intended to protect the environment should lead to accountability in appropriate circumstances for those individuals who intentionally set out to significantly damage the environment.[99] As a consequence, the intentional targeting of the environment as an act of warfare can *and should* give rise to individual criminal responsibility at the international level. Such acts can constitute an international crime that 'shocks the conscience'[100] of humankind as a whole. Indeed, the Universal Declaration of Human Rights affirms that 'disregard and contempt for human rights have resulted in barbarous acts which have outraged the conscience of mankind'.[101]

---

96 Parsons, *supra* note 37, 448.

97 Kenneth F. McCallion, 'International Environmental Justice: Rights and Remedies' (2003) 26 *Hastings International and Comparative Law Review* 427, 427.

98 Kenneth F. McCallion and H. Rajan Sharma, 'Environmental Justice Without Borders: The Need for an International Court of the Environment to Protect Fundamental Environmental Rights' (2000) 32 *George Washington Journal of International Law and Economics* 351, 354.

99 Robert McLaughlin, 'Improving Compliance: Making Non-State International Actors Responsible for Environmental Crimes' (2000) 11 *Colorado Journal of International Environmental Law and Policy* 377, 386.

100 *Reservations to the Convention on the Prevention and Punishment of the Crime of Genocide* (Advisory Opinion) [1951] ICJ Rep 16, 23.

101 United Nations General Assembly Resolution 217A (III) (10 December 1948) UN Doc A/810 on the Universal Declaration of Human Rights, preamble para 2.

Within this context, this book examines how acts that intentionally target the environment during the conduct of armed conflict give rise to individual responsibility under international law, and specifically under international criminal law. To reach this point, however, it is also necessary to consider existing international humanitarian law (*jus in bello*), and elements of international environmental law. This study is limited to those acts that utilise a method or means of warfare, in circumstances where there is an intention to target the natural environment as a victim, or to use it as a weapon, in the sense that these concepts are referred to above. It is submitted that the heinous nature of such acts is particularly characterised by this specific intent.[102]

To reiterate, therefore, this book does not seek to create an *environmental* mechanism, but rather to examine and, if necessary adapt, international *criminal* law to address the issue, although there will clearly be the expectation that additional environmental protection may also be achieved in this way.[103]

1.4. THE NEED FOR LEGAL REGULATION

The focus of this book is thus on appropriate *legal* regulation in response to acts of intentional environmental destruction during armed conflict. A fundamental characteristic of such regulation is its binding character supported, importantly, by associated enforcement mechanisms.

Law has traditionally been used to regulate acts that may affect the environment. In addition to those *jus in bello* treaties that may be relevant to environmental destruction during armed conflict,[104] there are many hundreds of multilateral treaties and bilateral instruments[105] that have been concluded by States in an effort to regulate the state of the environment, primarily in times of peace. To the extent, however, that some of the principles in those instruments might also apply during armed conflict, some parallel wartime environmental protections may emerge.[106]

The conclusion of many of these environmental instruments has coincided with the 'rise of normative environmental consciousness'[107] that emerged in the 1960s

---

102 The issue of intent is discussed in detail in chapter 5.
103 See also Michael G. Faure and Marjolein Visser, 'How to Punish Environmental Pollution? Some Reflections on Various Models of Criminalization of Environmental Harm' (1995) 3 *European Journal of Crime, Criminal Law and Criminal Justice* 316, 317.
104 These are discussed in detail in chapter 2.
105 McCallion and Sharma, *supra* note 98, 355.
106 Baker, *supra* note 28, 354. This issue is discussed in chapter 2.
107 Schmitt, *supra* note 63, 317.

and 1970s, when it was recognised that international law could (and should) assume practical relevance in terms of environmental issues.[108] This coincided with a growing awareness of the potential for future 'risks' to human health and the environment, and the uncertainties as to their effects, that could arise from the (in)action of States.[109] This trend has since continued, as illustrated, for example, by the increasing efforts over the past decade to implement legally binding standards to curb the effects of climate change. This is, as those efforts demonstrate, a complex, difficult and constantly evolving process. In some circumstances, even those international environmental agreements that historically have been 'successful' may no longer continue to remain effective without further refinement and expansion.[110]

In addition to legal regulation, there are a number of other mechanisms that may contribute to greater protection of the environment, perhaps even in armed conflict situations. Examples include important guidelines established through an 'increasing resort to the so-called "soft law" approach' to international environmental law,[111] as well as greater education of military personnel as to the long-term effects of environmental damage. However, although it has been argued by one commentator that the use of binding legal sanctions and standards actually exacerbates environmental damage, by entrenching an expectation that *some* degradation is acceptable or anticipated,[112] it would be naive to assume that a system comprising primarily of non-compulsory 'regulation', without the concurrent application of legally binding (and enforceable) standards of behaviour, would be sufficient to establish appropriate sanctions directed against the intentional targeting of the environment. This is particularly so in the circumstances of an armed conflict, where, in the mindset of those involved,

108 Barend van Niekerk, 'Environmental Pollution – The New International Crime' (1976) 93 *South African Law Journal* 68, 73.

109 Jacqueline Peel, *The Precautionary Principle in Practice: Environmental Decision-Making and Scientific Uncertainty* (2005), 1.

110 For example, it has been suggested that the terms of the Montreal Protocol on Substances that Deplete the Ozone Layer (opened for signature 16 September 1987, 26 ILM 154 (entered into force 1 January 1989)), described as 'one of the world's most successful multilateral environmental agreements', must be reconsidered in light of the threats posed by climate change: see Donald Kaniaru, Rajendra Shende, Scott Stone and Durwood Zaelke, 'Strengthening the Montreal Protocol: Insurance Against Abrupt Climate Change' (2007) 7(2) *Sustainable Development Law and Policy Journal* 3.

111 Patricia W. Birnie and Alan E. Boyle, *International Law and the Environment* (1st ed, 1992), 10. See also Steven Freeland, 'For Better or For Worse? The Use of 'Soft Law' within the International Legal Regulation of Outer Space' (2011) XXXVI *Annals of Air and Space Law* 409.

112 See, for example, Mark Halsey, 'Environmental Crime: Towards an Eco-Human Rights Approach' (1996) 8 *Current Issues in Criminal Justice* 217, who (at 235) suggests that there is a '*structural embeddedness* of environmental harm' implicit in the current general approach to environmental crime (emphasis in original).

(real or perceived) military and strategic issues would almost certainly outweigh most (if not all) environmental concerns.

The principal ways in which law has been utilised to regulate the protection of the environment have been through the use of national law, as well as, as indicated above, international law instruments, either on a multilateral or bilateral basis. Some attempts have also been made on a regional basis to standardise laws relating to environmental protection. These are each discussed below.

1.4.1. NATIONAL LAW

The emergence of an active environmental movement in many countries during the 1970s raised public consciousness in relation to the exploitation and destruction of the environment.[113] Environmentalists and civil society called for government action to combat threats to the natural environment.[114] Since then, many States have instituted systems of domestic environmental legal regulation, sometimes quite complex in structure. Environmental law in the United States, for example, has developed significantly since the 1970s, and now comprises an extensive body of statute and case law,[115] as well as a functioning enforcement mechanism.[116] Within these domestic frameworks, the concept of environmental *criminal* enforcement has slowly emerged, as policy makers 'became increasingly dissatisfied with [the] 'administrative dependence' of environmental criminal law'.[117] Nevertheless, even as late as the mid-1990s, environmental 'crimes' were still regarded as a relatively 'new' concept in criminal law lexicon.[118]

More recently, the issues of environmental crime and enforcement have been the subject of increasing discussion among national and international environmental law experts.[119] It has been argued that the development of national systems

113 Keith P. McManus, 'Civil Liberty for Wartime Environmental Damage: Adapting the United Nations Compensation Commission for the Iraq War' (2006) 33 *Boston College Environmental Affairs Law Review* 417, 419.

114 See, for example, J.H. Huebert and Walter Block, 'Space Environmentalism, Property Rights, and the Law' (2007) 37 *University of Memphis Law Review* 281, 283 and the corresponding footnotes.

115 Elizabeth Glass Geltman, *Modern Environmental Law: Policy and Practice* (1997), 1.

116 See, for example, Agence France-Presse, 'Clean-burning fools: BP fined $413 million', *The Sydney Morning Herald* (Sydney), 27–28 October 2007, 44.

117 Faure and Visser, *supra* note 103, 316.

118 Kathleen F. Brickey, 'Environmental Crime at the Crossroads: The Intersection of Environmental and Criminal Law Theory' (1996) 71 *Tulane Law Review* 487, 487.

119 Hans Addink, Bas Arts and Arthur Mol, 'Climate change policy in changing contexts: globalization, political modernization and legal innovation', in Ekko C. van Ierland, Joyeeta Gupta and Marcel T.J. Kok (eds), *Issues in International Climate Policy: Theory and Policy* (2003), 87.

of environmental criminal law have generally become more effective as they have evolved from rather simplistic models based on administrative licence schemes to models intended to protect human health and the environment.[120] Yet, existing national environmental legislation in various jurisdictions is neither consistent nor universal. This is even the case in circumstances where States are implementing the same international environmental treaty, since the implementation process, as well as the content of any implementing legislation, still falls within the sovereign discretion of each State.[121]

This does not, of course, deny the value in having binding domestic rules regulating specific environmental standards. An increasing number of States have, for example, even introduced the crime of ecocide into their national laws;[122] however, there have been very few prosecutions based on these provisions. Moreover, in many cases, unlike international criminal law, national environmental law lacks a punitive character. It is not uncommon for environmental lawyers to regard domestic environmental crimes not as 'real' crimes, but rather 'quasi-criminal regulatory offences'.[123] The penalties imposed under national law are often insufficient to deter environmental offenders, particularly where there are strong financial incentives to break the law.[124] Even where the terms of national law do appear on their face to provide for stringent environmental standards, enforcement in practice may be compromised, in order to encourage inward investment of capital.[125]

Those national laws that have been developed, in deference to the traditional principles of territoriality, do not adequately regulate *international* environmental misconduct,[126] and are thus typically not appropriate to deal with the transboundary consequences of intentional environmental destruction. In addition, they are unlikely to expressly apply to situations of armed conflict and, even if they were applicable, a party to the conflict is not likely to be concerned

120 Faure and Visser, *supra* note 103, 357.

121 For example, New Zealand and Australia, two neighbouring and very similar countries, have enacted very different terms in their respective implementing legislation of the Rome Statute: see Treasa Dunworth, 'The International Crimes and International Criminal Court Act 2000 (NZ): A Model for the Region?', in Neil Boister and Alberto Costi (eds), *Regionalising International Criminal Law in the Pacific* (2006), 145.

122 See Jean-Marie Henckaerts and Louise Doswald-Beck (eds), *Customary International Humanitarian Law* (Volume II: Practice, Part 1, 2005), 883–7 and the corresponding footnotes.

123 Michael Watson, 'Environmental Offences: The Reality of Environmental Crime' (2005) 7 *Environmental Law Review* 190, 190.

124 *Ibid*, 192.

125 See, for example, Rob White, 'Environmental Crime in Global Context: Exploring the Theoretical and Empirical Complexities' (2004) 16 *Current Issues in Criminal Justice* 271, 277.

126 Byung-Sun Cho, 'Emergence of an International Environmental Criminal Law?' (2001) 19 *UCLA Journal of Environmental Law and Policy* 11, 11.

about the domestic laws of the other parties, nor would it necessarily have reason to fear sanction under its own domestic laws.[127]

In sum, therefore, reliance on national law as the only (or principal) mechanism to address intentional environmental destruction during armed conflict is not appropriate. The reality is that national environmental law regimes are not designed, and have not been used for that purpose.

1.4.2. REGIONAL LAW

There is scope for some common principles of law to be developed at a *regional* level. At times, this may be more effective than national efforts regarding the protection of the environment, given the nature of some of the problems that may arise. The European Parliament and Council of the European Union have, for example, recognised that the protection of the environment 'cannot be sufficiently achieved by the Member States and can therefore ... be better achieved at Community level'.[128]

In the area of human rights protection, for example, we have seen the regionalization of regulatory systems, at least in some parts of the world. In the European, American and African regions, well-developed systems have emerged under widely accepted treaty regimes dealing with human rights and fundamental freedoms, reinforced with a structured judicial enforcement mechanism.[129] There is also an Arab Charter of Human Rights,[130] as well as some (albeit slow) movement towards the development of a regional human rights regimen in the Asian region.[131]

---

127 Leibler, *supra* note 5, 80.

128 European Parliament and Council of the European Union, 'Directive 2008/99/EC of the European Parliament and of the Council on the protection of the environment through criminal law', 19 November 2008, preamble para 14. This Directive came into force in December 2008.

129 European Convention for the Protection of Human Rights and Fundamental Freedoms, opened for signature 4 November 1950, 213 UNTS 221 (entered into force 3 September 1953), American Convention on Human Rights, opened for signature 22 November 1969, 1144 UNTS 123; 9 ILM 673 (entered into force 18 July 1978), and African Charter on Human and People's Rights, adopted 27 June 1981, 1520 UNTS 217; 21 ILM 58 (entered into force 21 October 1986).

130 See Arab Charter on Human Rights, adopted by the Council of the League of Arab States on 15 September 1994, reprinted in (1997) 18 *Human Rights Law Journal* 151 (not in force). This version was not ratified by any State and, in 2004, a revised text was adopted by an Arab Summit meeting in Tunisia, and entered into force in 2008: see Council of the League of Arab States Resolution 5437.

131 For further details, see Steven Freeland, 'International Criminal Justice in the Asia-Pacific Region – The Role of the International Criminal Court Treaty Regime' (2013) 11(5) *Journal of International Criminal Justice* 1029.

Nonetheless, the use of regional systems of law to address important issues could exacerbate *differences* between the regions,[132] rather than reinforce those *universal* standards that may be necessary to deal with issues such as intentional environmental destruction during armed conflict. In fact, regional legal systems are often designed to achieve different goals and take various forms,[133] precisely due to the unique characteristics of a particular region.

Of course, there may well be some advantages to regional legal regulation in relation to 'lesser' environmental crimes. In 1998, the Council of Europe opened for signature the Convention on the Protection of the Environment Through Criminal Law, which required States Parties to adopt 'appropriate measures as may be necessary to establish as criminal offences' certain acts of environmental damage.[134] Even though it recognised that criminal law was a secondary measure in relation to the prevention of destruction of the environment and was not directly binding for natural or legal persons,[135] this was the first regional instrument designed to address environmental protection through the use of criminal law.[136]

However, this initiative was not widely accepted. The definition of environmental crimes still differed among the domestic law of the European States, and the applicable sanctions were generally inadequate, with some countries continuing to apply civil rather than criminal penalties.[137] The Commission of the European Union (EU) unveiled a draft proposal for a Directive in 2007[138] that sought to create a series of 'green crimes' enforceable across the European Union (EU),[139] following an important decision of the European Court of Justice (ECJ)

---

132 See Steven Freeland, 'The Internationalization of Justice – A Case for the Universal Application of International Criminal Norms' (2007) 4 *New Zealand Yearbook of International Law* 45.

133 Gary W. Florkowski, *Managing Global Legal Systems* (2006), 71.

134 European Convention on the Protection of the Environment through Criminal Law, opened for signature 4 November 1998, 38 ILM 259 (not in force), article 2(1). As at January 2015, only 1 State (Estonia) had ratified this instrument, and it requires 3 ratifications to come into force: <http://conventions.coe.int/Treaty/Commun/ChercheSig.asp?NT=172&CM=1&DF=&CL=ENG> at 19 January 2015.

135 Thomas Richter, 'Transboundary environmental crimes: an analysis of Chinese and European Law', in Michael Faure and Song Ying (eds), *China and International Environmental Liability: Legal Remedies for Transboundary Pollution* (2008), 253, 254.

136 John P. Grant and J. Craig Barker, *International Criminal Law Deskbook* (2006), 164.

137 For example, at the time, Spain and Greece applied civil law penalties to the illegal shipping of waste, even though that was deemed to be an environmental crime under the draft directive: Fiona Harvey and Andrew Bounds, 'Regulatory regime put to the test at borders', *Financial Times* (London), 7 February 2007, 2.

138 Proposal of 9 February 2007 for a Directive of the European Parliament and of the Council on the protection of the environment through criminal law.

139 Andrew Bounds, 'Brussels in push to lay down 'green crime' laws', *Financial Times* (London), 7 February 2007, 1. An earlier Framework Decision 2003/80 on the Protection of the

confirming the power of the EU to take measures relevant to the national laws of Member States in respect of criminal penalties to combat serious environmental offences.[140]

This draft proposal was followed by the 2008 Directive, which allowed the EU to overrule Member States' competence in relation to criminal justice in this area, so as to ensure a minimum standard of environmental protection under criminal law, and consistency of approach in relation to environmental damage.[141] However, the crimes that it established, although serious, do *not* purport to apply to intentional environmental destruction during armed conflict, which is, of course, understandable given that a principal *raison d'être* for the establishment of what has become the EU was, in fact, to minimise the risk of future armed conflict within Europe.[142]

1.4.3. INTERNATIONAL LAW, ARMED CONFLICT AND THE ENVIRONMENT

The inter-relationship between intentional environment destruction, human security and armed conflict requires that effective enforcement measures exist at the international level. As indicated, both national law and regional law can only play a very limited role in addressing such acts. This book will therefore focus on the application of *international* law to the issue of intentional environmental damage during armed conflict. In doing so, it is acknowledged that any suggestion of *new* 'supranational' environmental regulation will likely encounter diplomatic opposition.[143] It is therefore necessary to first examine the existing principles and, only then, determine whether, and if so in what way, additional standards may be necessary, bearing in mind that these should, to the greatest

---

Environment had created intentional and negligence offences. This was, however, annulled on 13 September 2005.

140 *Commission of the European Communities v Council of the European Union,* Case C-176/03, Grand Chamber, 13 September 2005, (2005) EC Rep I-07879. The ECJ held that the power to require Member States to impose criminal penalties for certain environmental offences fell under pillar one of the EU structure, highlighting the importance of environmental protection within the European regulatory system: see Helle Krunke, 'Introduction: European Law as a Process between Constitutionalization and Democratization', in Hanne Petersen, Anne Lise Kjaer, Helle Krunke and Mikael Rask Madsen (eds), *Paradoxes of European Legal Integration* (2008), 9, 11: Jane Holder and Maria Lee, *Environmental Protection, Law and Policy: Text and Materials* (2nd ed, 2007), 414 footnote 70.

141 See *supra* note 128.

142 See the 1957 Treaty of Rome (Treaty Establishing the European Economic Community, opened for signature 25 March 1957, 298 UNTS 11 (entered into force 1 January 1958)), which followed on from the 1951 Treaty of Paris (Treaty Establishing the European Coal and Steel Community, opened for signature 18 April 1951, 261 UNTS 140 (entered into force 24 July 1952, expired 23 July 2002)).

143 Cho, *supra* note 126, 24.

extent possible, be consistent with existing practice and law, as well as utilizing already established enforcement mechanisms.[144]

International law – and more particularly international criminal law – is particularly suited to regulating such acts since, even though some international crimes such as genocide and crimes against humanity can (theoretically) occur in times of 'peace', the principles of international criminal law were developed principally within the context of armed conflict.[145] This is so even where the armed conflict may be non-international (as opposed to international) in character, since the consequences of such damage do not necessarily end at the borders of any one country. There is no overarching moral or legal justification for treating perpetrators of very serious 'international' crimes in internal conflicts differently from those engaged in international armed conflicts.[146] For example, during the conflict in Rwanda during the 1990s, culminating in the genocide that took place during 1994,[147] significant amounts of agricultural land were destroyed in order to force groups of people to move, both within the country and beyond its borders.[148] Moreover, many of the more recent conflicts have been internal in nature.[149]

It is true that the distinction between non-international and international armed conflicts is still important in the context of various international crimes,[150] a consequence of the traditional adherence of the 'non-intervention' principle that still pervades international law and international relations.[151] However, it is equally clear that international criminal law already does apply to 'armed

144 This is one important reason why intentional damage to the environment during armed conflict should be addressed by the ICC under the Rome Statute, rather than, potentially, through the establishment of a new judicial institution.

145 Orellana, *supra* note 82, 689.

146 See Jochen A. Frowein, Book Review – Theodor Meron, 'The Humanization of International Law' (2007) 101 *American Journal of International Law* 680, 681.

147 The ICTR has held that the conflict in Rwanda was an internal armed conflict under international law: Judgement, *Prosecutor v Akayesu*, Case No ICTR-96-4-T, Trial Chamber, 2 September 1998, para 627.

148 Weinstein, *supra* note 8, 700.

149 See Carl E. Bruch, 'All's Not Fair in (Civil) War: Criminal Liability for Environmental Damage in Internal Armed Conflict' (2001) 25 *Vermont Law Review* 695, 698 and the corresponding footnotes.

150 Certain international crimes within the jurisdiction of the international criminal tribunals can only be committed where the conflict is characterised as an international conflict: compare, for example, articles 8(2)(a) and 8(2)(b)) of the Rome Statute with articles 8(2)(c) and 8(2)(e). See also Decision on Defence Motion for Interlocutory Appeal on Jurisdiction, *Prosecutor v Duško Tadić*, Case No. IT-94-1, Appeals Chamber, 2 October 1995. This distinction is discussed further in chapters 4 and 5.

151 See also United Nations Charter, article 2(7).

conflict not of an international character',[152] as do the *jus in bello*.[153] In circumstances where intentional environmental damage during armed conflict is sufficiently serious to warrant criminalisation,[154] the same imperatives should apply irrespective of the legal nature of the armed conflict. Yet, the existing war crime relating to damage to the natural environment in the Rome Statute only applies in international armed conflict.[155]

Thus, unless the context requires it, the discussion that follows in this book does not distinguish between armed conflict of an international and non-international character, and the examination of the legal principles regulating crimes against the environment will include both types of armed conflict. In fact, there have been calls from time to time to 'unify' the *jus in bello*, so that they are equally applicable 'in all forms' of armed conflict.[156]

As will be discussed in chapter 2, there is already a body of *jus in bello* treaties that deal with certain aspects of intentional environmental damage during armed conflict. These instruments must be analysed to determine whether they adequately reflect the heinous nature associated with the *intention* to destroy the environment. On the other hand, the United Nations Charter, which is concerned with the maintenance of international peace and security,[157] makes no mention of the problem of environmental degradation. This goes some way towards explaining why the environment has not been of particular focus to the main organs of the United Nations, particularly the Security Council.[158] This is not particularly surprising given that, at the time that the United Nations Charter was drafted, the issue of environmental protection was not regarded at all as a priority.[159]

As a result, legal regulation aimed at promoting criminal law to enforce important international environmental principles has only been a relatively recent phenomenon.[160] International law has been even 'slower' to recognize the

152 See, for example, Rome Statute, articles 8(2)(c) and 8(2)(e).
153 See, for example, 1977 Additional Protocol II, as well as article 3 common to the four Geneva Conventions of 12 August 1949 for the Protection of War Victims (Common Article 3).
154 The appropriate 'threshold' level of damage warranting criminal culpability for crimes against the environment is discussed in chapters 4 and 5.
155 See Rome Statute, article 8(2)(b)(iv). This issue will be addressed in chapters 4 and 5.
156 See, for example, Bruch, *supra* note 149, 738–43 and 750 and the corresponding footnotes.
157 See United Nations Charter, article 1(1).
158 Of course, there are some United Nations agencies that are directed towards the environment, such as UNEP.
159 Van Niekerk, *supra* note 108, 70.
160 Martin Hedemann-Robinson, *Enforcement of European Union Environmental Law: Legal Issues and Challenges* (2007), 516.

need to address environment destruction during armed conflict situations.[161] It was not until 1977, following the television broadcasts of the environmental destruction wrought during the Vietnam War, that the concept of the 'natural environment' appeared in the treaty principles that set out the rules of international humanitarian law, although the overriding principle of 'military necessity' still generally applies in any assessment of environmental damage caused during armed conflict. There are some *jus in bello* instruments that do not specifically refer to military necessity in this context,[162] but it is necessary to critically consider the level of environmental damage required before a breach of those treaties is achieved, described by Carl Bruch as 'extreme'.[163]

In terms of many international environmental agreements that have emerged since the 1970s, most do not refer to armed conflict, although some may be applicable in wartime.[164] However, while many of these treaties – and the relevant principles of customary international law – may provide for the *prohibition* of certain forms of environmental destruction, even during armed conflict, they have been wholly ineffective in terms of the *prevention* of such action. With some exceptions, the general focus of these international environmental agreements has instead been in relation to principles between States *inter se* – the breach of which gives rise to principles of State responsibility under the traditional rules of general international law. As the International Court of Justice has noted:[165]

> in the field of environmental protection, vigilance and *prevention* are required on account of the often irreversible character of damage to the environment and of the limitations inherent in the very mechanism of reparation of this type of damage

What appears to be lacking is a mechanism that provides sufficient *deterrence* and 'facilitates condemnation and accountability' in a more effective manner.[166] There have been virtually no prosecutions of individuals for intentional

161 Hans-Peter Gasser, 'For Better Protection of the Natural Environment in Armed Conflict: A Proposal for Action' (1995) 89 *American Journal of International Law* 637, 638.

162 See Protocol I Additional to the Geneva Conventions of August 12, 1949, and relating to the Protection of Victims of International Armed Conflicts Opened for signature 8 June 1977, 1125 UNTS 3, 16 ILM 1391 (entered into force 7 December 1978) (1977 Additional Protocol I); Convention on the Prohibition of Military or Any Other Hostile Use of Environmental Modification Techniques Opened for signature 10 December 1976, 1108 UNTS 151; 16 ILM 88 (entered into force 5 October 1978) (ENMOD Convention). Both of these instruments are discussed in chapter 2.

163 Bruch, *supra* note 149, 703.

164 The *jus in bello* treaties, as well as the relevant international environmental agreements, are discussed in chapter 2.

165 *Gabcíkovo-Nagymaros Project (Hungary v. Slovakia)* (Judgment) [1997] ICJ Rep 7, para 140 (emphasis added).

166 Sharp, *supra* note 93, 5.

environmental destruction in such circumstances,[167] notwithstanding the inclusion in the *jus in bello* of the so-called 'Martens clause', which was first incorporated into the 1899 Hague Convention II with Respect to the Laws and Customs of War on Land,[168] and has been included in a number of subsequent treaties.[169] This book will explore the reasons for this and determine what (if any) 'solutions' may be necessary and appropriate, particularly bearing in mind that the prosecution of those who commit international crimes helps to preserve a common belief in the importance of implementing an 'international legal order', as well as providing (an element of) deterrence to those who may otherwise contemplate committing such crimes in the future.[170]

There have, in the past, been general calls – in non-binding instruments – for the *absolute* protection of the environment during times of war. The 1982 World Charter for Nature declares that '[n]ature *shall* be secured against degradation caused by warfare or other hostile activities',[171] and that '[m]ilitary activities damaging to nature *shall* be avoided'.[172] Ten years later, the Rio Declaration on Environment and Development[173] called for the protection of the environment during armed conflict, with principles 23 and 24 providing *inter alia*:[174]

> [t]he environment and natural resources of people under oppression, domination and occupation *shall* be protected (Principle 23);
>
> [w]arfare is inherently destructive of sustainable development. States *shall* therefore respect international law providing protection for the environment in times of armed conflict and cooperate in its further development, if necessary (Principle 24)

It has been somewhat optimistically asserted by one commentator that principle 23 establishes 'an absolute right to environmental protection',[175] thus sidestepping 'the pitfall of military necessity'.[176] Yet, less than six months after

---

167 One reason for this is that 'environmental crimes [have been] overshadowed by other atrocities': Weinstein, *supra* note 8, 698.

168 Hague Convention II with Respect to the Laws and Customs of War on Land, opened for signature 29 July 1899, 26 Martens (2nd) 949 (entered into force 4 September 1900).

169 See further the discussion in chapter 2.

170 Alberto Costi, 'Hybrid Tribunals as a Viable Transitional Justice Mechanism to Combat Impunity in Post-Conflict Situations' (2006) 22 *New Zealand Universities Law Review* 213, 215.

171 United Nations General Assembly Resolution 37/7 (28 October 1982) UN Doc A/RES/37/7, Annex, principle 5 (emphasis added). There were 110 States who voted in favour of this resolution, 1 (United States) against, with 18 abstentions.

172 *Ibid*, principle 20 (emphasis added).

173 Declaration of the United Nations Conference on Environment and Development (14 June 1992) UN Doc A/CONF.151/26 (Volume 1), 31 ILM 874 (Rio Declaration).

174 *Ibid*, principles 23 and 24.

175 Laurent R. Hourcle, 'Environmental Law of War' (2001) *Vanderbilt Law Review* 653, 678.

176 McManus, *supra* note 113, 428.

the Rio Declaration was adopted, the United Nations General Assembly, while taking note of the Declaration, stressed that 'destruction of the environment, *not justified by military necessity* and carried out wantonly, is clearly contrary to existing international law'.[177] A few years later, the International Court of Justice affirmed that the *jus in bello* prohibited 'harm greater than that unavoidable to achieve *legitimate military objectives*'.[178]

The Rome Statute currently continues along the same path, incorporating military advantage into a consideration of whether a war crime involving damage to the environment has been committed.[179] Some support this approach, arguing that it 'remedies what many have seen as a flaw' in conventional law dealing with environmental damage during armed conflict.[180] However, the impact of the military necessity factor upon the current international criminal legal regime regulating intentional destruction of the environment during armed conflict, as well as the existing thresholds of permissible damage, warrant a more circumspect analysis. The risks associated with allowing for weak accountability mechanisms can only become graver as weapons technology continues to evolve.[181]

1.5. INTERNATIONAL CRIMINAL LAW AS A REGULATORY MECHANISM

This book proceeds on the basis that international criminal law can play a leading role in addressing intentional destruction of the environment during armed conflict. To do so, the regulatory framework must be structured such that it does not, as is the current position, require a 'balancing test stacked heavily against the environment'.[182] As a precursor to this discussion, however, the following chapters will first analyse how far the existing conventional and customary principles of international law go towards achieving these goals. Should these, when complementing the current Rome Statute regime, be shown to be inadequate to deal with the intentional destruction of the environment

177 United Nations General Assembly Resolution 47/37 on the Protection of the environment in times of armed conflict (25 November 1992) UN Doc A/RES/47/37, preamble para 5 (emphasis added).

178 *Legality of the Threat or Use of Nuclear Weapons* (Advisory Opinion) [1996] ICJ Rep 242, para 78 (emphasis added).

179 Rome Statute, article 8(2)(b)(iv).

180 Schmitt, *supra* note 63, 282.

181 In this regard, see, for example, Commentary by the International Law Commission on Article 19 'Articles on State Responsibility for Internationally Wrongful Acts' [1976] 2 *Yearbook of the International Law Commission* Part II, 96, UN Doc A/CN.4/SER.A/1976/Add.l (Part 2), 108.

182 Sharp, *supra* note 93, 241.

during armed conflict, a further issue would arise as to how these flaws might best be rectified.

International criminal accountability on this issue results in the 'individualisation' of responsibility for intentional destruction of the environment during armed conflict, in that a guilty person(s) is held culpable without entire groups, or the State itself, deemed to be responsible.[183] This may aid in seeking external assistance to help remediate the environmental damage. International criminal law is also an appropriate form of regulation in this area because it more directly expresses the outrage that such acts gives rise to, as well as the imperative of global compliance with appropriate standards of behaviour.[184]

That said, the concept of an 'environmental crime' has not, until quite recently, been a point of focus in the otherwise rapid evolution of international criminal law. However, it is now accepted that international criminal law *is* an appropriate mode to deal with this issue, and it is already addressed in one provision of the Rome Statute, which is regarded by many as the most up-to-date codification of the nature and scope of international crimes, even though it does not reflect existing customary international law in every respect.[185] Article 8(2)(b)(iv) of the Rome Statute provides that a war crime within the context of an international armed conflict and 'within the established framework of international law' may have been committed in the event that an accused:[186]

> [i]ntentionally launch[es] an attack in the knowledge that such attack will cause ... widespread, long-term and severe damage to the environment which would be clearly excessive in relation to the concrete and direct overall military advantage anticipated

This specific provision draws on principles also found in the 1977 Additional Protocol I, which proscribe methods or means of warfare that are intended, or may be expected to cause 'widespread, long-term and severe' environmental damage. At the time that the 1977 Additional Protocol I was concluded, those

---

183 Naomi Roht-Arriaza and Javier Mariezcurrena (eds), *Transitional Justice in the Twenty-First Century: Beyond Truth versus Justice* (2006), 6.

184 Cho, *supra* note 126, 13–4.

185 A Trial Chamber of the ICTY has described the Rome Statute as follows:
'Depending on the matter at issue, the Rome Statute may be taken to restate, reflect or clarify customary rules or crystallise them, whereas in some areas it creates new law or modifies existing law. At any event, the Rome Statute by and large may be taken as constituting an authoritative expression of the legal views of a great number of States': Judgement, *Prosecutor v. Furundzjia*, Case No. IT-95-17/1, Trial Chamber, 10 December 1998, para 227. More recently, see Judgement, *Prosecutor v. Nikola Sainovic, Nebojsa Pavkovic, Vladimir Lazarevic and Sreten Lukic*, Case No. IT-05-87-A, Appeals Chamber, 23 January 2014, para 1648.

186 Rome Statute, article 8(2)(b)(iv). See chapter 4 for an analysis of this provision.

provisions were regarded as an important step forward in acknowledging the issue of environmental damage during conflict. Article 8(2)(b)(iv) of the Rome Statute has already been considered by the respective Offices of the Prosecutor of both the ICTY and the ICC,[187] but, for various reasons, was not used as a basis of any charges in either case. In analysing claims relating to the use of cluster bombs in the situation in Iraq, Luis Moreno-Ocampo, the then Prosecutor of the ICC, concluded *inter alia* that, on the basis of the information available to his office at that time, there was 'a lack of information indicating clear excessiveness in relation to military advantage'.[188]

Tara Weinstein has suggested that the environmental pollution wrought by Saddam's forces in 1991 would probably not have been sufficient to constitute a breach of article 13(b)(5) of the Statute of the Iraqi Special Tribunal, which, as noted, is identical with article 8(2)(b)(iv) of the Rome Statute.[189] The applicability of the existing legal regime to such acts will be the subject of detailed analysis in the remaining chapters of this book. In undertaking this examination, it will do well to bear in mind the words of the International Court of Justice:[190]

> [...] one of the most effective ways of preventing criminal acts, in general, is to provide penalties for persons committing such acts, and to impose those penalties effectively on those who commit the acts one is trying to prevent

1.6. COULD STATE CRIMINAL RESPONSIBILITY APPLY?

Before examining whether and how intentional destruction of the environment during armed conflict may attract *individual* criminal responsibility under international law, there is a preliminary, but important, threshold question to resolve; can the relevant State *also* be held criminally responsible for such acts (by individuals) where there is significant State involvement in the deliberate destruction?

---

187 See, for example, Final Report to the Prosecutor of the International Criminal Tribunal for the former Yugoslavia by the Committee Established to Review the NATO Bombing Campaign Against the Federal Republic of Yugoslavia, 13 June 2000, 39 ILM 1257, para 21; Office of the Prosecutor, The International Criminal Court, 'Iraq Response', (9 February 2006), 5 <www.iccnow.org/documents/OTP_letter_to_senders_re_Iraq_9_February_2006.pdf> at 19 January 2015.

188 'Iraq Response', *ibid*, 6.

189 Weinstein, *supra* note 8, 710.

190 *Case Concerning the Application of the Convention on the Prevention and Punishment of the Crime of Genocide (Bosnia and Herzegovina v. Serbia and Montenegro)* (Judgment) [2007] ICJ Rep 43, para 426 (*Application of the Genocide Convention Case*).

1.6.1. STATE CRIMINAL RESPONSIBILITY?

States bear a responsibility towards the environment, which extends also to situations of armed conflict.[191] In its Advisory Opinion in the *Legality of the Threat or Use of Nuclear Weapons Case*, the International Court of Justice affirmed that:[192]

> [t]he existence of the general obligation of States to ensure that activities within their jurisdiction and control respect the environment of other States or of areas beyond national control is now part of the corpus of international law relating to the environment

However, the Court did not prescribe any principles of criminal responsibility for a breach of this obligation.[193] Indeed, few, if any, international instruments give rise to specific legally binding criminal sanctions against States for the deliberate degradation of the environment. While issues of intentional environmental damage are subject to various 'non-criminal' legal processes at the international level, these generally do not reflect the magnitude of the destruction that may result, or the intentional nature of such acts. This regime more generally prescribes that criminal processes be undertaken at the domestic level based on traditional national jurisdiction principles,[194] thus giving rise to the problems of ineffective implementation and enforcement noted earlier.

States can incur 'classical (civil)'[195] responsibility for acts that constitute international crimes committed by individuals, or even by simply failing to provide for adequate national legislation making such acts a criminal offence, under the traditional international law principles of State responsibility.[196] This may arise if an organ of that State, or a person or group whose acts are attributable to the State, commits an international crime. State responsibility

---

191 Refer, for example, to the sentiments underpinning principle 24 of the Rio Declaration.

192 *Legality of the Threat or Use of Nuclear Weapons* (Advisory Opinion) [1996] ICJ Rep 242, para 29.

193 A Chamber for Environmental Matters was established in the International Court of Justice in 1993, but was not extended beyond 2006, since it had not been utilized by States, even in those cases that had raised important environmental issues.

194 See, for example, the United Nations Convention on the Law of the Sea (UNCLOS) (10 December 1982) 1833 UNTS 3; 21 ILM 1261, articles 213–222.

195 T.D. Gill, 'The "Genocide" case: Reflections on the ICJ's Decision in Bosnia-Herzegovina v Serbia', *The Hague Justice Portal* (The Netherlands), 26 March 2007 <www.haguejusticeportal.net/index.php?id=7266> at 19 January 2015.

196 One example of the failure of a State to properly incorporate an international crime (in this case genocide) into its national law is the domestic implementation of the Genocide Convention by Australia. Even though it ratified the Convention in 1949, the crime of genocide was not incorporated into Australian domestic law until 2002.

may exist even if no individual has been convicted of the crime by a competent court.[197]

However, this degree of State accountability represents a different level of culpability from the possibility that the State itself may be *criminally responsible*. This can be seen, for example, in relation to the accountability of a State for a violation of the Genocide Convention. In the discussions leading to the finalization of that instrument, there was considerable debate over the inclusion of a notion of State criminality.[198] However, this idea was, in the end, not incorporated into the treaty, which instead entails a civil liability regime for States. In the *Application of the Genocide Convention Case*, the International Court of Justice, when considering the various 'crimes' specified in article III of the Genocide Convention, and in particular the crime of 'complicity in genocide',[199] drew this distinction by stating:[200]

> [i]t would however not be in keeping with the object and purpose of the Convention to deny that the international responsibility of a State – *even though quite different in nature from criminal responsibility* – can be engaged through one of the acts, other than genocide itself, enumerated in Article III

This 'duality of responsibility'[201] is further illustrated by article 25(4) of the Rome Statute, which states that '[n]o provision in this Statute relating to individual criminal responsibility shall affect the responsibility of States under international law'.[202]

As noted by the International Court of Justice, article 58 of the International Law Commission's (ILC) Articles on the Responsibility of States for Internationally Wrongful Acts affirms 'the other side of the coin',[203] by emphasizing that '[t]hese articles are without prejudice to any question of the individual responsibility under international law of any person acting on behalf of a State.'[204] The distinction is more than a question of semantics – it demonstrates that, irrespective of the degree of involvement by the 'machinery' of a State, its

---

197 *Application of the Genocide Convention Case*, para 182. The International Court of Justice held that a State could be internationally responsible under the Genocide Convention for genocide and complicity to commit genocide, 'without an individual being convicted of the crime or an associated one'.

198 William A. Schabas, *Genocide in International Law* (2000), 419, 444.

199 Genocide Convention, article III(e).

200 *Application of the Genocide Convention Case*, para 167 (emphasis added).

201 *Ibid*, para 173.

202 Rome Statute, article 25(4).

203 *Application of the Genocide Convention Case*, para 173.

204 International Law Commission, Articles on the Responsibility of States for Internationally Wrongful Acts, annex to United Nations General Assembly Resolution 56/83 (12 December 2001) UN Doc A/RES/56/83, article 58 (ILC Articles on State Responsibility).

culpability for actions that constitute international crimes, even where they result in very serious consequences for humans or the environment, is to be regarded differently – one could assert with less 'stigma' – than the standards by which we judge those individuals who actually 'commit' the crime.

For example, the four Geneva Conventions of 1949[205] specify the scope of what now represent a number of war crimes – which can only be committed by individuals – associated with 'grave breaches' of those treaties[206] while, at the same time, 'merely' providing for State responsibility for a State Party where one of those crimes is committed by an individual who is the national of that State.[207]

Yet, in the period 1976–2001, the ILC,[208] which in 1949 was asked by the United Nations General Assembly to formulate the draft articles, advanced the notion of an international crime committed by States. In the Special Rapporteur's fifth report,[209] the ILC adopted *inter alia* draft article 19 into the document,[210] which distinguished between two different types of internationally wrongful acts giving rise to the responsibility of a State under international law – international crimes and international 'delicts'. The distinction rested on the subject matter

205 Geneva Convention for the Amelioration of the Condition of the Wounded and Sick in Armed Forces in the Field, opened for signature 12 August 1949, 75 UNTS 31 (entered into force 21 October 1950) (1949 Geneva Convention I), Geneva Convention for the Amelioration of the Condition of the Wounded, Sick and Shipwrecked Members of Armed Forces at Sea, opened for signature 12 August 1949, 75 UNTS 85 (entered into force 21 October 1950) (1949 Geneva Convention II), Geneva Convention Relative to the Treatment of Prisoners of War, opened for signature 12 August 1949, 75 UNTS 135 (entered into force 21 October 1950) (1949 Geneva Convention III), and 1949 Geneva Convention IV.

206 See 1949 Geneva Convention I, article 50; 1949 Geneva Convention II, article 51; 1949 Geneva Convention III, article 130; 1949 Geneva Convention IV, article 147. See also ICTY Statute, article 2 and Rome Statute, article 8(2)(a).

207 See Antonio Cassese, *International Criminal Law* (2003), 19 footnote 4. Also compare *R v. Bow Street Metropolitan Magistrate and others, Ex Parte* Pinochet Ugarte (Amnesty International and others intervening) (No 3) (1999) 2 All ER 97 (HL), with *Al-Adsani v. United Kingdom* (2001) 34 EHRR 273: in the former, the effect of the judgment was that an *individual* (Pinochet) alleged to have committed torture should face trial in a Spanish court, despite a claim of Head of State immunity whilst, in the latter, the majority of the European Court of Human Rights concluded that the defence of *State* immunity was sufficient to prevent the English courts from exercising jurisdiction in a civil case against a State (Kuwait) for alleged torture.

208 The ILC was established following the adoption by the United Nations General Assembly of Resolution 174 (II) (21 November 1947) UN Doc A/519, which approved the Statute of the International Law Commission (ILC Statute). Article 1(1) of the ILC Statute provides that the objects of the ILC are the 'promotion of the progressive development of international law and its codification.'

209 Fifth Report on State Responsibility, Roberto Ago, Special Rapporteur [1976] 2 *Yearbook of the International Law Commission* Part I, Chapter III, sections 1–4, pp. 3 *et seq.*, UN Doc A/CN.4/291 and Add. 1–2.

210 Report of the International Law Commission to the General Assembly on its work of the thirty-second session, [1980] 2 *Yearbook of the International Law Commission* Part II, 32, UN Doc A/CN.4/SER.A/1980/Add.l (Part 2).

of the international obligation that had been breached. An 'international crime' related to a 'limited category comprising particularly serious wrongs'[211] and was defined as:[212]

> [a]n internationally wrongful act which results from the breach by a State of an international obligation so essential for the protection of fundamental interests of the international community that its breach is recognized as a crime by that community as a whole

The draft article then specified a non-exhaustive list of acts that may constitute an international crime, which included:[213]

> a serious breach of an international obligation of essential importance for the safeguarding and preservation of the human environment, such as those prohibiting massive pollution of the atmosphere or of the seas

The other specified acts involved serious breaches of what can be regarded as peremptory rules or *jus cogens* norms of international law,[214] thus indicating the gravity accorded acts involving the destruction of the environment referred to in draft article 19(3)(d).[215] The ILC referred on several occasions to the concept of *jus cogens* norms of international law, and to article 53 of the VCLT, in explaining why it proposed the separate regimes of international responsibility in draft article 19.[216] By contrast, draft article 19(4) dealt with the failure to fulfil 'an obligation of lesser and less general importance',[217] and provided that '[a]ny internationally wrongful act which is not an international crime in accordance with paragraph 2, constitutes an international delict'. Supplementing draft article 19, draft articles 52 and 53 provided for the consequences of the commission by a State of an international crime, including the possibility of collective sanctions.

The question of whether States could be criminally responsible had also been raised by various commentators,[218] and draft article 19 gained some support

211 Commentary by the International Law Commission on Article 19 'Articles on State Responsibility for Internationally Wrongful Acts' [1976] 2 *Yearbook of the International Law Commission* Part II, 96, 97, UN Doc A/CN.4/SER.A/1976/Add.l (Part 2).
212 Draft article 19(2).
213 Draft article 19(3)(d).
214 See Article 53 of the Vienna Convention on the Law of Treaties (opened for signature 23 May 1969, 1155 UNTS 331; 8 ILM 679 (entered into force 27 January 1980)) (VCLT).
215 See draft article 19(3)(a), (b) and (c).
216 See, for example, paragraphs 16–18 of Commentary by the International Law Commission on Article 19 'Articles on State Responsibility for International Wrongfully Acts' [1976] 2 *Yearbook of the International Law Commission* Part II, 96, 102, UN Doc A/CN.4/SER.A/1976/Add.l (Part 2).
217 *Ibid.*
218 See, for example, Ian Brownlie, *International Law and the Use of Force by States* (1963), 154.

at the time, mainly from developing and Eastern European States. In its commentary on the draft article, the ILC noted that:[219]

> [c]ontemporary international law has reached the point of condemning outright the practice of certain States in ... acting ... gravely to endanger the preservation and conservation of the human environment ... [T]hese acts genuinely constitute "international crimes", that is to say international wrongs which are more serious than others and which as such, should entail more severe legal consequences

On the other hand, draft article 19 provoked controversy among some other States, as well as commentators,[220] and various members of the ILC itself, who argued that it promoted the idea of collective responsibility of the entire population of a State for the actions of their leaders, as well as the notion of collective punishment.[221] In the end, draft article 19 (and draft articles 52 and 53) was not included in the version of the draft Articles adopted by the ILC and noted by the General Assembly in 2001.[222]

Indeed, as confirmed by the International Court of Justice in the *Application of the Genocide Convention Case*, the concept of the international *criminal* responsibility of a State does not currently represent either customary international law or the general view and practice of States. This deferential view that international law takes with respect to State sovereignty is reflected in the Joint Declaration of Judges Shi and Koroma in that case, where they affirm that:[223]

> [The Genocide Convention] does not impose criminal responsibility on the State as a State. Indeed, it could not have done so at the time it was adopted given that the notion of crime of State was not part of international law and *even today general international law does not recognize the notion of the criminal responsibility of the State*

---

219 Commentary by the International Law Commission on Article 19 'Articles on State Responsibility for Internationally Wrongful Acts' [1976] 2 *Yearbook of the International Law Commission* Part II, 96, 109, UN Doc A/CN.4/SER.A/1976/Add.l (Part 2).

220 See, for example, Giorgio Gaja, 'Should All References to International Crimes Disappear from the ILC Draft Articles on State Responsibility?' (1999) *European Journal of International Law* 365; Alain Pellet, 'Can a State Commit a Crime? Definitely, Yes!' (1999) *European Journal of International Law* 425.

221 D.J. Harris, *Cases and Materials on International Law* (5th ed, 1998) 489, referring to the comments of Mr Rosenstock, the United States member of the ILC, reported at (1995) 89 *American Journal of International Law* 390, 393.

222 United Nations General Assembly Resolution 56/83 (12 December 2001) UN Doc A/RES/56/83 on the Responsibility of States for Internationally Wrongful Acts.

223 Joint Declaration of Judges Shi and Koroma, *Application of the Genocide Convention Case* [2007] ICJ Rep 279, para 1 (emphasis added).

Similar views had earlier been expressed by the Appeals Chamber of the ICTY in the *Blaskic* case, when it stated that '[u]nder present international law it is clear that States, by definition, cannot be the subject of criminal sanctions akin to those provided for in national criminal systems'.[224]

However, notwithstanding these statements, it can be argued that the sentiments enunciated in draft article 19(3)(d) may eventually reflect an emerging trend in relation to the law on environmental damage resulting from deliberate State policy. Articles 40 and 41 of the draft articles that were ultimately adopted by the ILC in 2001 are expressed to apply in the case of a serious breach by a State of 'an obligation arising under a peremptory norm of general international law.'[225] These provisions and the associated ILC commentary make no specific reference to damage to the environment, although the commentary does not purport to be exhaustive.[226] However, it is noteworthy to observe that, at its 65th session in 2013, the ILC decided to include the topic of 'Protection of the Environment in relation to Armed Conflicts' in its program of work, on the basis of the recommendations contained in a working group report.[227]

Notwithstanding the absence of a concept of *criminal* responsibility for States for intentional environmental destruction during armed conflict, there are various other enforcement mechanisms that have been instituted. Following the environmental damage occasioned in both Kuwait and Saudi Arabia by the Iraqi regime in the period during and immediately following the invasion of Kuwait, the United Nations Security Council passed a Resolution which, in part, provided that Iraq was 'liable under international law for any direct loss, damage – including environmental damage and the depletion of natural resources'.[228] Iraq eventually accepted, although under protest, the terms of that Resolution.[229] A compensation fund, administered by the United Nations Compensation

224 Judgement on the Request of the Republic of Croatia for Review of the Decision of Trial Chamber II of 18 July 1997, *Prosecutor v. Tihomir Blaskic*, Case No. IT-95-14-AR, Appeals Chamber, 29 October 1997, para 25.

225 International Law Commission, Draft Articles on Responsibility of States for International Wrongfully Acts contained in Report of the International Law Commission to the General Assembly on its work of the fifty-third session, [2001] 2:2 *Yearbook of the International Law Commission* 1, UN Doc A/CN.4/SER.A/2001/Add.1 (Part 2), 112.

226 Philippe Sands, *Principles of International Environmental Law* (2nd ed, 2005), 895.

227 International Law Commission, 'Protection of the environment in relation to armed conflicts' (5 June 2013) <www.un.org/law/ilc/index.html> at 19 January 2015. For further background, see Report of the International Law Commission to the General Assembly on its work of the sixty-third session, [2011], General Assembly Official Records, Sixty-sixth Session, Supplement No. 10, UN Doc A/66/10, Annex E, 'Protection of the environment in relation to armed conflicts'.

228 United Nations Security Council Resolution 687 (3 April 1991) UN Doc S/RES/687/1991, para 16.

229 See Koppe, *supra* note 13, 247 and the corresponding footnotes.

Commission (UNCC),[230] was subsequently established, and is still operating, to deal with *civil* claims for damage caused as a direct result of Iraq's invasion and subsequent occupation of Kuwait.[231]

Thus, although there may exist the possibility of legal redress against a State for intentional environmental destruction during armed conflict, this has not been extended to include criminal responsibility at the international level.

1.6.2. INDIVIDUAL CRIMINAL RESPONSIBILITY

International criminal law clearly applies to individuals. The judgment of the Nuremberg International Military Tribunal represents the traditional view, rejecting as it did the argument that individuals could not be held responsible for 'acts of state'.[232] In an oft-quoted passage, the Tribunal stated that, although:[233]

> international law imposes duties and liabilities upon individuals as well as upon States ... [c]rimes against international law are committed by men, not abstract entities, and only by punishing individuals who commit such crimes can the provisions of international law be enforced

This view was also repeated in various war crimes trials conducted by the United States Military Tribunal in Nuremberg.[234] The notion of individual criminal responsibility for international crimes is also reflected in the jurisdictional mandates of all subsequently established international(ised) criminal tribunals. These courts are not designed to prosecute non-natural entities, particularly States. The ICC, for example, only has jurisdiction over 'natural persons'.[235] Given that international criminal law is not in the position to hold a State culpable for such acts, this book will focus on how individuals, who themselves bear a responsibility towards the environment,[236] are regulated for acts of intentional environmental damage during armed conflict.

---

230 United Nations Security Council Resolution 687 (3 April 1991) UN Doc S/RES/687/1991 on Iraq-Kuwait, para 18.

231 United Nations Security Council Resolution 692 (20 May 1991) UN Doc S/RES/692/1991 on Iraq-Kuwait, para 3. See also *supra*, note 57.

232 Nina H.B. Jørgensen, *The Responsibility of States for International Crimes* (2000), 19.

233 *Judgment of the International Military Tribunal, Trial of the Major War Criminals* 1 October 1946, reprinted in (1947) 41 *American Journal of International Law* 172, 221.

234 See, for example, *United States v. Flick and Others* (Judgment), United States Military Tribunal, Nuremberg, (1949) IX War Crimes Reports 1, 18: *United States. v. Krupp and Others* (Judgment), United States Military Tribunal, Nuremberg, (1949) X War Crimes Reports 69, 150.

235 Rome Statute, article 25(1).

236 Patricia W. Birnie and Alan E. Boyle, *International Law and the Environment* (2nd ed, 2002) 282.

In this regard, there is the need to affirm such criminality at the international level, particularly in light of the 'traditional reluctance' of States to prosecute persons accused of international crimes within their domestic legal systems.[237] Given the well-accepted presumption that international crimes represent an affront to the community as a whole, it is argued in this book that intentionally targeting the environment during armed conflict can in certain circumstances meet this level of abhorrence. Warfare has been shown to have contributed to the large scale depletion of natural resources and serious degradation of the environment.[238] The international legal regime must therefore adequately provide for accountability in relation to such acts.

Seen in this context, it will be argued that the most appropriate mechanism of international criminal justice to prosecute such acts is the permanent international criminal court, the ICC, acting in accordance with the terms of the Rome Statute. Moreover, this may also 'incentivise' States – at least those that have ratified the Rome Statute – to 'upgrade' their domestic laws, in order to ensure that their respective national courts would have jurisdiction to deal with any alleged act that constituted an international crime within the mandate of the Court, thus minimising the risk of falling within the 'unable' criteria for complementarity set out in article 17 of the Rome Statute.[239]

Although there are, undoubtedly, some uncertainties as to precisely how aspects of the complementarity principle may apply in practice, it does represent a safeguard for those States who would otherwise be concerned that its nationals would face trial before the Court.[240] Even though a number of State Parties to the Rome Statute have yet to pass appropriate implementing legislation, as the international system of criminal justice evolves and the acceptance of the ICC broadens, national governments will come under increasing pressure to recognise the international crimes within their own legal systems. As has been suggested, the Rome Statute itself plays 'a powerful trendsetting role'[241] in articulating and shaping the actions of States.

---

237 Antonio Cassese, 'The Role of Internationalized Courts and Tribunals in the Fight Against International Criminality', in Cesare P.R. Romano, André Nollkaemper and Jann K. Kleffner (eds), *Internationalized Criminal Courts: Sierra Leone, East Timor, Kosovo, and Cambodia* (2004), 3, 3.

238 Sands, *supra* note 226, 308.

239 Under the system of complementarity established by the Rome Statute, a case is inadmissible *inter alia* where it 'is being investigated or prosecuted by a State which has jurisdiction over it, unless that State is unwilling or unable genuinely to carry out the investigation or prosecution': Rome Statute, article 17(1)(a). See also chapter 4.

240 See Alexis Goh and Steven Freeland, 'Australia and the International Criminal Court', in Gabriel A. Moens and Rudolphe Biffot (eds), *The Convergence of Legal Systems in the 21st Century – An Australian Approach* (2002), 285.

241 Mark A. Drumbl, *Reimagining Child Soldiers in International Law and Policy* (2012), 121.

1.7. KEY QUESTIONS

Following on from the observations above, the following chapters of this book focus on the following key questions:

(a) To what extent do treaty and customary principles of international law currently address the intentional destruction of the environment during armed conflict?
(b) Is the current position sufficient to allow for criminal accountability for such acts in appropriate circumstances?
(c) What function should the mechanisms of justice under international criminal law, in particular the ICC, play in addressing this issue?
(d) Is it therefore necessary to amend the Rome Statute in order to more appropriately deal with the intentional destruction of the environment during armed conflict?[242]

---

[242] As suggested in chapter 5, such actions might be termed as 'crimes against the environment', an expression that has previously been used by some commentators: see, for example, Stephen C. McCaffery, 'Crimes Against the Environment' (1986) 1 *International Criminal Law* 541.

CHAPTER TWO

REGULATION OF THE INTENTIONAL DESTRUCTION OF THE ENVIRONMENT DURING WARFARE UNDER TREATY LAW

2.1. INTRODUCTION – TREATIES UNDER INTERNATIONAL LAW

The previous chapter demonstrated the imperative for appropriate mechanisms of accountability to address the intentional destruction of the environment during armed conflict. This requires an analysis of the current treaty provisions, in order to assess whether they are sufficient and adequate in this regard. Most of the relevant conventional principles that do apply are specified in the treaties that set out the basic norms of international humanitarian law – the *jus in bello* ('rules in war'). There may also be other applicable general principles in a number of multilateral international environmental law instruments. The development of both of these categories of treaties coincides with two trends – the advances in military technology on the one hand, and the rise of environmental consciousness on the other.[1] An examination of these treaties provides a collective framework upon which a subsequent analysis of the Rome Statute of the International Criminal Court[2] – can (at least partially) be based.

It is first important to determine the nature of these instruments. The Vienna Convention on the Law of Treaties defines a treaty for the purposes of that instrument as follows:[3]

1 Christopher D. Stone, 'The Environment in Wartime: An Overview', in Jay E. Austin and Carl E. Bruch (eds), *The Environmental Consequences of War: Legal, Economic, and Scientific Perspectives* (2000), 16, 17–8.

2 Opened for signature 17 July 1998, 2187 UNTS 3; 37 ILM 999 (entered into force 1 July 2002) (Rome Statute).

3 Vienna Convention on the Law of Treaties, opened for signature 23 May 1969, 1155 UNTS 331; 8 ILM 679 (entered into force 27 January 1980) (VCLT), article 2(1)(a).

> an international agreement concluded between States in written form and governed by international law, whether embodied in a single instrument or in two or more related instruments and whatever its particular designation

Treaties are binding on States Parties, who are under an obligation to perform the terms of the relevant treaty in good faith in accordance with the principle of *pacta sunt servanda*.[4] The definition of a treaty in the VCLT does not necessarily encompass the complete range of international treaties, which also include those concluded by 'non-States'. For example, an international organisation like the United Nations can enter into treaties, since it possesses 'a large measure of international personality and the capacity to operate upon an international plane'.[5] However, for the purposes of this chapter, the definition in the VCLT suffices to describe the *jus in bello* and international environmental instruments that are discussed.

Treaties and custom represent the two principal 'sources' of international law.[6] The other sources, as set out in article 38(1) of the Statute of the International Court of Justice, are general principles of law, and judicial decisions and the writings of the most highly qualified publicists.[7] This book will focus primarily on international law principles under treaties (this chapter) or by virtue of customary international law (chapter 3). With regard to general principles of law, as discussed in chapter 1, there are no relevant and sufficiently widely accepted principles to be found within the national laws of States as to constitute a comprehensive 'source' of international law dealing with the intentional destruction of the environment during warfare. In the course of this analysis, the decisions of international(ised) criminal tribunals and the International Court of Justice, as well as other judicial decisions, will also be considered to the extent they are relevant.

---

4 *Ibid*, article 26.

5 *Reparation for Injuries Suffered in the Service of the United Nations* (Advisory Opinion) [1949] ICJ Rep 174, 179. See also Vienna Convention on the Law of Treaties between States and International Organizations or between International Organizations, opened for signature 21 March 1986 25 ILM 543 (not in force), preamble para 1.

6 It is generally asserted that article 38(1) of the Statute of the International Court of Justice, opened for signature 26 June 1945, 1 UNTS 16, 59 Stat. 1031 (entered into force 24 October 1945) (ICJ Statute), lists the so-called 'sources' of international law: see, for example, Georg Schwarzenberger, *International Law* (3rd ed, Vol 1, 1957), 21–22; Antonio Cassese, *International Law* (2nd ed, 2005), 156.

7 It should be noted that articles 21(1) and (2) of the Rome Statute codify the 'Applicable law' relevant to the judicial activities of the International Criminal Court.

2.2. INTERNATIONAL HUMANITARIAN LAW (*JUS IN BELLO*) – RELEVANT TREATY PROVISIONS

One (among many) tragic aspect of armed conflict is that innocent civilians are often (and increasingly) the victims. In 1900, it was estimated that civilians represented approximately five per cent of the casualties of conflict. By the time of the Second World War, this figure had risen to approximately 65 per cent. Human rights groups now calculate that approximately 90 per cent of all casualties in more recent armed conflicts[8] are made up of civilians, of which 40 per cent are children.[9] Such casualties arise through direct injury or death caused by military action, as well as other causes, including the destruction of the environment, which will often have immediate and longer-term effects on the security of human populations.

Over time, the disastrous consequences of armed conflict led to a general consensus that legal rules should be introduced and implemented in an effort to alleviate human suffering during hostilities. This process has rather astutely been described as:[10]

> a concerted attempt on the part of nations and individuals to bring the tools of rational analysis to bear upon a field of endeavour that often appears to be incorrigibly rooted in irrationality

This has seen the emergence of various legal principles limiting the methods and means of warfare, and prescribing the rights and protections both of civilians and non-civilians in times of hostilities, a distinction lying 'at the very heart of the law of armed conflict'.[11] These rules are regarded as 'the most essential' of 'all of the law of nations', in view of the broad and long-lasting effects of warfare.[12] The laws and customs of war had their origins in the customary practices of armies on the battlefield and have existed in various forms almost

8 Kristin Dorman, 'Proportionality and Distinction in the International Criminal Tribunal for the Former Yugoslavia' (2005) 12 *Australian International Law Journal* 83, 83.

9 Stephanie H. Bald, 'Searching for a Lost Childhood: Will the Special Court of Sierra Leone Find Justice for Its Children?' (2002) 18 *American University International Law Review* 537, 543, citing 'Women and Children Bear Brunt of War' *Saigon Times Daily* (Ho Chi Minh City), 3 May 2002. In the period 1986–1996, for example, it has been estimated that as many as two million children alone were killed in armed conflict: Nienke Grossman, 'Rehabilitation or Revenge: Prosecuting Child Soldiers for Human Rights Violations' (2007) 38:2 *Georgetown Journal of International Law* 323, 359.

10 Peter J. Richards and Michael N. Schmitt, 'Mars Meets Mother Nature: Protecting the Environment During Armed Conflict' (1999) 28 *Stetson Law Review* 1047, 1047.

11 W. Michael Reisman, 'Editorial Comment: Holding the Center of the Law of Armed Conflict' (2006) 100 *American Journal of International Law* 852, 856.

12 Dana Zartner Falstrom, 'Can International Law Survive the 21st Century? Yes, with Patience, Persistence and a Peek at the Past' (2007) 8 *San Diego International Law Journal* 291, 326. That

since antiquity.[13] Even in ancient civilizations, examples can be found of prohibitions on such things as the poisoning of water sources and the razing of forests.[14]

The first significant attempt to codify the laws of war came in 1863, when American President Lincoln issued a General Order authorising the so-called 'Lieber Instructions' or 'Lieber Code',[15] which introduced a set of legal guidelines to be applied at the time of the American Civil War. Although these rules were only binding on United States military forces, they corresponded with several important general aspects of the laws of war that existed at the time.[16] Following on from the Lieber Code, a number of other States drafted military manuals during the second half of the 19th Century.[17]

In 1864, the Geneva Convention for the Amelioration of the Condition of the Wounded in Armies in the Field was concluded, following a conference convened by the Swiss Federal Council.[18] This was the first international agreement that purported to specify rules of warfare, and was in part inspired by Henri Dunant's account of what he had witnessed on the Solferino battlefield.[19] As well as specifying principles relating to the relief for wounded,[20] the neutrality of medical personnel,[21] and the special status of the sign of the 'red cross on a white ground',[22] the 1864 Geneva Convention laid the groundwork for more detailed multilateral treaties on the laws of war.[23]

---

author opines (at 340) that war is sometimes 'necessary' to achieve the goals of international law – to 'uphold the peace and stability of the entire globe'.

13 For a discussion of the historical evolution of the regulation of armed conflict going back some 5,000 years, see M. Cherif Bassiouni (ed), *A Manual on International Humanitarian Law and Arms Control Agreements* (2000), 5–15.

14 Royal Swedish Ministry for Foreign Affairs, 'Conventional Weapons, their Deployment and Effects from a Humanitarian Aspect: Recommendations for the Modernization of International Law' (1973), 11.

15 Instructions for the Government of Armies of the United States in the Field, promulgated as General Orders No. 100, 24 April 1863.

16 Dietrich Schindler and Jiri Toman (eds), *The Law of Armed Conflicts* (4th ed, 2004), 3.

17 These included Prussia (1870), The Netherlands (1871), France (1877), Russia (1877 and 1904), Serbia (1878), Argentina (1881), Great Britain (1883 and 1904) and Spain (1893): see Michael A. Newton, Symposium, 'Modern Military Necessity: The Role and Relevance of Military Lawyers' (2007) 12 *Roger Williams University Law Review* 877, 886 and the corresponding footnote.

18 Opened for signature 22 August 1864, 18 Martens (1st) 607 (entered into force 22 June 1865) (1864 Geneva Convention).

19 See Henry Dunant, *A Memory of Solferino* (reprinted 1986).

20 1864 Geneva Convention, article 6.

21 *Ibid*, articles 1 and 2.

22 *Ibid*, article 7.

23 Schindler and Toman, *supra* note 16, vi. Traditionally, the law regulating the methods and means of warfare was referred to as the 'law of war' (*jus in bello*). It has subsequently also come to be known as the 'law of armed conflict.' Non-military organizations, in particular

Over time, important conventional norms – some of which also now represent rules of customary international law[24] – have emerged to regulate a broader range of military activities conducted during the course of armed conflict. There are now more than 100 instruments that comprise the *jus in bello*.[25] The scope of these norms extends beyond traditional State combatants, applying to non-State actors and even, in certain circumstances, to United Nations Peacekeeping Forces.[26] As a general observation, however, there remains significant tension between those restrictions imposed on the methods and means of warfare and the military goals sought through such acts, even though it has been argued that military considerations should not be the sole, or indeed primary, determining factor in how warfare is to be conducted.[27]

In its early codification, the *jus in bello* rules focused principally on the protection of medical personnel and the prohibition of certain types of weapons.[28] However, even in the absence of provisions limiting a specific aspect of warfare, the evolution of the *jus in bello* stemmed largely from a recognition of fundamental humanitarian values of a 'social, religious, political, moral, military and scientific' nature.[29] Indeed, Jean Pictet has defined international humanitarian law as:[30]

> a branch of public international law which owes its inspiration to a feeling for humanity and which is centered on the protection of the individual

---

the International Committee of the Red Cross, prefer the name 'international humanitarian law.' In the more recent *jus in bello* treaties, the term 'armed conflict' is used rather than 'war' or 'warfare': Michael H. Hoffman, 'The Legal Status And Responsibilities of Private Internet Users under the Law of Armed Conflict: A Primer for the Unwary on the Shape of Law to Come' (2003) 2 *Washington University Global Studies Law Review* 415, 417. These terms are used interchangeably in this book with no intended difference in meaning, unless otherwise expressly stated.

24 The relevant principles of customary international law are discussed in chapter 3.

25 Jaume Saura, 'Lawful Peacekeeping: Applicability of International Humanitarian Law to United Nations Peacekeeping Operations' (2007) 58 *Hastings Law Journal* 479, 488.

26 See Daphna Shraga, 'UN Peacekeeping Operations: Applicability of International Humanitarian Law and Responsibility for Operations-Related Damage' (2000) 94 *American Journal of International Law* 406.

27 See, for example, Richard A. Falk, *Law, Morality, and War in the Contemporary World* (1984), 9–12.

28 William A. Schabas, *Genocide in International Law* (2000), 15.

29 René Provost, *International Human Rights and Humanitarian Law* (2002), 5.

30 Jean Pictet, quoted in Yves Beigbeder, *Judging War Criminals: The Politics of International Justice* (1999), 1.

This is explicitly reflected in the Martens Clause,[31] which, as noted, first appeared in 1899, and which provided that the following broad principles were applicable to the conduct of warfare:[32]

> the principles of the law of nations, as they result from the usages established among civilized peoples, the laws of humanity, and the dictates of public conscience

There has been some conjecture over the years as to the precise significance of the Martens clause. Although it is probably the case that principles of humanity are not regarded as independent 'sources' of international law,[33] the International Court of Justice has indicated that they do express customary international law.[34] Theodor Meron suggests that the sentiments of the Martens Clause play an important role in the 'humanization of the law of war'.[35] The International Court of Justice has, on several occasions, emphasized 'elementary considerations of humanity' when interpreting principles of international law.[36]

In terms of its applicability to environmental damage, one commentator has suggested that '[i]n modern international law, there is no reason why [the principles included within the Martens clause] should not encompass environmental protection'.[37] At a minimum, it plays an important role in that, in terms of the conduct of warfare, even if an action is not expressly prohibited, it does not follow that it is 'necessarily permitted'.[38] Moreover, the Martens clause suggests the existence of a 'belief in common values',[39] by introducing a

---

31 Professor Friedrich (Fyodor) Martens was a delegate of Tsar Nicholas II at the Hague Peace Conferences of 1899 and 1907: Dieter Fleck (ed), *The Handbook of Humanitarian Law in Armed Conflicts* (1999), 28.

32 See, for example, the preamble of the Hague Convention IV Respecting the Laws and Customs of War, opened for signature 18 October 1907, 3 Martens (3rd) 461 (entered into force 26 January 1910).

33 Judgement, *Prosecutor v. Zoran Kupreskic, Mirjan Kupreskic, Vlatko Kupreskic, Drago Josipovic, Dragan Papic and Vladimir Santic*, Case No. IT-95-16-T, Trial Chamber, 14 January 2000, para 525. See also Knut Dörmann, *Elements of War Crimes under the Rome Statute of the International Criminal Court: Sources and Commentary* (2004), 168.

34 *Legality of the Threat or Use of Nuclear Weapons* (Advisory Opinion) [1996] ICJ Rep 226, para 84.

35 See Jochen A. Frowein, Book Review – Theodor Meron, 'The Humanization of International Law' (2007) 101 *American Journal of International Law* 680, 680.

36 See, for example, *Corfu Channel Case (United Kingdom of Great Britain and Northern Ireland v. Albania)* (Merits) (Judgment) [1949] ICJ Rep 4, 22; *Military and Paramilitary Activities in and against Nicaragua (Nicaragua v. United States of America)* (Merits) (Judgment) [1986] ICJ Rep 14, para 215.

37 Philippe Sands, *Principles of International Environmental Law* (2nd ed, 2005), 311.

38 Saura, *supra* note 25, 490. In relation to suggestions of an express permissive rule in general international law, see the decision of the Permanent Court of International Justice in *The Case of the S.S. 'Lotus' (France v. Turkey)* (Judgment) [1927] PCIJ Rep, Series A, No. 10, 18–9.

39 Dinah Shelton, 'Centennial Essay: In Honor of the 100th Anniversary of the AIJL and the ASIL: Normative Hierarchy in International Law' (2006) 100 *American Journal of International Law* 291, 295.

'public conscience' standard into the application of international humanitarian law to 'real-life' occurrences, although this may still be too vague to be of great assistance.[40] Yet, a consideration of several of the *jus in bello* treaties discussed below suggests that the rules of international humanitarian law are based on a balance between military necessity and principles of humanity.[41]

The central objectives of international humanitarian law are the protection of those affected by armed conflict, and the regulation of the conduct of hostilities. This reflects the need to establish a system of regulation to minimise the inevitable horrors rendered on 'people, property and *the environment*' by war.[42] The principles of international humanitarian law and the 'humanization that it brings to modern warfare' are now important elements in contemporary international law,[43] setting out minimum international standards that regulate *how, with what* and *against whom* armed hostilities can be conducted. This description is, of course, somewhat simplistic, but as has been noted by Judge Koroma, international humanitarian law is:[44]

> formulated and codified to restrict the use of various weapons and methods of warfare, [and] is intended to limit the terrible effects of war

In the same case, Judge Weeramantry explained the significance of international humanitarian law in the following way:[45]

> Humanitarian law and custom have a very ancient lineage. They reach back thousands of years. They were worked out in many civilizations ... many religious and philosophical ideas have been poured into the mould in which modern humanitarian law has been formed. They represented the effort of the human conscience to mitigate in some measure the brutalities and dreadful sufferings of war

The *jus in bello* rules are not directly concerned with the legality of a belligerent's resort to force – this is regulated by the *jus ad bellum* ('rules to war').[46] Rather,

---

40 Fleck, *supra* note 31, 29.

41 Anthony E. Cassimatis, 'International Humanitarian Law, International Human Rights Law, and Fragmentation of International Law' (2007) 56 *International and Comparative Law Quarterly* 623, 628–9.

42 L.C. Vohrah, Kelly Dawn Askin and Daryl A. Mundis, 'Contemporary Law Regulating Armed Conflict at Sea', in Nisuke Ando, Edward McWhinney and Rüdiger Wolfrum (eds), *Liber Amicorum: Judge Shigeru Oda* (Volume 2, 2002), 1523, 1541 (emphasis added).

43 Saura, *supra* note 25, 494.

44 Dissenting Opinion of Judge Koroma in *Legality of the Threat or Use of Nuclear Weapons* (Advisory Opinion) [1996] ICJ Rep 226, 558.

45 Dissenting Opinion of Judge Weeramantry in *Legality of the Threat or Use of Nuclear Weapons* (Advisory Opinion) [1996] ICJ Rep 226, 443.

46 The central core of the existing principles of the *jus ad bellum* is the prohibition of the threat or use of force, which is also reflected in article 2(4) of the United Nations Charter, opened

they deal with the methods and means employed during armed conflict, as well as the rights of those individuals involved in the hostilities. The Appeals Chamber of the International Criminal Tribunal for the former Yugoslavia (ICTY) has described international humanitarian law as:[47]

> a realistic body of law, grounded on the notion of effectiveness and inspired by the aim of deterring deviation from its standards to the maximum extent possible

However, even though the ICTY has provided a now well-accepted definition of what constitutes an armed conflict,[48] the precise nature of armed conflict continues to change, more particularly in recent decades, coinciding with the development of new and potentially devastating methods and means of warfare. For example, it is evident that recent armed conflicts differ from the type of war envisaged by the drafters of the 1949 Geneva Conventions[49] and the 1977 Additional Protocols,[50] not the least due to the much broader range of weaponry and technologically advanced 'hardware' and 'software' available for use by combatants,[51] as well as the general shift of the battlefield towards civilian

---

for signature 26 June 1945, 1 UNTS 16; 59 Stat. 1031 (entered into force 24 October 1945) (United Nations Charter). Although this book does not focus on the *jus ad bellum*, as noted in chapter 1, there have been suggestions made that an adaptation of those rules may emerge over time so as to permit a State to meet environmental threats to its security with force: Michael K. Murphy, 'Note: Achieving Economic Security with Swords as Ploughshares: The Modern Use of Force to Combat Environmental Degradation' (1999) 39 *Virginia Journal of International Law Association* 1181.

47 Judgement, *Prosecutor v Duško Tadić*, Case No. IT-94-1, Appeals Chamber, 15 July 1999, para 96.

48 See Decision on Defence Motion for Interlocutory Appeal on Jurisdiction, *Prosecutor v Duško Tadić*, Case No. IT-94-1, Appeals Chamber, 2 October 1995, para 70.

49 Geneva Convention for the Amelioration of the Condition of the Wounded and Sick in Armed Forces in the Field, opened for signature 12 August 1949, 75 UNTS 31 (entered into force 21 October 1950) (1949 Geneva Convention I), Geneva Convention for the Amelioration of the Condition of the Wounded, Sick and Shipwrecked Members of Armed Forces at Sea, opened for signature 12 August 1949, 75 UNTS 85 (entered into force 21 October 1950) (1949 Geneva Convention II), Geneva Convention Relative to the Treatment of Prisoners of War, opened for signature 12 August 1949, 75 UNTS 135 (entered into force 21 October 1950) (1949 Geneva Convention III), Geneva Convention Relative to the Protection of Civilian Persons in Time of War, opened for signature 12 August 1949, 75 UNTS 287 (entered into force 21 October 1950) (1949 Geneva Convention IV).

50 Protocol I Additional to the Geneva Conventions of August 12, 1949, and relating to the Protection of Victims of International Armed Conflicts, opened for signature 8 June 1977, 1125 UNTS 3, 16 ILM 1391 (entered into force 7 December 1978) (1977 Additional Protocol I) and Protocol II Additional to the Geneva Conventions of August 12, 1949, and relating to the Protection of Victims of Non-International Armed Conflicts, opened for signature 8 June 1977, 1125 UNTS 609, 16 ILM 1442 (entered into force 7 December 1978) (1977 Additional Protocol II).

51 The *jus in bello* treaties place a responsibility upon States to determine whether the use of a 'new' weapon represents a violation of applicable international law: see 1977 Additional Protocol I, article 36. See also James D. Fry, 'Contextualized Legal Reviews for the Methods

population centres. Additionally, many contemporary conflicts now involve non-State actors as belligerents.

As a result, the long-term threats and potential for death and destruction from armed conflict are now significantly graver than previously. It therefore follows that the minimum standards of what might once have been regarded as 'acceptable' behaviour by belligerents during armed conflict may also have changed, thus requiring closer scrutiny and more comprehensive legal regulation. To meet (some of) these challenges, the scope of the *jus in bello* rules has, over time, expanded on the questions of protection of property and of people and, more recently, express reference has also been made to the protection of the environment, at least in certain circumstances.[52] As Judge Lachs of the International Court of Justice has observed:[53]

> the great acceleration of social and economic change, combined with that of *science and technology*, have confronted law with a serious challenge: one it must meet, lest it lag even farther behind events than it has been wont to do.

It is therefore important to note that the interpretation of the principles outlined in the *jus in bello* treaties, as well as those customary international law principles (see chapter 3), are not to be regarded as 'static'. This is particularly so given that the methods, means, and underlying context of warfare are constantly changing and, even more broadly, that armed conflicts continue to occur despite any efforts by the broader international community to prevent them.[54]

As a consequence, the applicable law must adapt to meet changing circumstances. The mere fact, for example, that weapons technology has developed in ways that may not have been contemplated by the drafters of the relevant instruments does not necessarily mean that the principles of international humanitarian law are not applicable to their use. As the International Court of Justice noted in the *Legality of the Threat or Use of Nuclear Weapons* Advisory Opinion, adopting the written statement submitted by New Zealand:[55]

---

and Means of Warfare: Cave Combat and International Humanitarian Law' (2006) 44 *Columbia Journal of Transnational Law* 453, 468–90.

52 David Hunter, James Salzman and Durwood Zaelke, *International Environmental Law and Policy* (2nd ed, 2002), 1379.

53 Dissenting Opinion of Judge Lachs, *North Sea Continental Shelf Cases (Federal Republic of Germany v. Denmark and Federal Republic of Germany v. The Netherlands)* [1969] ICJ Rep 3, 230 (emphasis added).

54 Provost, *supra* note 29, 1.

55 *Legality of the Threat or Use of Nuclear Weapons* (Advisory Opinion) [1996] ICJ Rep 226, para. 86. One specific area where weapons technology has developed very significantly in recent years involves the military uses of outer space: see Steven Freeland, 'The Applicability of the *Jus in Bello* Rules of International Humanitarian Law to the Use of Outer Space' (2006) 49 *Proceedings of the Colloquium on the Law of Outer Space* 338; Michel Bourbonièrre, 'The

> international humanitarian law has evolved to meet contemporary circumstances, and is not limited in its application to weaponry of an earlier time. The fundamental principles of this law endure: to mitigate and circumscribe the cruelty of war for humanitarian reasons

In this way, the conventional *jus in bello* principles are, where appropriate, to be regarded as 'by definition evolutionary', as that phrase has been used by the International Court of Justice.[56] Indeed, the Court noted that:[57]

> an international instrument has to be interpreted and applied within the framework of the entire legal system prevailing at the time of the interpretation

On the other hand, the VCLT sets out clear principles for the interpretation of a treaty. The general rule of interpretation is set out in article 31, which provides *inter alia* that:[58]

> [a] treaty shall be interpreted in good faith in accordance with the ordinary meaning to be given to the terms of the treaty in their context and in the light of its object and purpose

These requirements are cumulative in nature and therefore should not be read in isolation.[59] The International Court of Justice has on several occasions confirmed that both article 31 and article 32 of the VCLT reflect customary international law,[60] and thus the rules that they set out might also apply to treaties that came into force prior to 1980.[61] Although the meaning of article 31(1) 'could conceivably change as a result of the practice of states with

---

Ambit of the Law of Neutrality and Space Security' (2006) 36 *Israel Yearbook on Human Rights* 205, 211.

56 *Legal Consequences for States of the Continued Presence of South Africa in Namibia (South West Africa) notwithstanding Security Council Resolution 276 (1970)* Advisory Opinion [1971] ICJ Rep 16, para 53.

57 *Id.*

58 VCLT, article 31(1).

59 James Harrison, 'The role of international conventions in solving transboundary pollution disputes', in Michael Faure and Song Ying (eds), *China and International Environmental Liability: Legal Remedies for Transboundary Pollution* (2008), 38, 41.

60 See, for example, *Territorial Dispute (Libyan Arab Jamahiriya v. Chad)* (Judgment) [1994] ICJ Rep 6, para 41; *Maritime Delimitation and Territorial Questions (Qatar v. Bahrain)* (Judgment) [1995] ICJ Rep 6, para 33; Case Concerning Oil Platforms (*Islamic Republic of Iran v. United States of America)* (Preliminary Objections, Judgment) [1996] ICJ Rep 803, para 23; *Legal Consequences of the Construction of a Wall in the Occupied Palestinian Territory* (Advisory Opinion) [2004] ICJ Rep 136, para 94.

61 The VCLT entered into force on 27 January 1980 and does not generally have a retroactive effect (article 4). As a consequence, it is generally regarded that the terms of the VCLT can, strictly speaking, only be applied to treaties that themselves come into force after that date.

regard to treaty interpretation',[62] in the absence of ambiguity in the terms of a treaty, it would be inappropriate to 'read into' a particular treaty rules relating to the intention to destroy the environment during armed conflict, if such rules do not appear from an interpretation based on the customary principles codified in the VCLT.[63]

This is so notwithstanding that it may be 'desirable' to infer rules into the applicable conventional provisions. This principle of interpretation was reaffirmed by Sir Percy Spender and Sir Gerald Fitzmaurice, when commenting on the decision of the International Court of Justice in the *International Status of South-West Africa* Advisory Opinion,[64] as follows:[65]

> [t]he principle of interpretation directed to giving provisions their maximum effect cannot legitimately be employed in order to introduce what would amount to a revision of those provisions

Thus, even though the meaning of a particular treaty may not necessarily be 'fixed' at the time of its conclusion[66] if, in the absence of ambiguity, the scope of a *jus in bello* treaty provision is clear in accordance with the well accepted principles of interpretation, then broader protections or prohibitions cannot simply be 'added' to the meaning of those provisions. This must be borne in mind when considering the treaties discussed below.

Another important issue is that of nomenclature. The earlier *jus in bello* treaties often refer to acts committed during 'war' or 'warfare'. As noted above, more recent treaties, particularly those concluded since the end of the Second World War, typically refer to 'armed conflict' or, occasionally, 'hostilities'. One reason for this is that a view gradually emerged in the first part of the 20th Century that

---

62 Michael Byers and Simon Chesterman, 'Changing the rules about rules? Unilateral humanitarian intervention and the future of international law', in J.L. Holzgrefe and Robert O. Keohane (eds), *Humanitarian Intervention: Ethical, Legal, and Political Dilemmas* (2003), 177, 187.

63 By way of illustration, in the *Interpretation of Peace Treaties Case* Advisory Opinion, the International Court of Justice refused to 'rewrite' a series of three peace treaties between the Allied Powers and Bulgaria, Hungary and Romania in such a way as to override the clear meaning of the text, affirming that '[i]t is the duty of the Court to interpret the Treaties, not to revise them': *Interpretation of Peace Treaties Case* (second phase) (Advisory Opinion) [1950] ICJ Rep 221, 229.

64 [1950] ICJ Rep 128.

65 As quoted in William A. Schabas, 'The Crime of Genocide in the Jurisprudence of the International Criminal Tribunals for the Former Yugoslavia and Rwanda', in Horst Fischer, Claus Kress and Sascha Rolf Lüder (eds), *International and National Prosecution of Crimes Under International Law: Current Developments* (2001), 447, 451.

66 See, for example, *Legal Consequences for States of the Continued Presence of South Africa in Namibia (South West Africa) notwithstanding Security Council Resolution 276 (1970)* (Advisory Opinion) [1971] ICJ Rep 16, para 53.

'war' was unlawful under international law. Indeed, in 1928, a highly symbolic instrument, usually referred to either as the 'Pact of Paris' or the 'Kellogg-Briand Pact',[67] renounced war 'as an instrument of national policy in their relations with one another.'[68]

Although this was a positive development in relation to the (il)legality of war, it became clear that it was an unrealistic and ineffective declaration.[69] Various States subsequently asserted that they would be exempted from the prohibitions when involved in a 'conflict' rather than a war,[70] leading to disagreement among commentators as to whether they applied to what came to be categorized as 'force short of war'.[71]

The illegality of war was subsequently reinforced by the adoption of the United Nations Charter, which expresses the determination of Members to 'save succeeding generations from the scourge of war'.[72] Moreover, uncertainties regarding the scope of the Kellogg-Briand Pact were resolved by the express prohibition of the threat or use of force specified in article 2(4) of the United Nations Charter, which provided for a more comprehensive conventional prohibition than had previously existed ('force' as opposed to 'war'). Nevertheless, the *jus in bello* rules are still sometimes referred to as the 'laws of war',[73] although as noted above, the expression 'international humanitarian law', is now generally preferred, since it highlights that the rules apply not just to States, but are also designed to protect individuals.[74]

With respect to the issue of intentional environmental destruction during *the course of* warfare, the most important of the *jus in bello* treaties are the Hague Conventions of 1899 and 1907, the four Geneva Conventions of 1949,

---

67 These were the names of the United States Secretary of State and the French Foreign Minister respectively who were involved in the negotiation of the treaty.

68 Treaty Providing for the Renunciation of War as an Instrument of National Policy (Kellog-Briand Pact), opened for signature 27 August 1928, 94 LNTS 57 (entered into force 24 July 1929), article I.

69 Beigbeder, *supra* note 30, 11.

70 Alex Conte, *Security in the 21st Century: The United Nations, Afghanistan and Iraq* (2005), 3.

71 See, for example, D.J. Harris, *Cases and Materials on International Law* (6th ed, 2004), 886–9 and the corresponding references.

72 United Nations Charter, preamble para 1.

73 See, for example, Adam Roberts and Richard Guelff, *Documents on the Laws of War* (3rd ed, 2005). However, the word 'war' is now used in many contexts, with recent usages including the 'war on drugs' and the 'global war on terror'. These might simply be examples of what could be regarded as poetic licence or, instead, may indicate that the definition of 'war' is expanding. Most commentators suggest that the term 'war' has a specific legal meaning: see, for example, Yoram Dinstein, *War, Aggression and Self-Defence* (3rd ed, 2001), 3. In any event, the international instruments that have been concluded in relation to these non-traditional 'wars' are not generally regarded as part of the *jus in bello*.

74 Michael Byers, *War Law: International Law and Armed Conflict* (2005), 9.

the Convention on the Prohibition of Military or Any Other Hostile Use of Environmental Modification Techniques of 1976,[75] and the Additional Protocols I and II to the Geneva Conventions of 1977. The importance of the obligations arising under The Hague and Geneva instruments has been reaffirmed by the United Nations Security Council in a number resolutions relating to the protection of civilians in armed conflict.[76]

While there are only a relatively small number of directly relevant treaty provisions, other conventional norms applicable in times of armed conflict, although not specifically directed towards environmental damage, may be pertinent in particular circumstances. In addition, several other treaties prohibit resort to certain weapons, the use of which may be highly destructive of the environment. The various *jus in bello* treaties that might be relevant to the issue of intentional destruction of the environment during warfare are discussed below in chronological order.

2.2.1. 1868 ST. PETERSBURG DECLARATION[77]

This important landmark in the evolution of international humanitarian law was the first modern treaty that sought to limit the type of weapons to be used during warfare.[78] It specified for the first time in an international instrument that 'the necessities of war ought to yield to the requirements of humanity'.[79] This is an early illustration of the customary international law 'principle of humanity',[80] which is discussed further in chapter 3. Not surprisingly, the St. Petersburg Declaration makes no mention of the environment, but does provide that:[81]

---

75 Opened for signature 10 December 1976, 1108 UNTS 151; 16 ILM 88 (entered into force 5 October 1978).

76 See, for example, United Nations Security Council Resolution 1674 (28 April 2006) UN Doc S/RES/1674/2006 on the Protection of Civilians in Armed Conflict, para 6; United Nations Security Council Resolution 1738 (23 December 2006) UN Doc S/RES/1738/2006 on the Protection of Civilians in Armed Conflict, preamble para 5 and 10.

77 Declaration Renouncing the Use, in Time of War, of Explosive Projectiles under 400 Grammes Weight, opened for signature 29 November 1868, 18 Martens (1st) 474 (entered into force 11 December 1868) (St. Petersburg Declaration).

78 Erik V. Koppe, *The Use of Nuclear Weapons and the Protection of the Environment During International Armed Conflict* (2006), 95. There have been earlier examples where specific weapons have been prohibited: for details, see Roberts and Guelff, *supra* note 73, 53.

79 St. Petersburg Declaration, para 1.

80 Michael N. Schmitt, 'Humanitarian Law and the Environment' (2000) 28 *Denver Journal of International Law and Policy* 265, 294.

81 St. Petersburg Declaration, para 3.

> the only legitimate object which States should endeavour to accomplish during war is to weaken the military forces of the enemy

These sentiments are reflected in more recent *jus in bello* instruments. On a literal reading of this provision, one could perhaps argue that *any* environmental damage occasioned during war by acts not intended to 'weaken the military forces of the enemy' would be prohibited. However, the issue of environmental damage was not at all a relevant factor in the drafting of the St. Petersburg Declaration; rather it was primarily directed against the use of 'exploding' bullets that shattered upon contact with hard and soft surfaces, such as those developed in the Indian city of Dum-Dum and used by the British Army in colonial India in the 19th Century.

The St. Petersburg Declaration expressly states that it does not apply to warfare in which a non-contracting State Party was a belligerent, even where a State Party was involved.[82] This so-called 'all participation' (*si omnes*) clause was a common term of several of the early *jus in bello* treaties. It was also included in the 1907 Hague Convention IV (see below), thus limiting the applicability of that treaty during the First World War, since a number of the major participants in that conflict were not States Parties.[83] By contrast, in the more recent international humanitarian law treaties – for example the 1949 Geneva Conventions (discussed below) – the treaty is applicable to all combatants who are States Parties, even where one (or more) of the combatants is not a State Party.[84]

2.2.2. 1899 HAGUE CONFERENCE CONVENTIONS AND DECLARATIONS

In 1874, a Diplomatic Conference was convened in Brussels at the initiative of Tsar Alexander II of Russia. The conference adopted an International Declaration on the Laws and Customs of War which, however, was not ratified, although it did have some influence in the negotiation of subsequent instruments.[85] As the 19th Century was drawing to a close, the major States of the time agreed to meet in The Hague to discuss possible restrictions as to the methods and means of warfare, with a particular focus on the limitation of armaments. A conference was held in 1899 to discuss these issues. Although there was little agreement in

82 *Ibid*, para 9 and 10.

83 Leslie C. Green, *The Contemporary Law of Armed Conflict* (2nd ed, 2000), 135.

84 Françoise Hampson, 'Other areas of customary law in relation to the Study', in Elizabeth Wilmshurst and Susan Breau (eds), *Perspectives on the ICRC Study on Customary International Humanitarian Law* (2007), 50, 50 and 71.

85 Hilaire McCoubrey, *International Humanitarian Law: Modern Developments in the Limitation of Warfare* (2nd ed, 1998), 19.

relation to arms limitations, the 26 participants concluded three conventions, the most important of which was the 1899 Hague Convention II with Respect to the Laws and Customs of War on Land.[86] This was the first comprehensive international instrument to codify those principles that had thus far developed from the practice of States.[87]

The 1899 Hague Conference also concluded with three declarations,[88] one of which required States to:[89]

> agree to abstain from the use of projectiles the sole object of which is the diffusion of asphyxiating or deleterious gases

It has been suggested that these restrictions were derived from then existing customary law principles.[90] They were embodied in the regulations annexed to the 1899 Hague Convention II and the 1907 Hague Convention IV Respecting the Laws and Customs of War on Land (see below). The wording from the St. Petersburg Declaration was also adopted into the 1899 Hague Declaration concerning expanding bullets,[91] and is widely assumed to represent customary international law, at least in relation to the broad prohibition of weapons that cause unnecessary suffering.[92]

---

86 Opened for signature 29 July 1899, 26 Martens (2nd) 949 (entered into force 4 September 1900) (1899 Hague Convention II). The other conventions were the Convention I for the Pacific Settlement of International Disputes, opened for signature 29 July 1899, 26 Martens (2nd) 720 (entered in force 4 September 1900) (1899 Hague Convention I), and Convention III for the Adaptation to Maritime Warfare of Principles of the Geneva Convention of 1864 opened for signature 29 July 1899, 26 Martens (2nd) 979 (entered in force 4 September 1900) (1899 Hague Convention III).

87 Bassiouni, *supra* note 13, 25.

88 These were the Declaration on the Launching of Projectiles and Explosives from Balloons, 29 July 1899, 32 Stat. 1839 (entered in force 4 September 1900) (1899 Hague Declaration I), the Declaration on the Use of Projectiles the Object of Which is the Diffusion of Asphyxiating or Deleterious Gases, date of signature 29 July 1899, 26 Martens (2nd) 998 (entered into force 4 September 1900) (1899 Hague Declaration II), and the Declaration on the Use of Bullets Which Expand or Flatten Easily in the Human Body, date of signature 29 July 1899, 26 Martens (2nd) 1002 (entered into force 4 September 1900) (1899 Hague Declaration III).

89 1899 Hague Declaration II, para 1.

90 Roberts and Guelff, *supra* note 73, 59.

91 Declaration on the Use of Bullets Which Expand or Flatten Easily in the Human Body, date of signature 29 July 1899, 26 Martens (2nd) 1002 (entered into force 4 September 1900).

92 See, for example, Robin Coupland and Dominique Loye, 'The 1899 Hague Declaration concerning Expanding Bullets: A treaty effective for more than 100 years faces complex contemporary issues' (2003) 85 *International Review of the Red Cross* 135, 135–7; Roberts and Guelff, *supra* note 73, 64.

2.2.3. 1907 HAGUE CONVENTION IV RESPECTING THE LAWS AND CUSTOMS OF WAR ON LAND[93]

The Final Act of the 1899 Conference called for a 'follow up' conference to address unresolved issues.[94] After some delay, primarily due to the Russo-Japanese War (1904–1905), this took place in The Hague in 1907. At this conference, 13 conventions were adopted by the 44 States that took part. Of these, the most relevant in relation to damage to the environment during warfare was the 1907 Hague Convention IV, which embodied some of the principles of the 1899 Hague Convention II. The regulations to the 1907 Hague Convention IV, annexed to the treaty, specified a quite comprehensive set of rules relating to the laws of war, including a number of principles that have subsequently come to represent rules of customary international law.

The 1907 Hague Convention IV and its regulations do not specifically address environmental destruction as such. However, article 22 of the regulations confirms that '[t]he right of belligerents to adopt means of injuring the enemy is not unlimited'. Some commentators have interpreted this provision as a basis for restrictions on the destruction of the environment during the conduct of warfare.[95] In addition, article 23(a) relates indirectly to the protection of the environment during the conduct of hostilities by prohibiting the use of 'poison or poisoned weapons'. Other provisions call for the protection of property, including article 23(g), which prohibits the destruction of 'the enemy's property', unless justified by the 'necessities of war'. While there are uncertainties as to what might constitute 'property' – it would not extend, for example, to 'objects' such as the climate or the atmosphere[96] – the prohibition probably does encompass tangible areas, whose destruction would constitute environmental damage.

The general prohibitions envisaged by article 23(g) of the 1907 Hague Convention IV formed the basis of charges brought against various German military and administrative officers following the conclusion of the Second World War. General Lothar Rendulic was accused before a United States Military Tribunal at Nuremberg of wanton destruction of property, based upon a 'scorched earth policy' implemented under his command in an attempt to hinder the advancing

---

93 Opened for signature 18 October 1907, 3 Martens (3rd) 461 (entered into force 26 January 1910) (1907 Hague Convention IV).

94 Schmitt, *supra* note 80, 294.

95 *Ibid*, 295, footnote 154.

96 Anthony Leibler, 'Deliberate Wartime Environmental Damage: New Challenges for International Law' (1992) 23 *California Western International Law Journal* 67, 105–6.

Russian forces in the Norwegian province of Finnmark.[97] In a decision that was, not surprisingly, widely criticised in Norway, the Tribunal, after considering the terms of article 23(g), dismissed the charges, concluding that Rendulic's actions were justified by military necessity founded upon his 'honest judgment on the basis of the conditions prevailing at the time', even though, as it transpired, his actions proved to be unnecessary in the circumstances.[98]

In emphasizing the significance of the military necessity justification expressed in article 23(g), the Tribunal stated:[99]

> [t]here is evidence in the record that there was no military necessity for this destruction and devastation. An examination of the facts in retrospect can well sustain this conclusion. But we are obliged to judge *the situation as it appeared to the defendant at the time.* If the facts were such as would justify the action by the exercise of judgment, after giving consideration to all the factors and existing possibilities, even though the conclusion reached may have been faulty, it cannot be said to be criminal. After giving careful consideration to all the evidence on the subject, we are convinced that the defendant cannot be held criminally responsible although when viewed in retrospect, the danger did not actually exist.

The conviction of Alfred Jodl, Chief of the Operations Staff of the High Command of the German Armed Forces, by the International Military Tribunal for the Trial of the German Major War Criminals at Nuremberg (Nuremberg Military Tribunal), was partly based on the Nazi regime's scorched earth policies.[100] The Nuremberg Military Tribunal found that, in October 1944, he had ordered the evacuation of all persons in Northern Norway and the burning of their houses, so they could not help the Russians. The Norwegian Government claimed that some 30,000 houses were damaged in this operation.[101]

The Nuremberg Military Tribunal rejected arguments by the defence that the *si omnes* clause in the 1907 Hague Convention IV,[102] as well as in the 1929

97 This case was one of 12 so-called 'Nuremberg follow-up trials' conducted by United States Military Tribunals pursuant to 'Law No. 10 on the Punishment of Persons Guilty of War Crimes, Crimes Against Peace and Against Humanity', issued by the Allied Control Council on 20 December 1945. These judgments are regarded by some commentators as still influencing international criminal law today: see, for example, Gerhard Werle, *Principles of International Criminal Law* (2005), 12–3.

98 *Trial of William List and Others* (The Hostages Trial) (Judgment), United States Military Tribunal, Nuremberg, (1949) VIII *Law Reports of Trials of War Criminals*, Case No. 47, 69.

99 *Ibid*, 68–9 (emphasis added).

100 Tara Weinstein, 'Prosecuting Attacks that Destroy the Environment: Environmental Crimes or Humanitarian Atrocities?' (2005) 17 *Georgetown International Environmental Law Review* 697, 704.

101 *Judgment of the International Military Tribunal, Trial of the Major War Criminals* 1 October 1946, reprinted in (1947) 41 *American Journal of International Law* 172, 316.

102 See 1907 Hague Convention IV, article 2.

Geneva Convention relative to the Treatment of Prisoners of War,[103] prevented prosecution on this issue, based on the fact that several of the belligerents were not States Parties. The Nuremberg Military Tribunal concluded that it was unnecessary to decide that question since:[104]

> the Convention expressly stated that it was an attempt 'to revise the general laws and customs of war,' which it thus recognised to be then existing, but by 1939 these rules laid down in the Convention were recognised by all civilized nations, and were regarded as being declaratory of the laws and customs of war which are referred to in Article 6 (b) of the Charter

In another Nuremberg trial, conducted by the United Nations War Crimes Commission, 10 German civil administrators who had worked at the Forestry Administration in the period 1939–1944 were accused of war crimes arising from their 'ruthless exploitation of Polish forestry', which involved widespread cutting down of Polish forests far in excess of what was appropriate to preserve the timber resources of the country. The United Nations War Crimes Commission concluded that a *prima facie* case of war crimes had been demonstrated, and nine of the accused were thus classified as war criminals.[105]

Some other provisions of the 1907 Hague Convention IV might also be relevant to the issue of intentional destruction of the environment. Article 55 imposes an obligation on an occupying State to 'safeguard' the capital value of various properties – including 'forests, and agricultural estates' – situated in occupied States, and to 'administer them in accordance with the rules of usufruct'.[106] These rules provide that an occupying State may use such properties, but must not destroy or permanently alter them.[107] A more general protection is afforded by article 23(e), which proscribes the use of 'arms, projectiles, or material calculated to cause unnecessary suffering'. This could possibly be applied to actions that damage the environment.[108] Moreover, as noted, the 1907 Hague

103 Opened for signature 27 July 1929, 118 LNTS 343 (entered into force 19 June 1931): see Theodor Meron, 'The Humanization of Humanitarian Law' (2000) 94 *American Journal of International Law* 239, 247–8 and the corresponding footnote.

104 *Judgment of the International Military Tribunal, Trial of the Major War Criminals* 1 October 1946, reprinted in (1947) 41 *American Journal of International Law* 172, 248–9.

105 Adam Roberts, 'Failures in protecting the environment in the 1990–91 Gulf War', in Peter Rowe (ed), *The Gulf War 1990–91 in International and English Law* (1993), 111, 124.

106 Usufruct is a Roman Law concept defined as 'The right of using and taking the fruits of something belonging to another': Shelia Bone (ed), *Osborn's Concise Law Dictionary* (9th ed, 2001), 396.

107 Marcos A. Orellana, 'Criminal Punishment for Environmental Damage: Individual and State Responsibility at a Crossroad' (2005) 17 *Georgetown International Environmental Law Review* 673, 679.

108 Major Walter G. Sharp, Sr., 'The Effective Deterrence of Environmental Damage During Armed Conflict: A Case Analysis of the Persian Gulf War' (1992) 137 *Military Law Review* 1, 10.

Convention IV also included a Martens clause, which was expressed to apply '[u]ntil a more complete code of the laws of war has been issued'.[109]

Nevertheless, it is clear that the 1907 Hague Convention IV does not expressly provide for extensive protection of the environment, or the prohibition of intentional destruction of the environment during hostilities. Rather, some of its provisions might be applicable to environmental destruction, if such damage results directly from actions specifically proscribed by the instrument. The 1907 Hague Convention IV also does not expressly provide for criminal liability arising from a violation of its terms.

The prohibitions included in the 1907 Hague Convention IV are considered to represent customary international law,[110] as are the annexed regulations.[111] As noted, in 1946, the Nuremberg Military Tribunal confirmed that:[112]

> [t]he rules of land warfare expressed in the [1907 Hague Convention IV] undoubtedly represented an advance over existing international law at the time of their adoption ... by 1939 these rules laid down in the Convention were recognised by all civilised nations, and were regarded as being declaratory of the laws and customs of war

2.2.4. 1925 GENEVA PROTOCOL FOR THE PROHIBITION OF THE USE IN WAR OF ASPHYXIATING, POISONOUS AND OTHER GASES, AND OF BACTERIOLOGICAL METHODS OF WARFARE

Despite the positive developments flowing from the 1899 and 1907 conferences in The Hague, the conclusion of further international agreements regulating the methods and means of warfare was stymied by the onset of the First World War. The then existing conventional provisions did not deter actions such as the widespread use, by both the Germans (at Ypres in 1915), and the British (at Loos in 1916), of mustard gas.[113] Indeed, in 1919, a British Commission had concluded:[114]

---

109 1907 Hague Convention IV, preamble para 8.

110 Patricia W. Birnie and Alan E. Boyle, *International Law and the Environment* (2nd ed, 2002), 148.

111 See, for example, *Legal Consequences of the Construction of a Wall in the Occupied Palestinian Territory* (Advisory Opinion) [2004] ICJ Rep 136, para 89.

112 *Judgment of the International Military Tribunal, Trial of the Major War Criminals* 1 October 1946, reprinted in (1947) 41 *American Journal of International Law* 172, 248–9.

113 James Carroll, 'Proving nuclear realists wrong', *The International Herald Tribune* (Paris), 23 June 2008 <www.iht.com/articles/2008/06/23/opinion/edcarroll.php> at 23 June 2008.

114 As quoted in *ibid.*

> [t]hat gas is a legitimate weapon of war [is beyond a] shadow of doubt ... for history shows that in no case has a weapon which has proved successful in war been abandoned

Nonetheless, at the end of the war, the 1919 Treaty of Versailles provided that:[115]

> [t]he use of asphyxiating, poisonous and other gases and all analogous liquids, materials or devices being prohibited, their manufacture and importation are strictly forbidden in Germany

Similar prohibitions were included in the 1922 Washington Treaty Relating to the Use of Submarines and Noxious Gases in Warfare,[116] concluded between France, Italy, Japan, the United Kingdom and the United States. However, France subsequently decided not to ratify this treaty, with the result that it never entered into force.

During the post-First World War period, there was relatively little enthusiasm for further codification of the laws of war. It was generally thought that a widespread revision of the *jus in bello*, even in the face of the obvious inadequacies that existed, would undermine confidence in the attempts of the newly-established League of Nations to prevent future wars.[117] Yet, whenever the League of Nations sought to promote disarmament discussions, it frequently met with widespread resistance, since the general international atmosphere in the late 1920s and 1930s was increasingly one of distrust.[118]

One of the relatively few steps forward during this period came in 1925, when an International Conference on the Control of the International Trade in Arms and in Implements of War, convened by the League of Nations, adopted the Geneva Protocol for the Prohibition of the Use in War of Asphyxiating, Poisonous and Other Gases, and of Bacteriological Methods of Warfare (1925 Geneva Protocol).[119] This instrument confirmed that the prohibition of the use of 'asphyxiating, poisonous or other gases, and of all analogous liquids, materials or devices'[120] was 'universally accepted as a part of International Law, binding alike the conscience and the practice of nations'.[121]

---

115 Treaty of Peace between the Allied and Associated Powers and Germany, opened for signature 28 June 1919, 112 BFSP 317 (entered into force 10 January 1920), article 171.

116 Roberts and Guelff, *supra* note 73, 155.

117 Schindler and Toman, *supra* note 16, vi.

118 David Armstrong, Lorna Lloyd and John Redmond, *From Versailles to Maastricht: International Organisation in the Twentieth Century* (1996), 46–7.

119 Opened for signature 17 June 1925, 26 Martens (3rd) 643 (entered into force 8 February 1928).

120 1925 Geneva Protocol, preamble para 1 and 2.

121 *Ibid*, preamble para 3.

The 1925 Geneva Protocol also extended the prohibition to 'bacteriological methods of warfare',[122] and was thus the first major convention that restricted the use of biological weapons,[123] a new legal development at that time and largely directed towards weapons of the future.[124] This instrument is widely accepted by States[125] (137 States Parties).[126] However, many States have entered reservations – although some of these were subsequently withdrawn[127] – when ratifying or acceding to the instrument, and there also have been several clear violations of its terms.[128]

The move towards broader acceptance of the prohibitions in the 1925 Geneva Protocol has also been encouraged by a number of United Nations General Assembly Resolutions. For example, in 1969, the General Assembly declared that the 1925 Geneva Protocol:[129]

> embodies the generally recognized rules of international law prohibiting the use in international armed conflicts of all biological and chemical methods of warfare, regardless of any technical developments

---

122 *Ibid*, para 1.

123 Ensign Florencio J. Yuzon, 'Deliberate Environmental Modification Through the Use of Chemical and Biological Weapons: "Greening" the Environmental Laws of Armed Conflict to Establish an Environmentally Protective Regime' (1996) 11 *American University Journal of International Law and Policy* 793, 827.

124 See generally Jack M. Beard, 'The Shortcomings of Indeterminacy in Arms Control Regimes: The Case of the Biological Weapons Convention' (2007) 101 *American Journal of International Law* 271.

125 For example, in 1989 alone, the 1925 Geneva Protocol was ratified, or acceded to, by Albania, Bangladesh, Cameroon, Equatorial Guinea, Grenada, Guinea-Bissau, North Korea, South Korea, Laos and Saint Kitts and Nevis.

126 International Committee of the Red Cross, 'State Parties to the Following International Humanitarian Law and Other Related Treaties' <www.icrc.org/ihl/%28SPF%29/party_main_treaties/$File/IHL_and_other_related_Treaties.pdf> at 19 January 2015.

127 Several States had sought to retain the right to retaliate in kind with weapons prohibited under the 1925 Geneva Protocol: Roberts and Guelff, *supra* note 73, 155. Subsequent treaties, such as the Convention on the Prohibition of the Development, Production and Stockpiling of Bacteriological (Biological) and Toxic Weapons and on their Destruction, opened for signature 10 April 1972, 1015 UNTS 163 (entered into force 26 March 1975) (1972 Biological Weapons Convention) and the Convention on the Prohibition of the Development, Production, Stockpiling and Use of Chemical Weapons and on Their Destruction, opened for signature 13 January 1993, 32 ILM 800 (entered into force 29 April 1997) (1993 Chemical Weapons Convention), have repudiated any such right, leading to the withdrawal of some of these reservations. For example, article I of the 1993 Chemical Weapons Convention prohibits the use 'under any circumstances' of the weapons covered by the instrument. See also *Legality of the Threat or Use of Nuclear Weapons* (Advisory Opinion) [1996] ICJ Rep 226, para 57.

128 See, for example, Roberts and Guelff, *supra* note 73, 156–7.

129 United Nations General Assembly Resolution 2603A (XXIV) (16 December 1969) UN Doc A/RES/2603(XXIV)A on the Question of Chemical and Bacteriological (Biological) Weapons, para 5.

The General Assembly has also declared that the use in international armed conflict of the following weapons is contrary to these rules:[130]

(a) Any chemical agents of warfare – chemical substances, whether gaseous, liquid or solid – which might be employed because of their direct toxic effects on man, animals or plants;
(b) Any biological agents of warfare – living organisms, whatever their nature, or infective material derived from them – which are intended to cause disease or death in man, animals or plants, and which depend for their effects on their ability to multiply in the person, animal or plant attacked

The war crime specified in the Rome Statute arising from the use of 'asphyxiating, poisonous or other gases, and all analogous liquids, materials or devices'[131] is derived from the 1925 Geneva Protocol.[132]

2.2.5. 1949 GENEVA CONVENTIONS

The Swiss city of Geneva has been the site of several landmark *jus in bello* agreements. In addition to the 1864 Geneva Convention and the 1925 Geneva Protocol, other conventions and protocols were concluded in 1868,[133] 1906[134] and 1929.[135] These were intended to revise and expand upon a number of the general principles of the 1864 Geneva Convention, as well as the rights of prisoners of war and other individuals.

The tragic events of the Second World War demonstrated that the codified rules of armed conflict at that time remained inadequate, particularly in relation to the protection of civilians and other non-combatants. The need to revise and extend the existing *jus in bello* treaty rules was obvious. A Diplomatic Conference for the Establishment of International Conventions for the Protection of Victims of War was held in Geneva from 21 April to 12 August 1949. The principal purpose

---

130 *Ibid*, para 8.

131 Rome Statute, article 8(2)(b)(xvii).

132 Knut Dörmann, 'War Crimes in the Elements of Crimes', in Horst Fischer, Claus Kress and Sascha Rolf Lüder (eds), *International and National Prosecution of Crimes Under International Law: Current Developments* (2001), 95, 128.

133 Additional Articles relating to the Condition of the Wounded in War, opened for signature 20 October 1868, 18 Martens (1st) 612. This instrument did not enter into force.

134 Convention for the Amelioration of the Condition of the Wounded and Sick in Armies in the Field, opened for signature 6 July 1906, 11 LNTS 440 (entered into force 9 August 1907). This instrument is no longer in force.

135 Convention for the Amelioration of the Condition of the Wounded and Sick in Armies in the Field, opened for signature 27 July 1929, 118 LNTS 303 (entered into force 19 June 1931); Convention Relative to the Treatment of Prisoners of War, opened for signature 27 July 1929, 118 LNTS 343 (entered into force 19 June 1931).

of the conference was to provide for a more comprehensive framework for the protection of victims of war. The conference set out to revise the 1907 Hague Convention X for the Adaptation to Maritime Warfare of the Principles of the 1906 Geneva Convention,[136] as well as the two 1929 Geneva Conventions, and to conclude a new convention directed towards the protection of civilians.

In the three-year period leading up to the conference, a series of meetings had taken place at which four draft conventions were prepared, ultimately to be adopted by the Geneva conference.[137] Each of the conventions deals with different, but related, aspects of the conduct of hostilities and the rights of, and protections to be afforded to victims.[138] In contrast to The Hague rules, which focused primarily on the methods and means of warfare, the principles codified in the Geneva rules were directed towards the 'humanitarian aspects of conflict', dealing specifically with the protection of both combatants and non-combatants.[139]

Nonetheless, both sets of rules reflect an overriding concern to mitigate the conduct and consequences of warfare, and thus there is a degree of overlap.[140] In this regard, the International Court of Justice has noted that:[141]

> [t]hese two branches of the law applicable in armed conflict have become so closely interrelated that they are considered to have gradually formed one single complex system, known today as international humanitarian law. The provisions of the Additional Protocols of 1977 give expression and attest to the unity and complexity of that law

As noted, unlike the 1907 Hague Conventions, the 1949 Geneva Conventions do not contain an 'all participation clause', and thus apply even where one of the belligerents is not a State Party.[142] This also reflects the fact that there has, over time, been a 'progressive trend towards the so-called 'humanisation'

---

136 Hague Convention X for the Adaptation to Maritime Warfare of the Principles of the 1906 Geneva Convention, opened for signature 18 October 1907, 3 Martens (3rd) 360 (entered into force 26 January 1910).

137 Roberts and Guelff, *supra* note 73, 195.

138 These were the wounded and sick in armed forces in the field (1949 Geneva Convention I), the wounded, sick and ship-wrecked in armed forces at sea (1949 Geneva Convention II), the treatment of prisoners at war (1949 Geneva Convention III) and the protection of civilians (1949 Geneva Convention IV).

139 See Yuzon, *supra* note 123, 813; Carl E. Bruch, 'All's Not Fair in (Civil) War: Criminal Liability for Environmental Damage in Internal Armed Conflict' (2001) 25 *Vermont Law Review* 695, 701.

140 McCoubrey, *supra* note 85, 2.

141 *Legality of the Threat or Use of Nuclear Weapons* (Advisory Opinion) [1996] ICJ Rep 226, para 75.

142 Green, *supra* note 83, 44.

of international legal obligations', with the result that traditional notions of reciprocity in the application of international humanitarian law have been eroded.[143] Instead, principles of international humanitarian law are now largely regarded as 'absolute' obligations, rather than being dependent on the conduct of opposing parties.

Each of the four 1949 Geneva Conventions has received almost universal acceptance, with over 190 States Parties and, together with the 1977 Additional Protocols, they form the core of the modern *jus in bello* rules.[144] The four treaties contain several common articles, both in terms of procedural aspects and substantive principles. One of the most important of these is common article 3, which specifies certain minimum standards of conduct during non-international armed conflicts,[145] based on principles that have been described by the International Court of Justice as reflecting 'elementary considerations of humanity'.[146] A breach of common article 3 is expressly defined as a crime within the competence of the International Criminal Tribunal for Rwanda (ICTR).[147]

143 Judgement, *Prosecutor v. Zoran Kupreskic, Mirjan Kupreskic, Vlatko Kupreskic, Drago Josipovic, Dragan Papic and Vladimir Santic*, Case No. IT-95-16-T, Trial Chamber, 14 January 2000, para 518.

144 Nsongurua J. Udombana, 'War is Not Child's Play! International Law and the Prohibition of Children's Involvement in Armed Conflicts' (2006) 20 *Temple International and Comparative Law Journal* 57, 72.

145 As noted, since the end of the Second World War, the majority of armed conflicts have been non-international (or internal) rather than international in nature: see Lindsay Moir, *The Law of Internal Armed Conflict* (2002), 1.

146 *Military and Paramilitary Activities in and against Nicaragua (Nicaragua v. United States of America)* (Merits) (Judgment) [1986] ICJ Rep 14, para 218, citing *Corfu Channel Case (United Kingdom of Great Britain and Northern Ireland v. Albania)* (Merits) (Judgment) [1949] ICJ Rep 4, 22.

147 See article 4 of the Statute of the International Criminal Tribunal for the Prosecution of Persons Responsible for Genocide and Other Serious Violations of International Humanitarian Law Committed in the Territory of Rwanda and Rwandan Citizens Responsible for Genocide and Other Such Violations Committed in the Territory of Neighbouring States, between 1 January 1994 and 31 December 1994, 33 ILM 1598 (ICTR Statute). The interpretation of common article 3 by the ICTR has, however, been the subject of some conjecture. For example, in 2001, the Belgian Government successfully requested to submit an *amicus curiae* brief before a Trial Chamber of that Tribunal, asserting that the provision had been 'too restrictively applied in the Tribunal's judgements in the cases of *Akayesu*, *Kayishema et al.*, and *Rutaganda*': Decision on the Kingdom Of Belgium's Application to File an *Amicus Curiae* Brief and on the Defence Application to Strike Out the Observations of the Kingdom of Belgium Concerning the Preliminary Response by the Defence, *Prosecutor v. Laurent Semanza*, Case No. ICTR-97-20-T, Trial Chamber, 9 February 2001, para 2. In considering the request, the ICTR was of the opinion that 'it may be useful to gather additional legal views on the scope of applicability of Article 3 common to the four Geneva Conventions and of Additional Protocol II' (para 10).

The 1949 Geneva Conventions were concluded shortly after the adoption by the United Nations General Assembly of the Universal Declaration of Human Rights (UDHR), which brought into focus the rights of individuals.[148] The UDHR influenced the subsequent codification of the laws of war, gradually transforming it into more of a 'human rights oriented law'.[149] Indeed, the 1949 Geneva Conventions refer to 'rights' of protected persons, as well as imposing obligations on belligerents. The 1949 Geneva Convention IV was the first *jus in bello* instrument specifically directed towards the protection of civilians in times of war,[150] a response to horrific events during the Second World War, which had seen civilians killed far in excess of the rate of any previous major conflict.[151]

Although it does not seek to replace the principles of The Hague Conventions, but rather is expressed to be 'supplementary' to those instruments, the 'civilian focus' in the 1949 Geneva Convention IV was a significant new development in the evolution of the *jus in bello*.[152] Despite the fact that it deals principally with the treatment of civilians held by an adversary, rather than focusing on the *effects* of a conflict, the 1949 Geneva Convention IV is the most important of the four conventions in relation to the issue of intentional environmental destruction during the course of armed conflict.

The 1949 Geneva Convention IV contains a general prohibition of 'pillage', which may be of some relevance.[153] Of greater significance, article 53 provides that:[154]

> [a]ny destruction by the Occupying Power of real or personal property belonging individually or collectively to private persons, or to the State [whose territory is being occupied], or to other public authorities, or to social or cooperative organizations, is prohibited, except where such destruction is rendered absolutely necessary by military operations

This provision only applies in relation to occupied territory,[155] and would therefore not be relevant to environmental destruction caused in locations that are not occupied but still are areas of armed conflict, as might be the case, for

148 Adopted by United Nations General Assembly Resolution 217A (III) (10 December 1948) UN Doc A/810 on the Universal Declaration of Human Rights.
149 Schindler and Toman, *supra* note 16, vii.
150 The 1899 Hague Convention II and 1907 Hague Convention IV both refer in general terms to civilians, but only in the context of enemy occupation.
151 International Committee of the Red Cross, 'What treaties make up international humanitarian law?' (October 2002) <www.icrc.org/Web/eng/siteeng0.nsf/html/5KZGE6> at 19 January 2015.
152 1949 Geneva Convention IV, article 154.
153 *Ibid*, article 33.
154 *Ibid*, article 53.
155 Section III of the 1949 Geneva Convention IV, in which article 53 appears, is headed 'Occupied Territories'.

example, with the aerial bombardment of an enemy country.[156] While this limits the applicability of the provision, it has been noted that occupied territory is 'particularly susceptible' to deliberate environmental damage not justified by military operations,[157] when compared to the territory of the Occupying Power itself. This was demonstrated, for example, by the Iraqi actions in Kuwait in 1991 referred to earlier. James Terry has suggested that, had the principles of the 1907 Hague Convention IV been followed by Iraq, 'there would have been no significant violation of the Kuwaiti environment'.[158]

As with article 23(g) of the 1907 Hague Convention IV, to the extent that article 53 does proscribe acts that constitute intentional destruction of the environment during armed conflict, this is subject to the overarching consideration of military necessity.[159] Although the provision purports to set a higher standard than the 1907 Hague Convention IV – by requiring that the act be *absolutely* necessary to avoid constituting a violation – it again reaffirms the viewpoint that military considerations take precedence over environmental destruction. That said, the 1949 Geneva Convention IV does regard as a 'grave breach' actions that amount to:[160]

> extensive destruction and appropriation of property, not justified by military necessity and carried out unlawfully and wantonly

Thus, if the destruction of property in an Occupied Territory is regarded as 'extensive', it may be deemed as a grave breach, in which case States Parties are obligated to 'provide effective penal sanctions', as well as search for alleged offenders and either try them before their own national courts, or hand them over to another State Party for trial, provided that a *prima facie* case has been made out.[161] It has been noted, however, that this *aut dedere aut judicare* obligation 'has not, in practice, been implemented' to any significant degree.[162]

---

156 Leibler, *supra* note 96, 106.

157 Schmitt, *supra* note 80, 298.

158 Colonel James P. Terry, 'The Environment and the Laws of War: The Impact of Desert Storm' (1992) 45 *Naval War College Review* 61, 63. Iraq was not a party to the 1907 Hague Convention IV at the time of its occupation of Kuwait.

159 See chapters 3 and 4.

160 1949 Geneva Convention IV, article 147.

161 *Ibid*, article 146. See also United Nations Security Council Resolution 1738 (23 December 2006) UN Doc S/RES/1738/2006 on the Protection of Civilians in Armed Conflict, preamble para 7. Acts that are grave breaches of the 1949 Geneva Conventions might also constitute war crimes, attracting individual criminal responsibility: see Statute of the International Tribunal for the Prosecution of Persons Responsible for Serious Violations of International Humanitarian Law Committed in the Territory of the Former Yugoslavia since 1991, 32 ILM 1159 (ICTY Statute), article 2, and Rome Statute, article 8(2)(a).

162 Beigbeder, *supra* note 30, 8–9.

2.2.6. 1954 HAGUE CONVENTION FOR THE PROTECTION OF CULTURAL PROPERTY IN THE EVENT OF ARMED CONFLICT[163]

Although earlier instruments had included general references to the protection of cultural property in times of war and occupation,[164] the 1954 Hague Convention represented the first comprehensive agreement to deal with the issue. It was a response to the widespread damage and destruction of such property during the Second World War. The convention, together with a Protocol,[165] was adopted under the auspices of the United Nations Educational, Scientific and Cultural Organization (UNESCO).[166] It sought to provide for a regime that protected cultural property not only during, but also prior to and after an armed conflict.[167]

The 1954 Hague Convention is largely directed towards protection of property that is not 'environmental' in nature; however the intentional destruction of the environment could place cultural objects at risk.[168] Obligations to protect cultural property specified in the convention have been supplemented by the 1977 Additional Protocols[169] and, as a general principle, represent rules of customary international law.[170] However, the obligations under the 1954 Hague Convention are, once again, subject to 'cases where military necessity imperatively requires ... a waiver'.[171] Moreover, any special protection placed

163 Opened for signature 14 May 1954, 249 UNTS 240 (entered into force 7 August 1956) (1954 Hague Convention).

164 See, for example, article 5 of the 1907 Hague Convention IX Concerning Bombardment by Naval Forces in Time of War, opened for signature 18 October 1907, 3 Martens (3rd) 604 (entered into force 26 January 1910).

165 1954 First Hague Protocol for the Protection of Cultural Property in the Event of Armed Conflict, opened for signature 14 May 1954, 249 UNTS 358 (entered into force 7 August 1956). This instrument deals principally with the exportation of cultural property from an occupied territory.

166 Article I(2)(c) of the UNESCO Constitution provides that the organization will 'maintain, increase and diffuse knowledge' by *inter alia* 'assuring the conservation and protection of the world's inheritance of books, works of art and monuments of history and science, and recommending to the nations concerned the necessary international conventions': UNESCO Constitution (adopted 16 November 1945) <http://portal.unesco.org/en/ev.phpURL_ID=15244&URL_DO=DO_TOPIC&URL_SECTION=201.html> at 19 January 2015. See also Juliana V. Campagna, 'War or Peace: It is Time for the United States to Ratify the 1954 Hague Convention for the Protection of Cultural Property in the Event of Armed Conflicts' (2005) 17 *Florida Journal of International Law* 271, 342–3.

167 Craig J.S. Forrest, 'The Doctrine of Military Necessity and the Protection of Cultural Property During Armed Conflicts' (2007) 37 *California Western International Law Journal* 177, 200.

168 Michael N. Schmitt, *supra* note 80, 299.

169 See the 1977 Additional Protocol I, article 53, and the 1977 Additional Protocol II, article 16.

170 Roberts and Guelff, *supra* note 73, 372.

171 1954 Hague Convention, article 2.

on refuges or centres that contain cultural property during times of armed conflict only applies provided they 'are not used for military purposes'.[172] The circumstances that constitute military necessity for the purposes of the protections afforded to cultural property are elaborated in the Second Protocol to the Hague Convention of 1954 for the Protection of Cultural Property in the Event of Armed Conflict (see below).[173]

2.2.7. 1972 BIOLOGICAL WEAPONS CONVENTION

This treaty expands upon the prohibitions specified in the 1925 Geneva Protocol, which had introduced several restrictions on the use of biological weapons. The 1972 Biological Weapons Convention provides that States Parties undertake:[174]

> never in any circumstances to develop, produce, stockpile or otherwise acquire or retain:
>
> 1. microbial or other biological agents, or toxins whatever their origin or method of production, of types and in quantities that have no justification for prophylactic, protective or other peaceful purposes;
> 2. weapons, equipment or means of delivery designed to use such agents or toxins for hostile purposes or in armed conflict

The 1972 Biological Weapons Convention provides for a conference to be held 'to review the operation of the Convention'. In 2006, the United Nations General Assembly 'welcomed' the Final Declaration of the Fourth Review Conference,[175] which reaffirmed:[176]

> that under all circumstances the use of bacteriological (biological) and toxin weapons and their development, production and stockpiling are effectively prohibited under article I of the Convention

The instrument also requires that State Parties 'destroy, or ... divert to peaceful purposes' all such agents, toxins, weapons, equipment and means of delivery,[177]

---

172 *Ibid*, article 8(1)(b).

173 Opened for signature 26 March 1999, 38 ILM 769 (entered into force 9 March 2004).

174 1972 Biological Weapons Convention, article I.

175 Final Declaration of the Fourth Review Conference of the Convention on the Prohibition of the Development, Production and Stockpiling of Bacteriological (Biological) and Toxic Weapons and on their Destruction, (6 December 1996) BWC/CONF.IV/9, part II.

176 United Nations General Assembly Resolution 61/102 (6 December 2006) UN Doc A/RES/61/102 on the Convention on the Prohibition of the Development, Production and Stockpiling of Bacteriological (Biological) and Toxic Weapons and on their Destruction, preamble para 4.

177 1972 Biological Weapons Convention, article II.

as well as restricting their transfer.[178] In implementing their obligations, States Parties are to observe 'all necessary safety precautions ... to protect populations and the environment'.

Some commentators argue that, strictly speaking, the 1972 Biological Weapons Convention is a 'disarmament' treaty, and thus does not form part of the *jus in bello*.[179] While this may be true, biological agents can be produced quickly and easily and have (in one form or another) been used in warfare for centuries, such as when, for example, plague-infected bodies were catapulted into cities under siege.[180] In their more contemporary form, the use of such weapons during armed conflict has the potential not only to kill large numbers of people, but also to destroy various elements of the environment. The 1972 Biological Weapons Convention therefore contributes to the body of treaty law that indirectly regulates the protection of the environment during hostilities.

The fact that biological weapons have only been used 'sporadically' during history in no way detracts from the potentially devastating impact that they may have.[181] Yet, there are several significant 'loopholes' in the 1972 Biological Weapons Convention – for example, the failure to prohibit 'biological weapons research'.[182] It should also be noted that the 1972 Biological Weapons Convention only applies to biological weapons, although it does make reference to the 'recognised objective' of prohibiting chemical weapons.[183] Both forms of weapons have the capability to contaminate soil, water and plants, and thus to devastate ecosystems and the natural environment, for periods far exceeding the damage caused by explosive munitions.[184] In 1971, a Conference on Disarmament decided to separate the negotiations regarding chemical and biological weapons, thus paving the way for the conclusion of the 1972 Biological Weapons Convention. Comprehensive international regulation of chemical

---

178 *Ibid*, article III.

179 See, for example, Schindler and Toman, *supra* note 16, 135.

180 Scott Spence, 'The Biological and Toxin Weapons Convention and its Implementation through National Law', speech delivered at the T.M.C. Asser Institute attended by the author, The Hague, The Netherlands, 25 November 2008.

181 For a description of the use of biological weapons during historical military campaigns, see Mark Wheelis, 'A Short History of Biological Warfare and Weapons', in Marie Isabelle Chevrier, Krzysztof Chomiczewski, Henri Garrigue, György Granasztói, Malcolm R. Dando and Graham S. Pearson (eds), *The Implementation of Legally Binding Measures to Strengthen the Biological and Toxin Weapons Convention* (2004), 15 *et seq*.

182 Michael P. Scharf, 'Clear and Present Danger: Enforcing the International Ban on Biological and Chemical Weapons Through Sanctions, Use of Force, and Criminalization' (1999) 20 *Michigan Journal of International Law* 477, 483.

183 1972 Biological Weapons Convention, article IX.

184 Barry Kellman, 'The Chemical Weapons Convention: A Verification and Enforcement Model for Determining Legal Responsibility for Environmental Harm Caused by War', in Jay E. Austin and Carl E. Bruch (eds), *The Environmental Consequences of War: Legal, Economic, and Scientific Perspectives* (2000), 579, 579.

weapons had to wait until the conclusion of the 1993 Chemical Weapons Convention (see below).

2.2.8. 1976 CONVENTION ON THE PROHIBITION OF MILITARY OR ANY OTHER HOSTILE USE OF ENVIRONMENTAL MODIFICATION TECHNIQUES[185]

Notwithstanding claims that it was done for 'a valid strategic and tactical purpose',[186] the widespread use by the United States of defoliants during the Vietnam War to destroy crops and forests as a deliberate military strategy led to significant concerns about the environmental effects of warfare, as well as the development of methods of environmental modification. These were further heightened by the revelation in 1972 that, during that war, the United States had attempted to manipulate weather patterns in the Indo-China region, in order to flood various strategic routes.[187] These actions highlighted quite starkly the need to introduce direct environmental protections into the body of law regulating the conduct of those engaged in armed conflicts.[188]

In July 1973, the United States Senate resolved to:[189]

> seek the agreement of other governments to a proposed treaty prohibiting the use of any environmental or geophysical modification activity as a weapon of war

Negotiations subsequently took place both within the United Nations General Assembly and under the auspices of the Conference of the Committee on Disarmament (CCD), including a series of bilateral discussions between the United States and the Soviet Union. In late 1974, the General Assembly accepted a Soviet Union initiative, which stressed the need to prohibit:[190]

---

185 Opened for signature 18 May 1977, 1108 UNTS 151; 16 ILM 88 (entered into force 5 October 1978) (ENMOD Convention). The instrument is dated 1976, since it was in that year that the text, together with a series of 'Understandings' relating to articles I, II, II and VIII of the convention (1976 Understandings), were adopted by the United Nations General Assembly and referred to all States for 'their consideration, signature and ratification': see United Nations General Assembly Resolution 31/72 (10 December 1976) UN Doc A/RES/31/72 on the Convention on the Prohibition of Military or any Other Hostile Use of Environmental Modification Techniques, preamble para I. The 1976 Understandings are not incorporated into the ENMOD Convention, but are generally thought to be part of the record of the negotiations leading to the conclusion of the instrument. However, there is not complete unanimity as to the legal status of the 1976 Understandings: see Koppe, *supra* note 78, 113–5.

186 Terry, *supra* note 158, 99.

187 Roberts and Guelff, *supra* note 73, 407.

188 Karen Hulme, *War Torn Environment: Interpreting the Legal Threshold* (2004), 71.

189 Roberts and Guelff, *supra* note 73, 407.

190 United Nations General Assembly Resolution 3264 (XXIX) (9 December 1974) UN Doc A/RES/3264(XXIX) on the Prohibition of Action to Influence the Environment and Climate for

> action to influence the environment and climate for military and other hostile purposes, which are incompatible with the maintenance of international security, human well-being and health

These developments culminated two years later with the conclusion of the ENMOD Convention, the main purpose of which is to prohibit the manipulation of the environment as a weapon of warfare. The core obligation requires that a State Party:[191]

> undertakes not to engage in military or any other hostile use of environmental modification techniques having widespread, long-lasting or severe effects as the means of destruction, damage or injury to any other State Party

The term 'environmental modification techniques' is defined as:[192]

> any technique for changing – through the deliberate manipulation of natural processes – the dynamics, composition or structure of the Earth, including its biota, lithosphere, hydrosphere and atmosphere, or of outer space

Article I of the ENMOD Convention is the first treaty obligation that directly addresses the issue of deliberate manipulation of the environment during hostilities. Indeed, it represents the first time that the word 'environment' appears in a *jus in bello* treaty.[193] The definition of environmental modification techniques appears to envisage a broad approach as to what might constitute the 'environment', although there are uncertainties as to the range of actions that might represent such a technique.[194] The terms of the ENMOD Convention are not limited to warfare or armed conflict, but are expressed to apply in broader circumstances – 'military or any other hostile use'. Thus, the ENMOD Convention may also apply in times of peace, although not in relation to environmental modification techniques undertaken 'for peaceful purposes'.[195]

Nor does the instrument differentiate between lawful and unlawful uses of force.[196] In this respect, the general ambit of the ENMOD Convention is wide, so

---

Military and Other Purposes, Incompatible with the Maintenance of International Security, Human Well-Being and Health.

191 ENMOD Convention, article I(1).

192 *Ibid*, article II.

193 Adam Roberts, 'The Law of War and Environmental Damage', in Jay E. Austin and Carl E. Bruch (eds), *The Environmental Consequences of War: Legal, Economic, and Scientific Perspectives* (2000), 47, 50.

194 See, for example, Hulme, *supra* note 188, 72, who poses the question as to whether 'the burning of a forest in order to kill its inhabitants' falls within the definition of 'environmental modification techniques'.

195 ENMOD Convention, article III.

196 Leibler, *supra* note 96, 82.

as to reflect the stated desire of the United Nations General Assembly at the time to avert the 'potential dangers' of environmental modification techniques.[197] It is also significant that the prohibitions specified in the convention are not expressed to be subject to any 'defence' of military necessity, which might otherwise (legally) justify the use of an environmental modification technique. A violation of the treaty thus occurs as soon as the threshold level of damage is caused through the use of such a technique.

Thus, in order to more clearly determine the scope of the general prohibition, it is necessary to ascertain the meaning of the qualifying words 'widespread, long-lasting or severe' as used in article I. These terms describe the requisite damage necessary to constitute a violation. They are expressed in the disjunctive – separated by the word 'or'. Consequently, an environmental modification technique that results in *any* of these effects is sufficient to (potentially) bring the prohibition into play. These three terms are defined, for the purposes of the ENMOD Convention, in the first of the 1976 Understandings as follows:[198]

> (a) 'widespread': encompassing an area on the scale of several hundred square kilometres;
> (b) 'long-lasting': lasting for a period of months, or approximately a season;
> (c) 'severe': involving serious or significant disruption or harm to human life, natural and economic resources or other assets

These definitions are:[199]

> intended exclusively for [the ENMOD Convention] and … not intended to prejudice the interpretation of the same or similar terms if used in connexion with any other international agreement

Notwithstanding the clarifications provided by the first of the 1976 Understandings, the use of these terms has been criticised as being too broad and vague, and thus not readily applicable in reality.[200] At the time that it became a signatory to the ENMOD Convention in May 1977, Turkey made an interpretive statement as follows:[201]

---

197 United Nations General Assembly Resolution 31/72 (10 December 1976) UN Doc A/RES/31/72 on the Convention on the Prohibition of Military or any Other Hostile Use of Environmental Modification Techniques, preamble.

198 1976 CCD Understanding Relating to Article I of ENMOD, 31 United Nations General Assembly Official Records Supp. No. 27 (A/31/27), Annex I.

199 *Id.*

200 See, for example, Roberts and Guelff, *supra* note 73, 408; Sharp, *supra* note 108, 20.

201 Interpretative Statement relating to the ENMOD Convention by the Government of Turkey, 18 May 1977, quoted in Roberts and Guelff, *supra* note 73, 415. As at January 2014, Turkey was still not a party to the ENMOD Convention: International Committee of the Red Cross, 'State Parties to the Following International Humanitarian Law and Other Related Treaties

> [i]n the opinion of the Turkish Government the terms "widespread", "long-lasting" and "severe effects" contained in the Convention need to be more clearly defined. So long as this clarification is not made, the Government of Turkey will be compelled to interpret itself the terms in question and consequently it reserves the right to do so as and when required

The fact that the definitions are expressly applicable only to the ENMOD Convention is important, since similar terms – 'widespread, long-term and severe' – are used in 1977 Additional Protocol I to qualify damage to the natural environment resulting from certain methods or means of warfare.[202] In addition, the crime of 'wilful and severe' damage to the environment was included in the first reading in 1991 of the Draft Code of Offences against the Peace and Security of Mankind (Draft Code of Mankind) prepared by the International Law Commission (ILC).[203] In reaction to the concerns raised by various governments, the Special Rapporteur subsequently proposed a 'slimmed down' version of the Draft Code of Mankind,[204] with the result that only four (international) crimes[205] were included in the text adopted by the ILC in 1996.[206]

---

as at 29 January 2014' <www.icrc.org/ihl/%28SPF%29/party_main_treaties/$File/IHL_and_other_related_Treaties.pdf> at 3 February 2014.

202 1977 Additional Protocol I, articles 35(3) and 55(1). These provisions are discussed in more detail later in this chapter.

203 The ILC was initially mandated by the United Nations General Assembly to prepare a Draft Code of Offences against the Peace and Security of Mankind in 1947: United Nations General Assembly Resolution 174 (II) (21 November 1947) UN Doc A/519 on the Establishment of an International Law Commission, para 2(b). The General Assembly then decided to delay further work on the draft, primarily due to difficulties relating to the definition of the crime of 'aggression': United Nations General Assembly Resolution 897 (IX) (4 December 1954) UN Doc A/RES/897/(IX) on the Draft Code of Offences against the Peace and Security of Mankind, para 3. The ILC recommenced its work on the draft in the early 1980s, culminating in the finalization of the 1996 text. In Judgement, *Prosecutor v. Furundzjia*, Case No. IT-95-17/1, Trial Chamber, 10 December 1998, the Trial Chamber of the ICTY described the 1996 text of the Draft Code of Mankind in the following terms (para 227):
'... the Draft Code is an authoritative international instrument which, depending upon the specific question at issue, may (i) constitute evidence of customary law, or (ii) shed light on customary rules which are of uncertain contents or are in the process of formation, or, at the very least, (iii) be indicative of the legal views of eminently qualified publicists representing the major legal systems of the world.'
Following the subsequent entering into force of the Rome Statute and the establishment of the International Criminal Court in 2002, the Draft Code of Mankind is generally thought to be of no further practical relevance: Rosemary Rayfuse, 'The Draft Code of Crimes Against the Peace and Security of Mankind: Eating Disorders at the International Law Commission' (1997) 8 *Criminal Law Forum* 43, 85.

204 See *ibid*, Rayfuse, 47 *et seq*.

205 These were the crimes of aggression, genocide, crimes against humanity and war crimes. The text also included the crime of 'Crimes against United Nations and associated personnel'.

206 Draft Code of Offences against the Peace and Security of Mankind, Report of the International Law Commission on Its Forty-eighth session, United Nations General Assembly Official Records 51st Sess., Supp. No. 10, 9 UN Doc A/51/10 (1996).

These did not include the previously proposed crime specifically relating to the environment.

These threshold terms are also relevant to individual criminal responsibility under article 8(2)(b)(iv) of the Rome Statute, which classifies as a war crime an attack launched:[207]

> in the knowledge that [it] will cause ... widespread, long-term and severe damage to the natural environment ...

The ENMOD Convention does not prohibit damage to the environment during the conduct of hostilities *per se*, even if it may be deliberate, unless such an act can be regarded as constituting an environmental modification technique. In this regard, the second of the 1976 Understandings sets out the following non-exhaustive list of phenomena that could be caused through the modification of the environment:[208]

> earthquakes; tsunamis; an upset in the ecological balance of a region; changes in weather patterns (clouds, precipitation, cyclones of various types, and tornadic storms); changes in climate patterns; changes in ocean currents; changes in the state of the ozone layer; and changes in the state of the ionosphere

The clarification goes on to state that each of these phenomena, when produced by an environmental modification technique:[209]

> would result, or could reasonably be expected to result, in widespread, long-lasting or severe destruction, damage or injury [to another State Party]

The ENMOD Convention calls for a process of consultation and cooperation in the event of a dispute. There is no reference either to criminal responsibility or civil liability in case of breach. The formal enforcement mechanism established under the instrument has been described as 'post hoc',[210] and 'political in nature'.[211] A State Party can lodge a complaint with the United Nations Security Council if it has reason to believe that another State Party is in violation of its terms. In the event that the Security Council considers that a violation of the convention that has harmed or is likely to harm a State Party has occurred, the other States Parties are required to provide assistance to that State Party, if it

---

207 Rome Statute, article 8(2)(b)(iv). This provision is discussed further in chapter 4.

208 1976 CCD Understanding Relating to Article II of ENMOD, 31 United Nations General Assembly Official Records Supp. No. 27 (A/31/27), Annex II.

209 *Id.*

210 Weinstein, *supra* note 100, 701.

211 Schmitt, *supra* note 80, 280.

should so request.[212] Any possible response by the Security Council to a violation is, of course, always subject to the power of veto being exercised by one of the five permanent members.[213]

Article VIII of the ENMOD Convention contemplated a review conference five years after the treaty entered into force. The First Review Conference took place in 1984 and the participants reaffirmed their support for the treaty. A Second Review Conference was held in September 1992, at which various differences of interpretation emerged regarding the scope of the instrument. This conference called for a meeting of a committee of experts to clarify these matters, but this did not eventuate.[214]

It should also be noted that there are only a relatively small number of States Parties to the ENMOD Convention (76 as at January 2015).[215] This probably reflects the fact that the use of environmental modification techniques as a method or means of warfare is not a strategy that is within the contemplation (or capability) of most States.[216] In this regard, the applicability of the ENMOD Convention is therefore relatively limited, given that it generally relates to the manipulation of the environment by way of extremely advanced technology.[217] Nevertheless, some commentators have suggested that the ENMOD provisions may now represent customary international law.[218]

2.2.9. 1977 ADDITIONAL PROTOCOL I

Even after the entry into force of the four 1949 Geneva Conventions, humankind has continued to witness a large number of conflicts 'affecting almost every continent'.[219] Although those treaties represented a very significant advance in

---

212 ENMOD Convention, article V(5).

213 See United Nations Charter, article 27(3).

214 Articles V(1) and V(2) of the ENMOD Convention enable a State Party to request that a 'Consultative Committee of Experts' be convened to discuss issues relating to the objectives or application of the treaty.

215 International Committee of the Red Cross, 'State Parties to the Following International Humanitarian Law and Other Related Treaties as at 29 January 2014' <www.icrc.org/ihl/%28SPF%29/party_main_treaties/$File/IHL_and_other_related_Treaties.pdf> at 19 January 2015.

216 Koppe, *supra* note 78, 107.

217 Major Richard M. Whitaker, 'Environmental Aspects of Overseas Operations' (1995) 4 *Army Lawyer* 27, 37.

218 See, for example, Roman Reyhani, 'Comment: Protection of the Environment During Armed Conflict' (2007) 14 *Missouri Environmental Law and Policy Review* 323, 325 and the corresponding footnote.

219 Jean-Marie Henckaerts, 'Study on customary international humanitarian law: A contribution to the understanding and respect for the rule of law in armed conflict' (2005) 87:857 *International Review of the Red Cross* 175, 175.

the development of the *jus in bello* rules, the changing nature of armed conflict highlighted various of their shortcomings. It became apparent that the rules regulating the methods and means of hostilities, as well as the rights of those affected by armed conflict, were no longer adequate to deal with the circumstances of every 'modern' conflict, and thus required revision and expansion.

In addition, the three decades following the conclusion of the 1949 Geneva Convention were characterised by a process of decolonization, primarily under the auspices of the United Nations,[220] with the result that many new States came into existence during that period.[221] The process of decolonization, and the associated recognition of the right of peoples to engage in a (armed) struggle in order to exercise their right of self-determination, also made a revision of the Geneva rules necessary, since armed conflict, hitherto primarily of an *inter*-State nature, was now increasingly becoming *intra*-State.[222]

Moreover, none of these new (decolonized) States had been involved in the finalization of the 1949 Geneva Conventions, and it was important that they had some input into the revised *jus in bello* provisions, so as to encourage a universality of acceptance.

In 1968, the United Nations General Assembly called for the adoption of:[223]

> additional humanitarian international conventions or for other appropriate legal instruments to ensure the better protection of civilians, prisoners and combatants in all armed conflicts and the prohibition and limitation of the use of certain methods and means of warfare

The International Committee of the Red Cross (ICRC) and the Swiss Federal Council subsequently convened a series of conferences. On 8 June 1977, less than a month after the ENMOD Convention was opened for signature, the participants at the fourth session of the Diplomatic Conference on the Reaffirmation and Development of International Humanitarian Law Applicable in Armed Conflicts adopted the Additional Protocols of 1977.[224] These

---

220 One of the expressed principal purposes of the United Nations is the 'self-determination of peoples': United Nations Charter, article 1(2). The International Court of Justice has also elaborated on the extent of this right in the *Western Sahara Case* (Advisory Opinion) [1975] ICJ Rep 12, para 54 *et seq.*

221 Cassese, *supra* note 6, 329.

222 Koppe, *supra* note 78, 117.

223 United Nations General Assembly Resolution 2444 (XXIII) (19 December 1968) UN Doc A/RES/2444(XXIII) on the Respect for Human Rights in Armed Conflicts, article 2(b).

224 A third Additional Protocol was adopted in 2005 – Protocol III Additional to the Geneva Conventions of August 12, 1949, and relating to the Adoption of an Additional Distinctive Emblem, date of adoption 8 December 2005, 45 ILM 558 (entered into force 14 January 2007).

instruments address international armed conflicts and non-international armed conflicts. The provisions relating to the former (in the 1977 Additional Protocol I) are much more extensive than those relating to non-international armed conflicts in the 1977 Additional Protocol II, which was the first treaty focusing solely on the regulation of such conflicts.[225] This is due partly to the fact that the negotiation of restrictions over 'civil wars' raised some difficult tensions between the (perceived) sovereign rights of States to deal with what they regarded as internal matters, and the need for the international community to respond to 'the new world of "internationalised" internal conflicts'.[226]

Yet, it was also clear that the rules elaborated in common article 3 of the 1949 Geneva Conventions had proven to be inadequate to deal with the increasing number (and range) of internal conflicts. In fact, the scope of law applicable to internal conflicts has since expanded quite significantly,[227] with every *jus in bello* treaty adopted since 1996 expressly applicable to both international and internal armed conflicts.[228] Although these provisions generally do not address damage to the environment directly, acts that constitute a violation of these restrictions may result in environmental degradation.

The 1977 Additional Protocol I significantly expands the ambit of 1949 Geneva Convention IV, and relates to the protection of the environment, both by setting standards as to what harm is permissible, and establishing rules as to what may constitute an acceptable target.[229] It contains a number of general provisions that expressly provide for protection of various 'environmental targets'.[230] More significantly, the treaty introduces two specific provisions that expressly apply to damage to the natural environment during the course of an international armed conflict.

---

225 International Committee of the Red Cross, 'Response of Jean-Marie Henckaerts to the Bellinger/Haynes Comments on Customary International Law Study' (2007) 46 *International Legal Materials* 959, 965.

226 Bill Bowring, *The Degradation of the International Legal Order? The Rehabilitation of Law and the Possibility of Politics* (2008), 78.

227 Roberts and Guelff, *supra* note 73, 483.

228 International Committee of the Red Cross, *supra* note 225, 965: see, for example, Amended Protocol II to the 1980 Convention on Certain Conventional Weapons on Prohibitions or Restrictions on the Use of Mines, Booby-Traps and Other Devices, date of adoption 3 May 1996, 35 ILM 1206 (entered into force 3 December 1998), articles 2 and 3; Ottawa Convention on the Prohibition of the Use, Stockpiling, Production and Transfer of Anti-Personnel Mines and on Their Destruction, opened for signature 3 December 1997, 36 ILM 1507 (entered into force 1 March 1999), article 1; Second Protocol to the Hague Convention of 1954 for the Protection of Cultural Property in the Event of Armed Conflict, opened for signature 26 March 1999, 38 ILM 769 (entered into force 9 March 2004), articles 3 and 22.

229 Philippe Sands and Paolo Galizzi (eds), *Documents in International Environmental Law* (2nd ed, 2004), 989.

230 Leibler, *supra* note 96, 107.

Dealing first with the more general provisions, article 54 sets out a non-exhaustive list of 'objects indispensable to the survival of the civilian population' that cannot be 'attack[ed], destroy[ed] or render[ed] useless'.[231] These prohibitions are subject to 'imperative military necessity' in the situation of the 'defence [by a State Party] of its national territory against invasion'.[232] As a consequence, article 54 has been described as providing 'very modest (at times, negligible) protection'.[233] The objects to which it applies include:[234]

> foodstuffs, agricultural areas for the production of foodstuffs, crops, livestock, drinking water installations and supplies and irrigation works

Article 56(1) of the 1977 Additional Protocol I prohibits attacks against certain 'works and installations containing dangerous forces', even in circumstances where they are regarded as 'military objectives'. These include 'dams, dykes and nuclear electrical generating stations'.[235] However, these protections cease to apply if the object is used:[236]

> in regular, significant and direct support of military operations and if such attack is the only feasible way to terminate such support

as well as, in the case of a dam or dyke, 'for other than its normal function'.[237] During the negotiations leading to the finalisation of the 1977 Additional Protocol I, another provision, relating to the protection of 'nature reserves', was discussed.[238] This proposal did not receive much support and was not included in the final version of the instrument.[239]

Turning to the more specific provisions, as noted, the 1977 Additional Protocol I was the first instrument to make direct reference to the issue of environmental damage during an international armed conflict. Although, as discussed below, these provisions fall well short of providing a comprehensive protection of the environment, they still represented a 'significant innovation'[240] for the time. Articles 35(3) and 55 expressly apply to damage caused to the 'natural

231 1977 Additional Protocol I, article 54 (2).
232 *Ibid*, article 54 (5).
233 Leibler, *supra* note 96, 108.
234 1977 Additional Protocol I, article 54 (2).
235 *Ibid*, article 56(1).
236 *Ibid*, articles 56(2)(a), (b) and (c).
237 *Ibid*, article 56(2)(a).
238 Draft article 48 *ter* provided that:
'Publicly recognized nature reserves with adequate markings and boundaries declared as such to the adversary shall be protected and respected except when such reserves are used specifically for military purposes'.
239 Hulme, *supra* note 188, 112.
240 Bassiouni, *supra* note 13, 28.

environment'. Unlike the ENMOD Convention, the 1977 Additional Protocol I does not specify what is encompassed by the natural environment, although the ICRC Commentary on the Additional Protocols suggests that:[241]

> [t]he concept of the natural environment should be understood in the widest sense to cover the biological environment in which a population is living. It does not consist merely of the objects indispensable to survival ... but also includes forests and other vegetation ... as well as fauna, flora and other biological or climatic elements

Article 35(3), which appears under the general heading of 'Methods and Means of Warfare, Combatant and Prisoner-of-War Status', specifies the following 'basic rule':[242]

> [i]t is prohibited to employ methods or means of warfare which are intended, or may be expected, to cause widespread, long-term and severe damage to the natural environment

On the other hand, article 55, headed 'Protection of the natural environment', specifies that:[243]

> 1. Care shall be taken in warfare to protect the natural environment against widespread, long-term and severe damage. This protection includes a prohibition of the use of methods or means of warfare which are intended or may be expected to cause such damage to the natural environment and thereby to prejudice the health or survival of the population.
> 2. Attacks against the natural environment by way of reprisals are prohibited

At first sight, these two articles appear to be quite similar. Both refer to the same threshold terms of environmental damage. Neither provision is expressed to be subject to any notion of military necessity, so that the existence of military advantage or imperative appears to be 'irrelevant' to their application.[244]

Yet there are some important differences between them. The provisions represent different forms of obligation,[245] and appear to operate from different perspectives. Whereas article 35(3) seeks to limit the methods and means of warfare, article 55 – which appears in a chapter headed 'Civilian Objects' – has been likened more to

241 International Committee of the Red Cross, 'Protocol Additional to the Geneva Conventions of 12 August 1949, and relating to the Protection of Victims of International Armed Conflicts (Protocol I), 8 June 1977 (ICRC Protocol I Commentary); Commentaries', para 2126 <www.icrc.org/ihl.nsf/COM/470-750070?OpenDocument> at 19 January 2015.
242 1977 Additional Protocol I, article 35(3).
243 *Ibid*, article 55.
244 Leibler, *supra* note 96, 109.
245 Hulme, *supra* note 188, 73.

a 'Geneva law', in that it 'creates a protected object, [being] the environment'.[246] The former prohibits acts that are intended (or may be expected) to cause a specific threshold of damage, while the latter seeks to impose a 'standard of care'.[247] Moreover, in its second sentence, article 55(1) refers (albeit not exclusively – 'includes') to harm to humans as a necessary element in its application. Whilst it is not clear whether this requirement must also be 'intended' and qualifies the duty of care set out in the first sentence,[248] this should be compared with article 35(3), which does not make any reference to human injury.

Some commentators argue that thc fact that article 55 refers to the 'health or survival of the population', whereas article 35(3) focuses on the environment without reference to human beings, gives rise to an 'intrinsic value paradigm' – where military commanders might be forced to choose between alternate acts that either cause human injury *or* damage environmentally fragile areas.[249] This raises complex (philosophical) questions as to the relative value of what might (initially) appear to be competing interests.[250]

Article 55(2) appears to provide for a blanket prohibition of *any* deliberate environmental damage that forms an act of reprisal. On its own terms, this represents a significant protection to the environment, at least in the relatively narrow context of reprisals. There is also some debate as to whether one of the articles sets a higher requisite standard than the other. For example, the ICRC Commentary suggests that the duty of 'care' elaborated in article 55(1) allows 'some latitude of judgment'.[251] On the other hand, some commentators argue that article 55(1) has a 'wider scope of application and entails a larger responsibility' than article 35(3).[252]

As noted, both provisions contain the terms 'widespread, long-term and severe' when specifying the quantum of harm to the environment. These terms are, of course, similar to those specified in article I(1) of the ENMOD Convention. However, as used in the 1977 Additional Protocol I, these terms relate to damage *to* the environment regardless of the weapons utilised – thus classifying the environment as a 'victim' – whereas in the ENMOD Convention, they apply to the effects of the manipulation *of* the environment in the sense that it has been utilized as a 'weapon'.[253]

---

246 Schmitt, *supra* note 80, 275.
247 *Ibid*, 276.
248 See, for example, Hulme, *supra* note 188, 74–8.
249 See, for example, Richards and Schmitt, *supra* note 10, 1084.
250 See *ibid*, 1082–6.
251 ICRC Protocol I Commentary, *supra* note 241, para 2133.
252 See, for example, Koppe, *supra* note 78, 123–5; Hulme, *supra* note 188, 80–8.
253 Roberts and Guelff, *supra* note 73, 408; Yoram Dinstein, *The Conduct of Hostilities under the Law of International Armed Conflict* (2004), 189.

Also as noted, the clarifications of those terms in the 1976 CCD Understanding Relating to Article I of the ENMOD Convention expressly apply only to that instrument. This is significant, given that the ENMOD Convention and the 1977 Additional Protocol I, although negotiated separately, were concluded at approximately the same time. It can reasonably be assumed that those involved in the drafting of each instrument would have been aware of the other.[254] Moreover, the *travaux préparatoires* of the two instruments indicate that the terms were intended to have different meanings in each of them.[255]

Anthony Leibler has suggested that the meaning of the terms as used in article 35(3) and 55(1) of the 1977 Additional Protocol I might be as follows:[256]

(1) 'Widespread': encompassing at least an entire region of several hundred kilometres;
(2) 'Long-term': lasting for at least several decades;
(3) 'Severe': causing death, ill-health or loss of sustenance to thousands of people, at present or in the future

Unlike article I(1) of the ENMOD Convention, in article 35(3) and 55(1), the terms are used in the conjunctive ('and') form, with the result that the relevant environmental harm must meet *each* of those descriptors before the provision is applicable. Even apart from the uncertainties as to the precise meaning of these terms, this indicates that an extremely high degree of harm must be proven before either provision will be relevant.

This threshold was considered by a committee (the Committee) established by the Prosecutor of the ICTY on 14 May 1999 to examine certain aspects of the North Atlantic Treaty Organization's (NATO) conduct during 'Operation Allied Force', a bombing campaign conducted against the (former) Federal Republic of Yugoslavia from 24 March to 9 June 1999.[257] The Committee had been asked to determine whether the Prosecutor should be advised to begin an *ex-officio* investigation, pursuant to article 18 of the ICTY Statute, into allegations

254 *Id*, Dinstein.

255 See Laurence Boisson de Chazournes and Philippe Sands (eds), *International Law, The International Court of Justice and Nuclear Weapons* (1999), 345 and the corresponding footnote.

256 Leibler, *supra* note 96, 111. Florencio Yuzon finds such attempts to characterize these terms as 'unavailing': Yuzon, *supra* note 123, 822. See also Jozef Goldblat, 'The Mitigation of Environmental Disruption by War: Legal Approaches', in Arthur Westing (ed), *Environmental Hazards of War: Releasing Dangerous Forces in an Industrialized World* (1990) 48, 52.

257 For a description of the background to the implementation of Operation Allied Force, see Steven Freeland, 'The Bombing of Kosovo and the Milosevic Trial: Reflections on Some Legal Issues' [2002] *Australian International Law Journal* 150, 156–9.

that NATO personnel had committed serious violations of international humanitarian law during that campaign.[258]

In June 2000, the Committee issued its final report,[259] which raised some interesting issues relating to international law and the principles of the *jus in bello*. Whilst recognizing that mistakes and 'errors of judgment' may have occurred,[260] the Committee concluded that, on the basis of evidence available to it at the time, it would recommend to no investigation be commenced. The Prosecutor accepted all of these recommendations, and did not proceed to investigate the matter further. Nevertheless, the reasoning and final conclusions in the Committee Report have attracted significant criticism.[261]

Within its analysis of NATO's actions, the Committee found it necessary to consider the possible applicability of articles 35(3) and 55 of the 1977 Additional Protocol I to the environmental damage caused by the bombing campaign, in particular, relating to the use of depleted uranium projectiles and cluster bombs.[262] The Committee concluded that the two provisions had a 'very high threshold of application,' stating that:[263]

> [t]he adjectives 'widespread, long-term, and severe' used in Additional Protocol I are joined by the word 'and', meaning that it is a triple, cumulative standard that needs to be fulfilled.
> Consequently, it would appear extremely difficult to develop a *prima facie* case upon the basis of these provisions, even assuming they were applicable.

---

258 See ICTY Statute, article 18. The recommendations of the Committee were not binding on the Prosecutor, who retained complete independence as to whether or not to initiate a formal investigation.

259 Final Report to the Prosecutor of the International Criminal Tribunal for the former Yugoslavia by the Committee Established to Review the NATO Bombing Campaign Against the Federal Republic of Yugoslavia, 13 June 2000, 39 ILM 1257 (Committee Report).

260 *Ibid*, para 90.

261 See, for example, Michael Cottier, 'Did NATO Forces Commit War Crimes During the Kosovo Conflict? Reflections on the Prosecutor's Report of 13 June 2000', in Horst Fischer, Claus Kress and Sascha Rolf Lüder (eds), *International and National Prosecution of Crimes Under International Law: Current Developments* (2001), 505; Freeland, *supra* note 257.

262 See Convention on Cluster Munitions, opened for signature 3 December 2008, 48 ILM 357 (entered into force 1 August 2010), which came into force despite opposition by major producers of cluster bombs, and those countries that stockpile them, including the United States, Russia and China: 'Cluster bomb ban treaty approved' (28 May 2008) *BBC News* <http://news.bbc.co.uk/2/hi/europe/7423714.stm> at 19 January 2015. As at January 2015, the United States, Russia and China have neither signed nor ratified this convention: International Committee of the Red Cross, 'State Parties to the Following International Humanitarian Law and Other Related Treaties' <www.icrc.org/ihl/%28SPF%29/party_main_treaties/$File/IHL_and_other_related_Treaties.pdf> at 19 January 2015.

263 Committee Report, para 15.

The Committee found that the required level of damage had not been reached, concluding that, while there was an obligation to avoid excessive long-term damage to the natural environment even when bombing legitimate military targets, what was 'excessive' in this context could not be clearly defined.[264] It also held that the real impact of the campaign at that time was 'unknown and difficult to measure.'[265] In essence, it has been argued, the Committee's interpretation of the provisions meant that they would have 'almost no applicability' in virtually every circumstance.[266]

The Committee's reference to a requirement of 'excessive' damage is of interest in two respects. First, there is, in fact, *no* reference to this threshold in the two articles of the 1977 Additional Protocol I, although it is used in other provisions of that instrument,[267] which suggests that the drafters of the two articles deliberately chose not to include the word. Notwithstanding this, it appears that the Committee found it appropriate to 'add' this extra requirement – contrary to the accepted principles of treaty interpretation – so that the provisions would accord with what it regarded as the general (customary) international law obligation. Yet, by immediately observing that what constitutes excessive damage is perhaps not always capable of clear definition, the Committee Report severely compromised the force of any such obligation. The Committee, in essence, tacked on a requirement that it then held was incapable of application to the events before it. Clearly, this was a flawed approach.

Secondly, the war crime defined in the Rome Statute requires that the:[268]

> widespread, long-term and severe damage to the natural environment [must be] clearly *excessive* in relation to the concrete and direct overall military advantage anticipated

The International Criminal Court has not yet been required to determine precisely what this provision might mean. Somewhat confusingly, given its finding that a determination of what is excessive is difficult, the Committee concluded that the use of the word 'clearly' in article 8(2)(b)(iv) 'ensures that criminal responsibility would be entailed only in cases where the excessiveness of the incidental damage was obvious.'[269]

---

264 *Ibid*, para 23.

265 *Id.*

266 Aaron Schwabach, 'NATO'S War in Kosovo and the Final Report to the Prosecutor of the International Criminal Tribunal for the Former Yugoslavia' (2001) 9 *Tulane Journal of International and Comparative Law* 167, 175.

267 See, for example, 1977 Additional Protocol I, article 57(2)(a)(iii).

268 Rome Statute, article 8(2)(b)(iv) (emphasis added).

269 Committee Report, para 21.

In the course of its findings, the Committee also referred to the *Legality of the Threat or Use of Nuclear Weapons* Advisory Opinion,[270] suggesting – incorrectly – that the International Court of Justice had cast doubt as to whether article 55 also represented customary international law.[271] In its decision, the Court concluded as follows:[272]

> [t]he Court notes further-more that Articles 35, para 3, and 55 of Additional Protocol I provide additional protection for the environment. Taken together, these provisions embody a general obligation to protect the natural environment against widespread, long-term and severe environmental damage; the prohibition of methods and means of warfare which are intended, or may be expected, to cause such damage; and the prohibition of attacks against the natural environment by way of reprisals. These are powerful constraints for all the States having subscribed to these provisions

Notwithstanding the Court's sentiments, it is apparent that the protections to the environment provided by articles 35(3) and 55(1) of the 1977 Additional Protocol I are inadequate, given both the uncertainties and ambiguities of the threshold terms included, as well as the 'extreme environmental damage'[273] required for either provision to be applicable. Moreover, in light of these uncertainties, as well as the large number of reservations lodged by States Parties to the 1977 Additional Protocol I, it is widely thought that these provisions do not represent customary international law,[274] although this view is not unanimous.[275] At the same time, however, it should be noted that the instrument does reflect the principles of the Martens clause by specifying that:[276]

> civilians and combatants remain under the protection and authority of the principles of international law derived from *established custom*, from the principles of humanity and from the dictates of public conscience

It has been suggested that, because they purport to prohibit environmental damage in circumstances where such acts might be justifiable under the principle of military necessity, articles 35(3) and 55(1) of the 1977 Additional Protocol I 'probably still have no customary equivalents'.[277] Certainly, the quote referred to above from the *Legality of the Threat or Use of Nuclear Weapons*

---

270 [1996] ICJ Rep 226.

271 Committee Report, para 15. See Freeland, *supra* note 257, 163.

272 [1996] ICJ Rep 226, para 31.

273 Bruch, *supra* note 139, 703.

274 See, for example, Dinstein, *supra* note 253, 185; Anthony Leibler, *supra* note 96, 112; Yuzon, *supra* note 123, 822.

275 Sands, *supra* note 37, 315.

276 1977 Additional Protocol I, article 1(2) (emphasis added). See also 1977 Additional Protocol II, preamble para 4.

277 Fleck, *supra* note 31, 118. That author also applies the same argument to the specific environmental prohibitions in the ENMOD Convention.

Advisory Opinion does not expressly contradict this view, seeming instead to emphasize how important the provisions are as *treaty* obligations, with the implied conclusion that non-States Parties to the convention are not bound by these provisions. It is relevant to note, however, that both of the 1977 Additional Protocols have been widely ratified,[278] although not by the United States.[279]

The terms of article 35(3) of the 1977 Additional Protocol I are repeated in the preamble to the 1981 Certain Conventional Weapons Convention.[280] France has lodged a reservation to that instrument in which it considered that the obligations codified in the paragraph only applied to 'States Parties to the Protocol', and thus by implication did not represent a principle of customary international law.[281]

The terms of the 1977 Additional Protocol I do not expressly categorise a violation of either article 35(3) or article 55 as a 'grave breach' of the instrument.[282] A proposal that a violation of article 55(1) would fall within the grave breaches regime of the Geneva rules was defeated in the discussions leading to the conclusion of the 1977 Additional Protocol I.[283] Nevertheless, there is some disagreement as to whether, and in what circumstances, a breach of these provisions may constitute a grave breach.[284] This further reflects uncertainties both as to the precise scope of each provision, as well as the extent to which they represent principles of customary international law.

Overall, although the inclusion of the prohibitions provided for in both articles 35(3) and 55 of the 1977 Additional Protocol I represented a step forward

278 As at January 2015, there were 174 States Parties to the 1977 Additional Protocol I and 168 States Parties to the 1977 Additional Protocol II: International Committee of the Red Cross, 'State Parties to the Following International Humanitarian Law and Other Related Treaties' <www.icrc.org/ihl/%28SPF%29/party_main_treaties/$File/IHL_and_other_related_Treaties.pdf> at 19 January 2015.

279 The United States became a signatory to both of the 1977 Additional Protocols on 12 December 1977.

280 Convention on Prohibitions or Restrictions on the Use of Certain Conventional Weapons Which May be Deemed to be Excessively Injurious or to Have Indiscriminate Effects, opened for signature 10 April 1981, 1342 UNTS 137 (entered into force 2 December 1983) (1981 Certain Conventional Weapons Convention), preamble para 4.

281 At the time of making this reservation (10 April 1981), France was not a State Party to the 1977 Additional Protocol I (it became a State Party on 11 April 2001).

282 See 1977 Additional Protocol I, articles 11 and 85, and 1949 Geneva Convention IV, article 147.

283 Hulme, *supra* note 188, 77.

284 Compare, for example, Rymn James Parsons, 'The Fight to Save the Planet: U.S. Armed Forces, "Greenkeeping," and Enforcement of the Law Pertaining to Environmental Protection During Armed Conflict' (1998) 10 *Georgetown International Environmental Law Review* 441, 452, with Rayfuse, *supra* note 203, 84.

in the *jus in bello* at the time,[285] they have not established a comprehensive regime of accountability for intentional environmental damage in times of war, particularly since the threshold requirements of damage are unclear and, in any case, are too high as to have significant practical application. They do have some value, however, highlighting the need for combatants in armed conflicts to consider the environment, a point that has also been emphasized by the International Court of Justice.[286]

2.2.10. 1981 CERTAIN CONVENTIONAL WEAPONS CONVENTION

This treaty, together with three Protocols,[287] was adopted at the second session of a United Nations Conference held in Geneva in 1980.[288] Already by this time, States Parties to the 1977 Additional Protocol I were, as noted in chapter 1, under the following obligation with respect to 'new weapons':[289]

> [i]n the study, development, acquisition or adoption of a new weapon, means or method of warfare ... to determine whether its employment would, in some or all circumstances, be prohibited by this Protocol or by any other rule of international law

The 1981 Certain Conventional Weapons Convention and its Protocols provide for several prohibitions and restrictions in respect of particular weapons. The obligation to ensure that weapons employed during an armed conflict

---

285 Dinstein, *supra* note 253, 185.

286 *Legality of the Threat or Use of Nuclear Weapons* (Advisory Opinion) [1996] ICJ Rep 226, para 30.

287 The most relevant of these for the purposes of protection of the environment during hostilities is Protocol III on Prohibitions or Restrictions on the Use of Incendiary Weapons to the, opened for signature 10 April 1981, 1342 UNTS 137 (entered into force 2 December 1983) (1981 Protocol III). The other two Protocols relate to Non-Detectable Fragments and Mines, Booby-Traps and Other Devices respectively. Of course, by restricting these forms of weapons in particular circumstances, those two protocols may also have indirect effects in terms of protecting the environment. This is particularly the case in relation to restrictions on the use of anti-personnel mines, the use of which can render large areas of land unusable. The Ottawa Convention on the Prohibition of the Use, Stockpiling, Production and Transfer of Anti-Personnel Mines and on Their Destruction, opened for signature 3 December 1997, 36 ILM 1507 (entered into force 1 March 1999), is discussed below.

288 United Nations Conference on Prohibitions or Restrictions of Use of Certain Conventional Weapons Which May be Deemed to be Excessively Injurious or to Have Indiscriminate Effects. For a description of the background to the negotiations leading to the conclusion of the 1981 Certain Conventional Weapons Convention, see Captain J. Ashley Roach, 'Certain Conventional Weapons Convention: Arms Control or Humanitarian Law?' (1984) 105 *Military Law Review* 3, 5–17.

289 1977 Additional Protocol I, article 36.

are legal was, of course, not new.[290] The underlying theme of the 1981 Certain Conventional Weapons Convention and its Protocols derives from fundamental principles applied in earlier *jus in bello* treaties – that the 'methods and means of warfare [are] not unlimited' and that, during the conduct of armed conflict, the use of the following is prohibited:[291]

> weapons, projectiles and material and methods of warfare of a nature to cause superfluous injury or unnecessary suffering

As noted, the preamble of the 1981 Certain Conventional Weapons Convention reiterates that:[292]

> it is prohibited to employ methods or means of warfare which are intended, or may be expected, to cause widespread, long-term and severe damage to the natural environment

For the purposes of its interpretation, and notwithstanding any reservations, this provision forms part of the 'context' of the treaty – material related to its conclusion[293] – even though it does not in and of itself create a legally binding obligation.[294] Moreover, the preamble of a treaty may also assist in the determination of its 'object and purpose',[295] assuming that it is not inconsistent with the main text of the treaty.[296]

In any event, the reaffirmation of a general prohibition of 'widespread, long-term and severe damage to the natural environment' in the preamble of the 1981 Certain Conventional Weapons Convention – terms similar to those found in

---

290 W. Hays Parks, 'Conventional Weapons and Weapons Reviews' (2005) 8 *Yearbook of International Humanitarian Law* 55, 57. In early 2013, an arms trade treaty that sought to impose restrictions on the trade and export of certain conventional weapons, including weapons that might be used for acts of genocide, crimes against humanity and war crimes, was adopted by the United Nations General Assembly: United Nations General Assembly Resolution 67/234 B (2 April 2013) UN Doc A/RES/67/234 B on the Arms Trade Treaty. The treaty subsequently entered into force in December 2014.

291 1981 Certain Conventional Weapons Convention, preamble para 3.

292 *Ibid*, preamble para 4. In similar vein to the French reservation in relation to this paragraph noted earlier, the United States lodged an 'understanding' to the effect that this paragraph: 'which refers to the substance of provisions of Article 35(3) and Article 55(1) of [the 1977 Additional Protocol I], applies only to States which have accepted those provisions'.

293 Anthony Aust, *Modern Treaty Law and Practice* (2nd ed, 2007), 234.

294 See VCLT, article 31(2).

295 Tim Hillier, *Sourcebook on Public International Law* (1998), 144. See, for example, *Golder v. United Kingdom* (1975) 1 EHRR 524, where the European Court of Human Rights (at 535–6) stated that:
'the preamble is generally very useful for the determination of the 'object' and 'purpose' of the instrument to be construed'.

296 Aust, *supra* note 293, 426.

the then recently concluded ENMOD Convention and the 1977 Additional Protocol I – reflected the growing concern of the international community for the protection of the environment during times of armed conflict. Antonio Cassese, while acknowledging that the general prohibition in the preamble does not reflect customary law, argues that it may already reflect 'a general consensus and thus is binding on all members of the world community',[297] notwithstanding the uncertainties regarding the precise meaning to be attributed to each of the descriptors.

Article 2(4) of the 1981 Protocol III also provides some protection for the environment, although this has been described as 'merely incidental'.[298] This instrument relates to the use of incendiary weapons,[299] reflecting international concern triggered by the frequent use of napalm by the United States Air Force during the Vietnam War,[300] and which were raised again in relation to the alleged use by Israel of white phosphorus weapons during its military action in Gaza in early 2009.[301] Under the heading of 'Protection of civilians and civilian objects', it provides as follows:[302]

> [i]t is prohibited to make forests or other kinds of plant cover the object of attack by incendiary weapons except when such natural elements are used to cover, conceal or camouflage combatants or other military objectives, or are themselves military objectives

A 'military objective' in this context is defined as:[303]

> any object which by its nature, location, purpose or use makes an effective contribution to military action and whose total or partial destruction, capture or neutralization, in the circumstances ruling at the time, offers a definite military advantage

'Civilian objects' are defined as all objects 'which are not military objectives'.[304] It has been suggested that a forest would be a military objective for these purposes 'in jungle warfare or if it were likely to be used by retreating or advancing forces as a cover in the future'.[305] In one sense, article 2(4) of the 1981

---

297 Cassese, *supra* note 6, 421–2.
298 Yuzon, *supra* note 123, 823.
299 Article 1(1) of the 1981 Certain Conventional Weapons defines an incendiary weapon for the purposes convention.
300 Roberts and Guelff, *supra* note 73, 517.
301 Rory McCarthy, 'Fatal burns never seen before Israel's war', *The Guardian* (London), 21 January 2009, 24.
302 1981 Protocol III, article 2(4).
303 *Ibid*, article 1(3).
304 *Ibid*, article 1(4).
305 Green, *supra* note 83, 141.

Protocol III could be said to provide for broader protection of the environment than either of articles 35(3) and 55(1) of the 1977 Additional Protocol I, in that it is not limited by any requisite threshold of damage. However, in other respects, it is narrower than articles 35(3) and 55(1), since it applies only to attacks by incendiary weapons causing damage to forests and other kinds of plant cover.

Moreover, unlike articles 35(3) and 55(1), article 2(4) of the 1981 Protocol III allows for an exception based on military necessity, thus making the provision more acceptable to States.[306] Erik Koppe suggests that this undermines the practical effect of article 2(4), arguing that the type of 'scorched earth tactics' contemplated by the provision would most likely only be undertaken where there *was* a military advantage to be gained.[307] Articles 35(3) and 55(1) of the 1977 Additional Protocol I might still apply in such circumstances, but then one is again faced with the difficulties of proving the requisite very high threshold level of environmental damage.

In 2003, the States Parties to the 1981 Certain Conventional Weapons Convention adopted a Protocol dealing with 'Explosive Remnants of War'.[308] This instrument was the first multilateral agreement to address the problem of unexploded and abandoned ordnances, which pose a serious and ongoing threat to civilians long after the end of an armed conflict, and render the affected environment as unusable. In adopting the 1981 Protocol V, the States Parties reiterated:[309]

> their determination to address the humanitarian, developmental and *environmental* impact of explosive remnants of war, including existing explosive remnants of war

The 1981 Protocol V came into force in late 2006 and introduces various obligations upon participants in armed conflict in relation to the clearance, removal or destruction of such ordnances in relation to territory under its control.[310]

---

306 Schmitt, *supra* note 80, 289.

307 Koppe, *supra* note 78, 153–4.

308 Protocol V on Explosive Remnants of War to the Convention on Certain Conventional Weapons on Prohibitions or Restrictions on the Use of Mines, Booby-Traps and Other Devices, date of adoption 28 November 2003, 2399 UNTC 100 (entered into force 12 November 2006) (1981 Protocol V). For a discussion of the negotiations leading to the finalization of the 1981 Protocol V, see Louis Maresca, 'A New Protocol on Explosive Remnants of War: The History and Negotiation of Protocol V to the 1980 Convention on Certain Conventional Weapons' (2004) 86:856 *International Review of the Red Cross* 815.

309 Declaration on the Occasion of the Entry into Force of the Protocol on Explosive Remnants of War (Protocol V) Adopted by the Third Review Conference of the Convention on Prohibitions or Restrictions on the Use of Certain Conventional Weapons Which May be Deemed to be Excessively Injurious or to Have Indiscriminate Effects on 13 November 2006, preamble para 8 (emphasis added) <www.unog.ch/80256EDD006B8954/%28httpAssets%29/682877B24ABFF1BFC12572810063FB74/$file/Declaration+on+EIF+Protocol+V.pdf> at 20 January 2015.

310 1981 Protocol V, article 3.

2.2.11. 1993 CHEMICAL WEAPONS CONVENTION

As noted, the 1972 Biological Weapons Convention only went part of the way to addressing some of the shortcomings of the 1925 Geneva Protocol, which had specified prohibitions on the use of chemical and biological weapons, but had been silent with regard to their production and possession.[311] It would be a further two decades before a comprehensive international treaty would be finalised relating to chemical weapons. In the meantime, chemical weapons had been used, to tragic effect, in various conflicts, including by both sides in the Iran–Iraq War in the 1980s,[312] and also by the Saddam Hussein regime against the Kurds in Halabja and elsewhere in 1988.[313]

The 1993 Chemical Weapons Convention was concluded after lengthy negotiations and welcomed by the United Nations General Assembly, which reiterated the:[314]

> urgent necessity of a total ban on chemical weapons, so as to abolish an entire category of weapons of mass destruction, and thus to eliminate the risk to mankind of renewed use of these inhumane weapons

While the preamble of the 1993 Chemical Weapons Convention refers specifically to the use of herbicides,[315] they are not specifically included in the definitions of 'chemical weapons' and a 'toxic chemical' in the substantive clauses of the instrument,[316] notwithstanding that their use can have very

---

311 Schindler and Toman, *supra* note 16, 239.

312 Schmitt, *supra* note 80, 289.

313 See United Nations Secretary-General, 'Report of the Mission Dispatched by the Secretary-General to Investigate Allegations of the Use of Chemical Weapons in the Conflict between The Islamic Republic of Iran and Iraq' (12 March 1986) UN Doc S/17922, para 58(a). In April 1991, the United Nations Security Council noted Iraq's 'prior use of chemical weapons' and 'invited' that country to reaffirm its obligations under the 1925 Geneva Protocol and the 1972 Biological Weapons Convention (this was prior to the conclusion of the 1993 Chemical Weapons Convention): United Nations Security Council Resolution 687 (3 April 1991) UN Doc S/RES/687/1991 on Iraq-Kuwait, preamble para 8 and para 11 respectively. In September 2007, the Iraqi High Criminal Tribunal confirmed the death sentence in the 'Anfal' trial against Ali Hassan al-Majid (known as 'Chemical Ali') and four others for crimes against humanity involving the deaths of approximately 100,000 Kurds during 1988, including approximately 5,000 Kurdish civilians who were killed by chemical weapons (mustard and nerve gas) in the northern Iraqi town of Halabja: "Chemical Ali' Sentence Confirmed' (4 September 2007) *BBC News* <http://news.bbc.co.uk/2/hi/middle_east/6977842.stm> at 20 January 2015.

314 United Nations General Assembly Resolution 47/39 (30 November 1992) UN Doc A/RES/47/39 on the Convention on the Prohibition of the Development, Production, Stockpiling and Use of Chemical Weapons and on Their Destruction, preamble para 5.

315 1993 Chemical Weapons Convention, preamble para 7.

316 *Ibid*, articles II(1) and (2).

serious ecological and environmental effects.[317] However, those definitions are broad in scope and, in any event, the United States – which had negotiated for the exclusion of herbicides within the definitional provisions – has formally renounced their 'first use ... in time of armed conflict'.[318]

Like the 1972 Biological Weapons Convention, the 1993 Chemical Weapons Convention is usually described as a 'disarmament agreement'[319] rather than an international humanitarian law instrument. However, the use of chemical weapons would inevitably have detrimental effects on the environment. The 1993 Chemical Weapons Convention comprehensively prohibits the use of chemicals that are likely to harm the environment,[320] and represents the first international agreement where a complete category of weapons is banned and their use prohibited. Article I(1) of the convention provides that the States Parties undertake 'never under any circumstances':[321]

> (a) To develop, produce, otherwise acquire, stockpile or retain chemical weapons, or transfer, directly or indirectly, chemical weapons to anyone;
> (b) To use chemical weapons;
> (c) To engage in any military preparations to use chemical weapons;
> (d) To assist, encourage or induce, in any way, anyone to engage in any activity prohibited to a State Party under this Convention.

In addition, States Parties are required to destroy their stockpiles of chemical weapons, as well as their production facilities,[322] and are prohibited from using 'riot control agents as a method of warfare'.[323] States Parties must also 'assign the highest priority to ... protecting the environment' during the transportation, sampling, storage and destruction of chemical weapons,[324] the destruction of weapons facilities,[325] the national implementation of obligations under the treaty,[326] and during production activities that are not prohibited.[327]

---

317 Royal Swedish Ministry for Foreign Affairs, *supra* note 14, 109.
318 See Dinstein, *supra* note 253, 188 and the corresponding footnote.
319 For a comprehensive analysis of the disarmament aspects of the 1993 Chemical Weapons Convention, see Eric P.J. Myjer (ed), *Issues of Arms Control Law and the Chemical Weapons Convention* (2001).
320 Schmitt, *supra* note 80, 291.
321 1993 Chemical Weapons Convention, article I(1).
322 *Ibid*, articles I(2), I(3) and I(4).
323 *Ibid*. A riot control agent is defined in article I(5).
324 *Ibid*, article IV(10).
325 *Ibid*, article V(11).
326 *Ibid*, article VII(3).
327 *Ibid*, Annex on implementation and verification, part VI(C)(7).

A specialised body – The Hague-based Organization for the Prohibition of Chemical Weapons (OPCW) – is established under the 1993 Chemical Weapons Convention, with the following specific tasks:[328]

> to achieve the object and purpose of this Convention, to ensure the implementation of its provisions, including those for international verification of compliance with it, and to provide a forum for consultation and cooperation among States Parties

While the 1993 Chemical Weapons Convention represents another step forward towards an international regime of weapons control, some commentators argue that its enforcement mechanisms fall short of what is required to ensure the prevention of the proliferation of chemical weapons.[329] Michael Scharf suggests that the success of the 1972 Biological Weapons Convention and the 1993 Chemical Weapons Convention will ultimately depend upon a widespread acceptance of a 'chemical and biological weapons taboo'.[330]

2.2.12. 1997 OTTAWA CONVENTION ON THE PROHIBITION OF THE USE, STOCKPILING, PRODUCTION AND TRANSFER OF ANTI-PERSONNEL MINES AND ON THEIR DESTRUCTION[331]

The indiscriminate use of anti-personnel mines and similar weapons can have serious detrimental environmental impacts, rendering otherwise valuable areas unusable and highly dangerous, and obstructing economic development.[332] In the 1980s and 1990s, their use became the subject of significant concern. A 1996 Report issued by the ICRC, citing the United States State Department, estimated that there were about 84 million 'uncleared' landmines in 64 countries around the world.[333] Despite the fact that there has been steady progress in the clearing of some mined areas – for example, in 2006, over 450 square kilometers of land

328 *Ibid*, article VIII(1).

329 See, for example, Kellman, *supra* note 184, 600–1.

330 Scharf, *supra* note 182, 510.

331 Opened for signature 3 December 1997, 36 ILM 1507 (entered into force 1 March 1999) (1997 Ottawa Convention).

332 Christopher Stone refers to contrary arguments that suggest that the mining of 'delicate landscapes' such as the Sinai or the Korean Demilitarized Zone might even represent a 'benefit to the environment by providing the natural terrain with a macabre reprieve from human incursion': Stone, *supra* note 1, 21.

333 International Committee of the Red Cross, 'Anti-personnel Landmines: Friend or Foe? – A Study of the Military Use and Effectiveness of Anti-personnel Mines' (1996) <www.icrc.org/eng/assets/files/other/icrc_002_0654.pdf> para 1, at 20 January 2015 (ICRC Landmines Report).

was cleared[334] – the task ahead looks formidable. Indeed, the ICRC Landmines Report estimated that there were a 100 million such mines stockpiled around the world,[335] and referred to United Nations projections that:[336]

> if the use of mines were stopped immediately it would take 1,100 years and $33 billion dollars to clear, at current rates, those already in place

As noted, Protocol II to the 1981 Certain Conventional Weapons Convention relates to mines, booby-traps and other devices.[337] While it did prohibit the use of such devices indiscriminately or when directed against civilians,[338] its practical effects were minimal. It did not restrict the use of such weapons *per se*, did not place any limits on their manufacture or transfer, and was not formally applicable to civil wars.[339] In 1996, the 1981 Protocol II was amended so as to rectify various shortcomings, but no agreement could be reached on a total prohibition of the use of anti-personnel mines.

Included within a landmark Resolution headed 'General and Complete Disarmament'[340] adopted by the United Nations General Assembly in late 1996, States were urged:[341]

> to pursue vigorously an effective, legally binding international agreement to ban the use, stockpiling, production and transfer of anti-personnel landmines ... as soon as possible

A series of meetings led by Canada eventually gave rise to the conclusion of the 1997 Ottawa Convention, which was expressed to be based upon:[342]

> the principle of international humanitarian law that the right of the parties to an armed conflict to choose methods or means of warfare is not unlimited, on the principle that prohibits the employment in armed conflicts of weapons, projectiles and materials and methods of warfare of a nature to cause superfluous injury

---

334 Stephanie Holmes, 'Delicate territory: the landmine ban' *BBC News*, 2 December 2007 <http://news. bbc.co.uk/2/hi/in_depth/7121419.stm> at 20 January 2015.

335 ICRC Landmines Report, para 103.

336 *Ibid*, para 1.

337 Protocol II on Prohibitions or Restrictions on the Use of Mines, Booby-Traps and Other Devices to the, opened for signature 10 April 1981, 1342 UNTS 137 (entered into force 2 December 1983) (1981 Protocol II).

338 *Ibid*, articles 3(2) and (3).

339 Roberts and Guelff, *supra* note 73, 645.

340 United Nations General Assembly Resolution 51/45 (10 December 1996) UN Doc A/RES/51/45 on General and Complete Disarmament.

341 *Ibid*, para 1.

342 1997 Ottawa Convention, preamble para 11.

or unnecessary suffering and on the principle that a distinction must be made between civilians and combatants

In terms similar to the general obligations under the 1972 Biological Weapons Convention and the 1993 Chemical Weapons Convention, States Parties to the 1997 Ottawa Convention undertake, 'never under any circumstances':[343]

(a) To use anti-personnel mines;
(b) To develop, produce, otherwise acquire, stockpile, retain or transfer to anyone, directly or indirectly, anti-personnel mines;
(c) To assist, encourage or induce, in any way, anyone to engage in any activity prohibited to a State Party under this Convention.

The 1997 Ottawa Convention provides for a complete ban on the use of anti-personnel mines, and thus provides for further (indirect) protection of the environment, as well as for remediation of their harmful effects. Clearly, however, the effectiveness of such treaty-based bans is dependent upon widespread ratification and practical implementation. Without universal acceptance of these prohibitions, 'awkward situations' may arise.[344]

Moreover, it is apparent that the scale of the existing problem of landmines poses significant challenges to the effective implementation of its principal goals. The 1997 Ottawa Convention contemplates that the process of destroying all stockpiles of anti-personnel mines in areas under a State's jurisdiction or control may take up to 10 years, with allowance for an extension of time based on various submitted reasons, including '[t]he humanitarian, social, economic, and environmental implications of the extension'.[345]

2.2.13. 1999 SECOND PROTOCOL TO THE HAGUE CONVENTION OF 1954 FOR THE PROTECTION OF CULTURAL PROPERTY IN THE EVENT OF ARMED CONFLICT[346]

As noted, the 1954 Hague Convention could be waived in certain cases of military necessity. This was seen as compromising the protections afforded by

343 *Ibid*, article 1.1.
344 See, for example, Byers, *supra* note 74, 124, where the author describes how, in 2001, Canadian Soldiers had been ordered by their American commander to lay mines around their camp in Afghanistan. When the Canadian soldiers refused to do so, American soldiers, who were not subject to the restrictions of the 1997 Ottawa Convention, laid the mines.
345 1997 Ottawa Convention, article 5(4)(c).
346 Opened for signature 26 March 1999, 38 ILM 769 (entered into force 9 March 2004) (Second Protocol to the 1954 Hague Convention).

that instrument, particularly after several of those States that had called for the waiver failed to ratify the treaty.[347] During the 1990s, there were renewed calls to supplement the provisions of the convention and to clarify the extent of its application. The fact that it was not applicable to non-international armed conflicts also raised some concerns, particularly in light of the fact that:[348]

> in many recent cases the destruction of the physical evidence of the existence of the national, ethnic and/or religious community under attack has been an integral part of the various types and levels of humanitarian abuse of the civilian peoples, through to the level of alleged genocide [in the then former Yugoslavia]

A Diplomatic Conference in The Hague in March 1999 adopted the Second Protocol to the 1954 Hague Convention. This instrument supplements the 1954 Hague Convention in a number of ways. In particular, it expressly applies to armed conflicts not of an international character[349] and, in addition, clarifies the concept of 'military necessity' for the purposes of determining whether a 'waiver' may be invoked under article 4(2) of the 1954 Hague Convention.[350] The instrument also specifies particular precautions that a State Party must take in the conduct of military operations, so as to minimize the possibility of 'incidental damage to cultural property' protected under article 4 of the 1954 Hague Convention. Included within these is the obligation to 'cancel or suspend an attack if it becomes apparent':[351]

> that the attack may be expected to cause incidental damage to cultural property protected under Article 4 of the [1954 Hague] Convention which would be excessive in relation to the concrete and direct military advantage anticipated

2.3. APPLICABILITY OF INTERNATIONAL ENVIRONMENTAL LAW TREATIES

As noted, as well as a consideration of the relevant *jus in bello* treaty instruments, it is pertinent to consider whether and, if so, to what extent, any of the existing international environmental law treaties might also regulate intentional environmental destruction during armed conflict.

---

347 Forrest, *supra* note 167, 210.

348 Patrick J. Boylan, *Review of the Convention for the Protection of Cultural Property in the Event of Armed Conflict: (The Hague Convention of 1954)* (1993), para A8.

349 Second Protocol to the 1954 Hague Convention, articles 3(1) and 22(1).

350 *Ibid*, article 6.

351 *Ibid*, article 7(d)(ii). The words in articles 7(b) and 7(d)(ii) of the Second Protocol to the 1954 Hague Convention are similar to those used in article 8(2)(b)(iv) of the Rome Statute, although the latter refers to damage that is 'clearly' excessive and requires a consideration of the 'overall' military advantage.

2.3.1. DEVELOPMENT OF INTERNATIONAL ENVIRONMENTAL LAW TREATIES

The body of law that is now categorized as international environmental law (IEL) has developed very rapidly over the past few decades, particularly following the 1972 Stockholm Declaration,[352] which, as noted, was the first significant statement of fundamental international principles relating to the protection of the environment.[353] The Stockholm Conference had been convened by the United Nations General Assembly, which was 'convinced' that there was:[354]

> the need for intensified action at national, regional and international level to limit and, where possible, to eliminate the impairment of the human environment and in order to protect and improve the natural surroundings in the interests of man

Shortly afterwards, the United Nations Environment Programme (UNEP) was established '[t]o provide leadership and encourage partnership in caring for the environment by inspiring, informing, and enabling nations and peoples to improve their quality of life without compromising that of future generations.'[355] Further momentum for the development of international environmental standards was provided by the 1992 United Nations Conference on Environment and Development.[356]

Even before these landmark events, however, there had been a number of important developments, including the judicial awards in the *Fur Seals Arbitration*[357] and the *Trail Smelter Arbitration*[358] – the latter being described

---

352 Declaration of the United Nations Conference on the Human Environment (16 June 1972) UN Doc A/CONF.48/14/Rev.1 (1972 Stockholm Declaration).

353 Parsons, *supra* note 284, 455.

354 United Nations General Assembly Resolution 2398 (XXIII) (3 December 1968) UN Doc A/RES/2398(XXIII) on the Problems of the Human Environment, preamble para 11.

355 United Nations Environment Programme, 'About UNEP: The Organization' <www.unep.org/Documents.Multilingual/Default.asp?DocumentID=43> at 20 January 2015.

356 This conference either directly resulted in, or provided the impetus for, the conclusion of a number of important IEL instruments, including the Declaration of the United Nations Conference on Environment and Development (14 June 1992) UN Doc A/CONF.151/26 (Volume 1), 31 ILM 874 (1992 Rio Declaration), the Non-Legally Binding Authoritative Statement of Principles for a Global Consensus on the Management, Conservation and Sustainable Development of All Types of Forests (14 June 1992) UN Doc A/CONF.151/26 (Vol. III), 31 ILM 881, the United Nations Framework Convention on Climate Change, opened for signature 4 June 1992, 31 ILM 849 (entered into force 21 March 1994) (UNFCCC), the Convention on Biological Diversity, opened for signature 5 June 1992, 31 ILM 822 (entered into force 29 December 1993), and the United Nations Conference on Environment and Development Report, 'Agenda 21' UN Doc A/CONF.151/26/Rev.1 (vol I) (1993).

357 *Bering Sea Fur Seals Fisheries Arbitration (Great Britain v. United States)* (1893) 1 Moore's International Arbitration Awards 755. For a discussion of the case, see Sands, *supra* note 37, 561–7.

358 *Trail Smelter Arbitration (United States v. Canada)* (1941) 3 RIAA 1905. This case arose from damage caused in Washington State by sulphur dioxide emissions from a smelter in Trail,

as the 'grandfather' of international environmental law[359] – as well as the conclusion of various treaties dealing with diverse environment-related issues, including nuclear testing,[360] oil pollution at sea,[361] and the dumping of radioactive waste.[362]

The use of nuclear weapons during the Second World War had begun a process of discussion regarding the environmental consequences of such weapons.[363] The 1972 Stockholm Declaration made specific reference to these weapons, principle 26 providing as follows:

> [m]an and his environment must be spared the effects of nuclear weapons and all other means of mass destruction. States must strive to reach prompt agreement, in the relevant international organs, on the elimination and complete destruction of such weapons

As noted in chapter 1, the 1982 World Charter for Nature contained a number of general principles relating to the environment and armed conflict, including that '[n]ature shall be secured against degradation caused by warfare or other hostile activities' and '[m]ilitary activities damaging to nature shall be avoided'.[364]

Ten years later, the Rio Declaration provided that:[365]

> [t]he environment and natural resources of people under oppression, domination and occupation shall be protected

and:

British Colombia. The Arbitral Tribunal concluded (at 1965) that '[n]o State has the right to use or permit the use of its territory in such a manner as to cause injury ... in or to the territory of another ...' This decision is regarded as establishing the general principle of State responsibility for environmental damage: Leibler, *supra* note 96, 70. Within the context of international armed conflict, the International Court of Justice has endorsed the *Trail Smelter* principle, affirming that it was 'every State's obligation not to allow knowingly its territory to be used for acts contrary to the rights of other States': *Corfu Channel Case (United Kingdom of Great Britain and Northern Ireland v. Albania)* (Merits) (Judgment) [1949] ICJ Rep 4, 22.

359 Kenneth F. McCallion, 'International Environmental Justice: Rights and Remedies' (2003) 26 *Hastings International and Comparative Law Review* 427, 429.

360 Treaty Banning Nuclear Weapons Tests in the Atmosphere, in Outer Space and under Water, opened for signature 5 August 1963, 480 UNTS 43 (entered into force 10 October 1963).

361 International Convention for the Prevention of Pollution of the Sea by Oil, opened for signature 12 May 1954, 327 UNTS 3 (entered into force 26 July 1958).

362 Antarctic Treaty, opened for signature 1 December 1959, 402 UNTS 71 (entered into force 23 June 1961), article 5(1).

363 John O'Brien, *International Law* (2002), 551.

364 United Nations General Assembly Resolution 37/7 (28 October 1982) UN Doc A/RES/37/7 on the World Charter for Nature, Annex, principles 5 and 20.

365 1992 Rio Declaration, principles 23 and 24.

> [w]arfare is inherently destructive of sustainable development. States shall therefore respect international law providing protection for the environment in times of armed conflict and cooperate in its further development, if necessary

The 1972 Stockholm Declaration and the 1992 Rio Declaration are important examples of the broad non-binding 'soft law' instruments that also relate to the environment. These instruments provide guidelines or standards of conduct that may often influence the actions of States in relation to environmental protection, but they do not have the legal 'force' of binding treaties.[366] Rather, it is only the reflection (if any) in treaties and customary law of the relevant concepts contained in these soft law instruments that gives rise to international legal obligations.[367]

As noted, the past three decades have also seen a large number of binding international instruments directed towards protecting the environment.[368] Perhaps not surprisingly, there is no single legal definition of the 'environment' consistently applied in these instruments;[369] rather, the meaning of that expression (if it is defined at all) will depend upon the context of the particular treaty.

Cherif Bassiouni has calculated that, in the period 1899–1996, there had been 55 instruments that have dealt with unlawful acts against certain internationally protected elements of the environment.[370] Whilst this list also includes the ENMOD Convention and the 1977 Additional Protocol I, several of the other more significant conventions had been negotiated and agreed under the auspices of the UNEP,[371] which, however, for most practical purposes lacks real authority to enforce international environmental law.[372]

---

366 For a discussion of the role of soft law instruments in various international law regimes, see Christine Chinkin, 'The Challenge of Soft Law: Development and Change in International Law' (1989) 38 *International and Comparative Law Quarterly* 850; Dinah Shelton, *Commitment and Compliance; The Role of Non-Binding Norms in the International Legal System* (2000): Steven Freeland, 'For Better or For Worse? The Use of 'Soft Law' within the International Legal Regulation of Outer Space' (2011) XXXVI *Annals of Air and Space Law* 409.

367 Pierre-Marie Dupuy, 'Soft Law and the International Law of the Environment' (1991) 12 *Michigan Journal of International Law* 420, 428.

368 Kenneth F. McCallion and H. Rajan Sharma, 'Environmental Justice Without Borders: The Need for an International Court of the Environment to Protect Fundamental Environmental Rights' (2000) 32 *George Washington Journal of International Law and Economics* 351, 355.

369 John H. Currie, Craig Forcese and Valerie Oosterveld, *International Law: Doctrine, Practice, and Theory* (2007), 719.

370 M. Cherif Bassiouni, *Introduction to International Criminal Law* (2003), 155–6 and 253–5.

371 These include the Montreal Protocol on Substances that Deplete the Ozone Layer, opened for signature 16 September 1987, 26 ILM 154 (entered into force 1 January 1989) (Montreal Protocol), the UNFCCC, and the Kyoto Protocol to the United Nations Framework Convention on Climate Change, opened for signature 16 March 1998, 37 ILM 22 (entered into force 16 February 2005).

372 Peggy Rodgers Kalas, 'International Environmental Dispute Resolution and the Need for Access by Non-State Entities' (2001) 12 *Colorado Journal of International Environmental Law and Policy* 191, 230–1.

Drawing from the *Trail Smelter Arbitration* principles, many of these IEL agreements give rise to obligations on States to exercise 'due diligence' to prevent environmental damage to other States – or, in some cases, to areas beyond the territorial jurisdiction of any State – arising from activities under their jurisdiction or control. It has therefore been argued that accountability arises not so much from the environmental harm *per se*, but rather from the failure to exercise the requisite standard of care.[373] These instruments, both soft law and binding agreements, are typically not punitive in nature.[374] Instead, the prevention of environmental damage is the 'underlying objective'.[375]

In the *Gabcíkovo-Nagymaros Project Case*, the International Court of Justice was asked to consider a dispute relating (in part) to the possible environmental effects of a system of locks diverting a section of the River Danube. Although it decided the dispute primarily on the basis of the relevant treaty[376] rather than on environmental principles, as noted previously, the Court affirmed that it was:[377]

> mindful that, in the field of environmental protection, vigilance and prevention are required on account of the often irreversible character of damage to the environment and of the limitations inherent in the very mechanism of reparation of this type of damage.

2.3.2. APPLICABILITY DURING WARFARE?

One could assert that environmental law and policy alone may not be sufficient to protect the environment in normal (peace time) circumstances.[378] Whereas IEL has traditionally been developed as an autonomous discipline,[379] increasingly diverse international (and national) human activities continue to have a greater impact upon the environment. This is even more the case in relation

---

373 Mark Allan Gray, 'The International Crime of Ecocide' (1996) 26 *California Western International Law Journal* 215, 242.

374 Robert McLaughlin, 'Improving Compliance: Making Non-State International Actors Responsible for Environmental Crimes' (2000) 11 *Colorado Journal of International Environmental Law and Policy* 377, 380.

375 Hulme, *supra* note 188, 52.

376 This was a bilateral treaty between Hungary and Czechoslovakia dated 17 September 1977. The Court held that Slovakia, as successor to Czechoslovakia, became a party to the treaty as from 1 January 1993: *Gabcíkovo-Nagymaros Project (Hungary v. Slovakia)* (Judgment) [1997] ICJ Rep 7, 83.

377 *Ibid*, para 140.

378 Avi Brisman, 'Toward a More Elaborate Typology of Environmental Values: Liberalizing Criminal Disenfranchisement Laws and Policies' (2007) 33 *New England Journal on Criminal and Civil Confinement* 283, 454.

379 Jane Holder and Maria Lee, *Environmental Protection, Law and Policy: Text and Materials* (2nd ed, 2007), xliv.

to environmental damage during armed conflict, where, as noted in chapter 1, experience has shown that the soft-law declarations and binding IEL agreements have largely been ineffective and, in any event, may not even be applicable.

Indeed, the development of international law regarding the protection of the environment during wartime has 'lagged' behind the development of environmental protection during peace time.[380] One reason for this is that, in general, most rules of international law, including many of the IEL principles – but with the exception of the *jus in bello*, and perhaps also human rights principles[381] – do not expressly extend to situations of armed conflict.[382] There are very few multilateral IEL treaties that specifically relate to environmental protection during armed conflict, and many are silent as to their applicability following the outbreak of hostilities.[383] There is no general or specific rule of international law that assumes that such agreements will continue to operate during times of hostilities, and there is significant disagreement among commentators as to what the correct position may be.[384] Moreover, there is 'insufficient uniformity of opinion' among States on the issue.[385] Previously, it had traditionally been assumed that all treaties between the belligerents in a war terminated *ipso facto* upon the outbreak of hostilities; however, it is now more generally thought that the question will depend on the type of treaty itself.[386]

The VCLT is not particularly helpful in this regard, simply stating that:[387]

> [t]he provisions of the present Convention shall not prejudge any question that may arise in regard to a treaty from … the outbreak of hostilities between States

380 See Andy Rich, 'The Environment: Adequacy of Protection in Times of War' (2004) 12 *Penn State Environmental Law Review* 445, 455 and the corresponding footnote.

381 In *Legal Consequences of the Construction of a Wall in the Occupied Palestinian Territory* (Advisory Opinion) [2004] ICJ Rep 136, the International Court of Justice (at para 106) considered that 'the protection offered by human rights conventions does not cease in case of armed conflict'. This repeats the view expressed by the Court in *Legality of the Threat or Use of Nuclear Weapons* (Advisory Opinion) [1996] ICJ Rep 226, para 25.

382 Hampson, *supra* note 84, 51.

383 Sands, *supra* note 37, 309.

384 See Silja Vöneky, 'Peacetime Environmental Law as a Basis of State Responsibility for Environmental Damage Caused by War', in Jay E. Austin and Carl E. Bruch (eds), *The Environmental Consequences of War: Legal, Economic, and Scientific Perspectives* (2000), 190, 193–4 and the corresponding footnotes.

385 Iain Scobbie, 'The approach to customary international law in the Study', in Elizabeth Wilmshurst and Susan Breau (eds), *Perspectives on the ICRC Study on Customary International Humanitarian Law* (2007), 15, 41. The author notes (at 41) the differing submissions made by the four States that addressed the issue in *Legality of the Threat or Use of Nuclear Weapons* (Advisory Opinion) [1996] ICJ Rep 226, also referred to in the 2005 ICRC 'Study on Customary International Law', which is discussed in chapter 3.

386 Vöneky, *supra* note 384, 197 and the corresponding footnotes.

387 VCLT, article 73.

In 1993, a panel of experts was convened under the auspices of the ICRC to draft 'Guidelines for Military Manuals and Instructions on the Protection of the Environment in Times of Armed Conflict'. Article 5 of this document provided that:[388]

> international environmental agreements and relevant rules of customary law may continue to be applicable in times of armed conflict to the extent that they are not inconsistent with the applicable law of armed conflict

In 1994, the ICRC Guidelines were submitted to the United Nations General Assembly which, without formally approving the draft, passed a general resolution relating to the 'Decade of International Law (1990–1999), in which it:[389]

> invite[d] all States to disseminate widely the revised guidelines for military manuals and instructions on the protection of the environment in times of armed conflict received from the International Committee of the Red Cross and to give due consideration to the possibility of incorporating them into their military manuals and other instructions addressed to their military personnel

However, this invitation has thus far not been widely taken up, although a number of States, including the United States (in its Army Operational Law handbook),[390] and Germany (in its Joint Services Regulations), have amended their respective military regulations in direct response the ICRC Guidelines, but even then not to the degree of environmental protection reflected in that document.[391]

In the end, whether a specific IEL treaty applies during periods of armed conflict will generally be determined by reference to the express terms of the treaty itself. There have been several suggested methodologies as to how this question

388 International Committee of the Red Cross, 'Guidelines for Military Manuals and Instructions on the Protection of the Environment in Times of Armed Conflict' (ICRC Environment Guidelines), article II(5), (1995) <www.icrc.org/eng/resources/documents/misc/57jn38.htm> at 20 January 2015. See also Hans-Peter Gasser, 'Guidelines for Military Manuals and Instructions on the Protection of the Environment in Times of Armed Conflict' (1996) 311 *International Review of the Red Cross* 230, appendix.

389 United Nations General Assembly Resolution 49/50 (9 December 1994) UN Doc A/RES/49/50 on the United Nations Decade of International Law, para 11.

390 In addition, the Uniform Code of Military Justice (UCMJ), which applies to all members of the United States Armed Forces, provides for prosecution in situations of environmental damage caused by United States military personnel. Some commentators believe that this is the most appropriate enforcement mechanism to deal with illegal environmental damage during wartime: see, for example, Eric Talbot Jensen and James J. Teixeira, Jr., 'Prosecuting Members of the U.S. Military for Wartime Environmental Crimes' (2005) 17 *Georgetown International Environmental Law Review* 651, 671.

391 Schmitt, *supra* note 80, 270–1.

might be determined in practice. Philippe Sands classifies the applicability (or otherwise) of different IEL treaties in times of armed conflict in the following ways:[392]

(a) treaties that expressly exclude their applicability in relation to damage that occurs as a result of war or armed conflict;
(b) treaties that allow for total or partial suspension at the instigation of one of the Parties;
(c) treaties that require the consequences of hostilities to influence subsequent decisions under the relevant treaty;
(d) treaties that expressly exclude their applicability to any military activities, even during times of peace;
(e) treaties that expressly apply to specific activities associated with the conduct of hostilities; or
(f) treaties that expressly or impliedly apply at all times.

Edith Brown Weiss, while acknowledging that most IEL treaties do not indicate whether they are applicable during an armed conflict, asserts that some treaties can only effectively fulfil their purposes if they do continue to apply during times of hostilities.[393]

In more general terms, Silja Vöneky asserts that State practice and legal doctrine commonly result in the following five categories of treaties continuing to bind States Parties even during times of international armed conflict:[394]

(a) treaties expressly providing for continuance during war;
(b) treaties that are compatible with the maintenance of war;
(c) treaties creating an international regime or status;
(d) human rights treaties; and
(e) *jus cogens* rules and obligations *erga omnes*.[395]

---

392 See Sands, *supra* note 37, 309–10, who provides examples of IEL treaties that fall within each of these categories.

393 Edith Brown Weiss, 'Opening the Door to the Environment and to Future Generations', in Laurence Boisson de Chazournes and Philippe Sands (eds), *International Law, The International Court of Justice and Nuclear Weapons* (1999), 338, 347–8. The author cites as examples the Convention for the Protection of the World Cultural and Natural Heritage, opened for signature 16 November 1972, 11 ILM 1358 (entered into force 17 December 1975), the Convention on International Trade in Endangered Species of Wild Fauna and Flora, opened for signature 3 March 1973, 993 UNTS 243 (entered into force 1 July 1975), and the Montreal Protocol.

394 Vöneky, *supra* note 384, 198.

395 This is not an entirely accurate statement, since a *jus cogens* rule or *erga omnes* obligation is a principle of customary international law, although it may also be included as a term(s) of a treaty. The existence of identical conventional and customary rules was recognized by the International Court of Justice in *North Sea Continental Shelf Cases (Federal Republic of*

Silja Vöneky provides a number of examples of IEL treaties that may fall within the first three categories specified.[396] In terms of human rights treaties, article 12 of the ICESCR recognizes 'the right of everyone to the enjoyment of the highest attainable standard of physical and mental health'.[397] The provision specifies that, in order to achieve the realization of this right, States Parties to the treaty shall take necessary steps for '[t]he improvement of all aspects of environmental and industrial hygiene'.[398] With respect to this provision, the Special Rapporteur of the (then) United Nations Commission on Human Rights asserted that:[399]

> in times of armed conflict, the deliberate causing of large-scale environmental damage which severely affects the health of a considerable proportion of the population concerned, or creates risks for the health of future generations, amounts to a serious violation of [article 12 ICESCR]

All in all, in the absence of specific provisions dealing with the point, the application of IEL and other significant multilateral treaties in times of armed conflict is often problematic. For example, the 1982 United Nations Convention on the Law of the Sea[400] was intended to regulate the use of the seas in times of peace.[401] However, the terms of the treaty may still be relevant in relation to various military uses of the seas.[402]

Germany v. Denmark and Federal Republic of Germany v. The Netherlands) (Judgment) [1969] ICJ Rep 3, para 71: see also *Military and Paramilitary Activities in and against Nicaragua (Nicaragua v. United States of America)* (Merits) (Judgment) [1986] ICJ Rep 14, para 177–8. Article 53 of the VCLT defines a 'peremptory norm', and this is often used as a definition of a *jus cogens* rule. An obligation *erga omnes* has been described by the International Court of Justice as an obligation owed by a State 'towards the international community as a whole ... the concern of all States ... [and that] all States can be held to have a legal interest in their protection': *Barcelona Traction, Light and Power Company, Limited (Belgium v. Spain)* (Judgment) [1970] ICJ Rep 3, para 33. The International Court of Justice has also made reference to the issue of *erga omnes* obligations in the context of environmental concerns: see *Nuclear Tests Case (Australia v. France)* (Judgment) [1974] ICJ Rep 253, para 50; *Nuclear Tests Case (New Zealand v. France)* (Judgment) [1974] ICJ Rep 457, para 52.

396 Vöneky, *supra* note 384, 198–203.

397 International Covenant on Economic, Social and Cultural Rights, opened for signature 16 December 1966, 993 UNTS 3 (entered into force 3 January 1976) (ICESCR), article 12(1).

398 *Ibid*, article 12(2)(b).

399 Commission on Human Rights, Report prepared by Special Rapporteur Walter Kälin in accordance with Human Rights Commission Resolution 1991/67, 'Report on the situation of human rights in Kuwait under Iraqi occupation' (16 January 1992) UN Doc E/CN.4/1992/26, para 208, quoted in Vöneky, *supra* note 384, 201.

400 Opened for signature 10 December 1982, 1833 UNTS 3; 21 ILM 1261 (entered into force 16 November 1994) (UNCLOS).

401 R.R. Churchill and A.V. Lowe, *The Law of the Sea* (2nd ed, 1988), 307.

402 There are a number of treaties within the *jus in bello* rules relating to the use of the seas, stemming primarily from the Hague Peace Conferences. In addition, the 1949 Geneva Convention II addresses the rights of the wounded, sick and shipwrecked members of armed forces at sea. For a discussion of the relevant rules relating to the military uses of the sea, see, for example, *ibid*, Churchill and Lowe, 307 *et seq*; Wolff Heintschel von Heinegg, 'The Law of

When presented with the opportunity to do so, the International Court of Justice has chosen not to 'categorize' the vast array of IEL treaties in terms of their possible applicability to times of armed conflict, but has instead adopted a different approach. In the *Legality of the Threat or Use of Nuclear Weapons* Advisory Opinion, the Court heard conflicting arguments as to whether certain IEL treaties – as well as some of the *jus in bello* treaties discussed above[403] – were applicable in times of armed conflict in general, and to the use of nuclear weapons in particular. In response, the Court somewhat 'side-stepped' the differing viewpoints,[404] instead concluding:[405]

> that the issue is not whether the treaties relating to the protection of the environment are or are not applicable during an armed conflict, but rather whether the obligations stemming from these treaties were intended to be obligations of total restraint during military conflict

Citing Principle 24 of the Rio Declaration, the Court continued:[406]

> [t]he Court does not consider that the treaties in question could have intended to deprive a State of the exercise of its right of self-defence under international law because of its obligations to protect the environment.

However, as noted in chapter 1, the Court subsequently affirmed that environmental considerations *were* relevant when assessing the acts designed to support military objectives.

In his report on the effects of armed conflicts on treaties, the ILC Special Rapporteur, whilst acknowledging that there was no consensus among States on the specific legal question, suggested that the comments of the International Court of Justice:[407]

> provide general and indirect support for the use of a presumption that environmental treaties apply in case of armed conflict

In reaching its conclusions, the Court affirmed that (customary) principles of international environmental law are relevant to the conduct of armed conflict, in that they can help to determine whether a particular action that has adverse

---

Armed Conflict at Sea', in Fleck (ed), *The Handbook of Humanitarian Law in Armed Conflicts* (1999), 405–83; Dinstein, *supra* note 253, 102–8.

403 These included the ENMOD Convention and the 1977 Additional Protocol I.

404 Sands, *supra* note 37, 310.

405 [1996] ICJ Rep 226, para 30.

406 *Id.*

407 First Report on the Effects of Armed Conflicts on Treaties, by Ian Brownlie, Special Rapporteur, (21 April 2005) UN Doc A/CN.4/552, para 91.

effects on the environment is reasonable and proportional in the context of what constitutes military necessity in the circumstances.[408] It has also been suggested that many of the arguments presented to the Court were premised on the continued applicability of the environmental treaties during an armed conflict involving the use of a nuclear weapon.[409]

Yet, even then, in the end, the Court could not categorically say that the threat or use of nuclear weapons – which would have dire environmental consequences – would in every circumstance constitute a violation of international law.[410] The irony of this finding was highlighted by Roger Clark, who asked the question:[411]

> [i]f it is absolutely forbidden by humanitarian law to use one dum-dum bullet on one person, how can it not be absolutely forbidden to use a nuclear weapon on millions?

Yet, as a result of the Court's ultimate decision – perhaps more accurately described as a 'non-decision' – if the use of such weapons may be lawful in specific circumstances, it follows that the consequential and undoubtedly severe environmental damage associated with such use would also be lawful.[412] At the same time, the findings of the Court highlight the fact that the IEL treaties in general do not *directly* address the issue of the use of nuclear weapons. This is indicative of the 'great caution shown by [S]tates',[413] when codifying principles in relation to criminal responsibility for environmental damage, particularly in the context of warfare.

2.3.3. LIABILITY AND STATE RESPONSIBILITY RATHER THAN CRIMINAL RESPONSIBILITY

As noted in chapter 1, to the extent that issues of accountability for environmental damage have been specifically addressed in IEL treaties, the tendency has been to design this as a regime of liability for States, based upon traditional principles

---

408 See also Betsy Baker, 'Legal Protections for the Environment in Times of Armed Conflict' (1993) 33 *Virginia Journal of International Law* 351, 354–5.

409 International Law Commission, 'The effect of armed conflict on treaties: an examination of practice and doctrine: Memorandum by the Secretariat' (1 February 2005) UN Doc A/CN.4/550, footnote 117.

410 On this issue the Court was divided equally, with the casting vote of President Bedjaoui deciding the matter: see ICJ Statute, article 55(2).

411 Roger S. Clark, 'International Court of Justice: Advisory Proceedings on the Legality of the Threat or Use of Nuclear Weapons (Question Posed by the General Assembly): The Laws of Armed Conflict and the Use or Threat of Use of Nuclear Weapons' (1996) 7 *Criminal Law Forum* 265, 288.

412 Karen Hulme, 'Natural Environment', in Elizabeth Wilmshurst and Susan Breau (eds), *Perspectives on the ICRC Study on Customary International Humanitarian Law* (2007), 204, 216.

413 Birnie and Boyle, *supra* note 110, 286.

of State responsibility,[414] rather than as a system encompassing criminal responsibility for individuals (or States). This is reflective of traditional notions of environmental justice, which are concerned with 'the distribution of benefits and burdens in the use and management of environmental resources'.[415] The application of these principles means that the liability that arises is 'remedied' by the making of reparations,[416] which may often constitute a monetary compensation – a traditional 'civil' remedy.[417] This approach has been used, for example, in relation to marine pollution and nuclear accidents; in both of these enforcement regimes, however, there is no express reference or applicability to armed conflict.

Although not 'just' an IEL treaty, UNCLOS does regulate the marine environment. As noted above, the extent to which it applies during times of armed conflict remains uncertain. Section 9 of UNCLOS deals with 'Responsibility and Liability' and provides that:[418]

> [S]tates are responsible for the fulfillment of their international obligations concerning the protection and preservation of the marine environment. *They shall be liable in accordance with international law*

In similar vein, a number of IEL treaties introduce a scheme of liability for damage caused by nuclear accidents.[419] While in certain circumstances these conventions impose liability not upon States, but rather upon the operator of a nuclear installation or ship (which may include the handler or carrier of nuclear material), they do not impose any criminal responsibility.

---

414 O'Brien, *supra* note 363, 559.

415 Gerard C. Rowe, 'Environmental Justice as an Ethical, Economic and Legal Principle', in Benjamin J. Richardson and Klaus Bosselmann (eds), *Environmental Justice and Market Mechanisms* (1999), 58, 58.

416 See *Case Concerning the Factory at Chorzów (Germany v. Poland)* (Jurisdiction) [1927] PCIJ Rep, Series A, No. 9, 21.

417 Parsons, *supra* note 284, 454.

418 UNCLOS, article 235(1) (emphasis added). See further Birnie and Boyle, *supra* note 110, 382–92.

419 The principal treaties dealing with damage caused by nuclear accidents are the Paris Convention on Third Party Liability in the Field of Nuclear Energy, opened for signature 29 July 1960, 956 UNTS 251 (entered into force 1 April 1968) and the Vienna Convention on Civil Liability for Nuclear Damage, opened for signature 29 May 1963, 1063 UNTS 255 (entered into force 12 November 1977), established under the auspices of the Organisation for Economic Development (OECD) and the International Atomic Energy Agency (IAEA) respectively. Two amending Protocols to these conventions were adopted in 1988 (Joint Protocol Relating to the Application of the Vienna Convention on Civil Liability for Nuclear Damage and the Paris Convention on Third Party Liability in the Field of Nuclear Energy, opened for signature 21 September 1988, 1672 UNTS 301 (entered into force 27 April 1992)) and 1997 (Protocol to amend the 1963 Vienna Convention on Civil Liability for Nuclear Damage, opened for signature 12 September 1997, 36 ILM 1462 (entered into force 4 October 2003)).

It has been asserted that neither of these liability regimes, nor other accountability systems provided for within the framework of the IEL treaties, are particularly effective in 'shap[ing] the behaviour of the polluter',[420] and that existing international dispute mechanisms relating to environmental destruction are 'inadequate to protect the rights of individuals and the environment in the 21st century.'[421] Indeed, it has been noted that, although there are certain IEL treaties that have successfully established a liability regime, this remains the exception, and setting up such regimes for other types of environmental damage 'has been much more difficult'.[422] Whether or not these criticisms underestimate the value and effect of the broad range of IEL conventions in operation, in general terms, to the extent that specific IEL treaties do apply during the course of armed conflict, they are weak and inadequate to provide for a comprehensive and stringent set of binding principles regulating acts of intentional environmental damage.

As noted, one reason for this is that the enforcement procedures and accountability regime they establish are not directed towards the criminalization of such acts, but rather are based on the concept of State responsibility. Even then, State responsibility in the context of international environmental damage has traditionally not often been raised, and a significant number of IEA treaties do not even contain specific environmental standards that could give rise to claims of a 'wrongful act' in a situation of violation.[423] There is, consequently, relatively little international case law concerning State responsibility for environmental damage, although this may now be changing.[424]

In the end, the enforcement of international environmental law 'rests with national interest'.[425] Given the sometimes highly politicized circumstances surrounding disputes between States involving environmental harm, as well

420 Malgosia Fitzmaurice, 'International Responsibility and Liability', in Daniel Bodansky, Jutta Brunnée and Ellen Hey (eds), *The Oxford Handbook of International Environmental Law* (2007), 1010, 1034.

421 Kalas, *supra* note 372, 192.

422 See Anthony Aust, *Handbook of International Law* (2005).

423 Elli Louka, *International Environmental Law: Fairness, Effectiveness, and World Order* (2006), 468–9.

424 For example, a number of disputes between States involving environmental issues have in recent times been brought before the International Court of Justice. These include disputes relating to the environmental impact of a proposed system of locks diverting part of the River Danube (*Gabcíkovo-Nagymaros Project (Hungary v. Slovakia)* (Judgment) [1997] ICJ Rep 7), and of two proposed pulp mills on the River Uruguay (*Case Concerning Pulp Mills on the River Uruguay (Argentina v. Uruguay)* (Judgment) [2010] ICJ Rep 14). More recently, that Court has adjudicated in a dispute between Australia and Japan regarding Japan's whaling activities in Antarctic waters: see *Whaling in the Antarctic (Australia v. Japan: New Zealand intervening)* (Judgment) 31 March 2014 <www.icj-cij.org/docket/files/148/18136.pdf>.

425 Ben Boer, Ross Ramsay and Donald R. Rothwell, *International Environmental Law in the Asia Pacific* (1998), 18.

as the broader ramifications that may arise when one State seeks to enforce an obligation against another State through a judicial (and highly public) process, there remains an inherent disincentive to undertake contentious adjudication, even though it may be legally justified under the principles of State responsibility.[426] This may be even more so in circumstances where the particular IEL treaty does not establish a comprehensive dispute resolution procedure. In this sense, the pressures of *realpolitik* may further limit the already inadequate environmental protections in times of armed conflict afforded by IEL treaties.

Of course, as noted in chapter 1, liability was attributed to Iraq following its deliberate acts of 'environmental terrorism'[427] during the 1990–1991 Gulf War. The United Nations Security Council resolved that Iraq was 'liable under international law' for *inter alia* 'environmental damage and the depletion of natural resources'. This represented the first formal international acknowledgement that deliberate acts of environmental destruction during wartime were compensable.[428] It was also the first time that international law was used as the tool by which parties sought compensation for wartime environmental damage.[429] Given the high 'profile' of the damage, an enforcement regime was instituted to coordinate and determine claims for compensation – principally through the establishment and operation of the United Nations Compensation Commission (UNCC).[430] This regime is significantly more comprehensive than most other mechanisms established under international law.

Yet, this was an *ad hoc* response that was implemented after the fact. What the events in Kuwait in 1990–1991 did demonstrate is that, for all practical purposes, such acts were not adequately dealt with by the existing IEL treaties. Certainly, the provisions of the IEL treaties to which Iraq was a State Party clearly did not deter it from carrying out the acts of environmental destruction.[431] In addition,

---

426 Orellana, *supra* note 107, 686.

427 This was a term used by the United States Pentagon at the time to describe the actions of the Iraqi forces: Andrew Rosenthal, 'War in the Gulf: The President; Bush Calls Gulf Oil Spill a 'Sick' Act by Hussein', *The New York Times* (New York), 26 January 1991 <www.nytimes.com/1991/01/26/world/war-in-the-gulf-the-president-bush-calls-gulf-oil-spill-a-sick-act-by-hussein.html> at 20 January 2015.

428 Luan Low and David Hodgkinson, 'Compensation for Wartime Environmental Damage: Challenges to International Law After the Gulf War' (1995) 35 *Virginia Journal of International Law* 405, 406.

429 Hunter, Salzman and Zaelke, *supra* note 52, 1392.

430 See United Nations Security Council Resolution 687 on Iraq-Kuwait (3 April 1991) UN Doc S/RES/687/1991, para 18.

431 Iraq was not in violation of either the ENMOD Convention, or the 1977 Additional Protocol I, in relation to such acts, since it was not a State Party to either instrument at the time (it became a State Party to the 1977 Additional Protocol I in April 2010: International Committee of the Red Cross, 'State Parties to the Following International Humanitarian Law and Other Related Treaties <www.icrc.org/ihl/%28SPF%29/party_main_treaties/$File/IHL_and_other_related_Treaties.pdf> at 19 January 2015. Moreover, at least in the opinion

due to the nature of these treaties, no individuals were found criminally responsible for such acts under international law.

Moreover, from a broader perspective, it has been argued that reliance on the principles of State responsibility for environmental destruction 'merely imposes a threat of [possible] post-war financial obligations', which a belligerent State will almost invariably 'overlook' in order to achieve its military objectives.[432] In most circumstances, the benefits of achieving military objectives during hostilities will outweigh the financial burdens associated with the protection of the environment.[433]

In any event, the degree of effectiveness of compensatory regimes focusing on the concept of reparation for environmental damage is even more complex and unsatisfactory given the difficulty with arriving at an appropriate evaluation (in monetary terms) of the damage caused to the environment,[434] as well as the severely limited powers of enforcement under which the secretariats established under the various IEL treaties will have to operate.[435] Of course, there may also be a considerable time lag before the full environmental effects of any such acts can be more accurately determined, thus potentially giving rise to considerable delays in relation to any system of sanctions based upon a (monetary) reparation for actual damage caused.

2.4. CONCLUSION

This chapter has reviewed the various relevant *jus in bello* instruments and broad IEL framework to demonstrate that no comprehensive treaty regime has been established to provide for accountability in respect of the intentional destruction of the environment during armed conflict. To the extent that conventional environmental limits do exist, compliance is not uniform and often relies upon the requisite political will within the international community.[436]

---

of some commentators, the environmental protection provisions of those two treaties did not represent customary international law at the time of such actions: see, for example, Dinstein, *supra* note 253, 193. However, the actions of the Iraqi forces may have constituted a violation of 1949 Geneva Convention IV: Louis Rene Beres, 'Iraqi Crimes and International Law: The Imperative to Punish' (1993) 21 *Denver Journal of International Law and Policy* 335, footnote 16, and may have infringed the necessity or proportionality requirements under customary international law (see further chapter 3) governing military activities: Birnie and Boyle, *supra* note 110, 149.

432 Leibler, *supra* note 96, 77.

433 Yuzon, *supra* note 123, 803.

434 Hulme, *supra* note 188, 60.

435 Kalas, *supra* note 372, 223.

436 Gillian D. Triggs, *International Law: Contemporary Principles and Practices* (2006), 829.

The IEL treaties also provide little direct relevance to the issue. Apart from the fact that it is not clear as to how far any particular IEL instrument may apply in times of war, to the extent that they are applicable, they do not provide a scheme for criminal responsibility for such actions. Rather, they institute a regime based on traditional principles of State responsibility, incorporating notions of liability and the obligation of a State to make reparation or provide compensation where there has been a violation, without addressing specifically the issue of individual criminal responsibility.[437]

It is true that the International Court of Justice has affirmed that environmental considerations are to be taken into account in times of armed conflict. However, this still leaves much room for uncertainty. While it may be effective on an *ex post facto* basis, in circumstances where there is a broad consensus that some form of enforcement must be implemented (such as occurred after the 1990–1991 Gulf War), in general terms the (possible) application of international environmental principles, and a regime based on liability as an enforcement mechanism, provides very little disincentive, and is too weak to represent an effective system of deterrence.

In terms of the *jus in bello*, it has been noted that, while those rules may have, over time, become more 'humanized', they have not shown any signs of being 'environmentalized' to any significant degree.[438] There are, as has been discussed, a number of *jus in bello* provisions that do refer either directly or indirectly to the environment, reflecting what has been described as the 'incipient convergence' of international humanitarian law and increasing concern for the environment.[439] However, for the most part, these provisions are either subject to overarching considerations of military necessity (for example in the case of article 2(4) of the 1981 Protocol III), subject to seemingly unattainable threshold standards (in the case of articles 35(1) and 55(1) of the 1977 Additional Protocol I), or capable of restrictive interpretation (the ENMOD Convention). It is also the case that only a limited range of intentional acts of environmental destruction would be covered by the more specific provisions in the 1977 Additional Protocol I or the ENMOD Convention.[440]

As noted, in response to the failure of the *jus in bello* principles to adequately address the deliberate environmental destruction caused by the Iraqi forces

437 Lyal S. Sunga, *The Emerging System of International Criminal Law: Developments in Codification and Implementation* (1997), 224.
438 Mark A. Drumbl, 'International Human Rights, International Humanitarian Law, and Environmental Security: Can the International Criminal Court Bridge the Gaps?' (2000) 6 *ILSA Journal of International and Comparative Law* 305, 306.
439 Richards and Schmitt, *supra* note 10, 1048.
440 Dinstein, *supra* note 253, 196.

in 1990–1991, a 'radical'[441] proposal emerged to conclude a 'Fifth Geneva' Convention on the Protection of the Environment in Time of Armed Conflict. A conference was convened in June 1991 in order to discuss what the provisions of such an instrument might cover, including prohibitions on the use of the environment as a weapon, of weapons aimed against the environment, and of 'indirect' damage to the environment.[442]

Participants at the conference acknowledged that the existing corpus of treaty law was highly deficient in the area of environmental protection, and the proposal also gained the backing of Greenpeace.[443] However, on the whole, this 'solution' was not more broadly supported, with several participants arguing that, instead of negotiating a new instrument, greater efforts should be made to encourage more effective implementation of the existing rules, notwithstanding their shortcomings.[444] One participant made the observation that 'creating an effective legal regime in relation to environmental protection will require more than a comprehensive international treaty'.[445]

Various other commentators have subsequently suggested that the proposed Fifth Geneva Convention would not properly differentiate between 'active' and 'passive' environmental damage caused during armed conflict,[446] and that it may not be an appropriate or adequate way to address the 'challenges of our time', since there is a 'need for a more complex and diversified approach' to regulating the *jus in bello*.[447]

The generally critical reaction to the proposal for a new treaty for the protection of the environment during war was perhaps indicative of the fact that the broad assumptions regarding the emergence of a 'New World Order' following the Gulf War were overly ambitious.[448] Certainly, however, it also reflected a reality

441 Gasser, *supra* note 388, 639.

442 Hunter, Salzman and Zaelke, *supra* note 52, 1395.

443 *Id.*

444 Günther Handl, Book Review – Glen Plant (ed), 'Environmental Protection and the Law of War: A "Fifth Geneva" Convention on the Environment in Time of Armed Conflict' (1993) 42:4 *The International and Comparative Law Quarterly* 976, 976.

445 Comments by Richard Falk in Glen Plant (ed), *Environmental Protection and the Law of War: A "Fifth Geneva" Convention on the Environment in Time of Armed Conflict* (1992), 95.

446 See, for example, Eric Talbot Jensen, 'The International Law of Environmental Warfare: Active and Passive Damage During Armed Conflict' (2005) 38 *Vanderbilt Journal of Transnational Law* 145, 181.

447 Carsten Stahn, 'The Future of *Jus Post Bellum*', in Carsten Stahn and Jann K. Kleffner (eds), *Jus Post Bellum: Towards a Law of Transition From Conflict to Peace* (2008), 231, 236.

448 Handl, *supra* note 444, 977. The idea of a 'New World Order' was raised by United States President George H.W. Bush in a speech in September 1990, when he said: 'Out of these troubled times ... a new world order ... can emerge: a new era – freer from the threat of terror, stronger in the pursuit of justice, and more secure in the quest for peace': George H.W. Bush, Address Before a Joint Session of Congress, 11 September 1990.

that States were reluctant to adequately address, through binding international legal instruments, situations where participants in an armed conflict intend to damage to the environment. As the review of the relevant treaties in this chapter has indicated, this remains the position today.

The following chapter therefore considers whether, instead, the principles of customary international law, particularly those relating to the *jus in bello*, offer any greater degree of appropriate legal regulation of, and accountability for such acts.

CHAPTER THREE

REGULATION OF THE INTENTIONAL DESTRUCTION OF THE ENVIRONMENT DURING WARFARE UNDER CUSTOMARY INTERNATIONAL LAW

3.1. INTRODUCTION – CUSTOMARY INTERNATIONAL LAW WITHIN INTERNATIONAL LAW

The previous chapter examined those treaty provisions in the *jus in bello* instruments, as well as in some international environmental law (IEL) agreements, which might, in specific circumstances, directly or indirectly apply to intentional environmental damage during armed conflict. As was seen, there are several treaty provisions of relevance which, in conformity with the general nature of the *jus in bello* rules, restrict the methods and means of warfare, but they do not comprehensively prohibit such acts. Indeed, at least within those two treaty regimes, there is no mechanism in place that would adequately assign culpability. In addition, some of the (potentially) applicable conventional rules place military necessity as an overarching consideration, thus significantly diluting their effectiveness in seeking to regulate intentional environmental destruction.

It is therefore necessary to consider whether there are any rules of international law of a non-conventional nature that may also be applicable. This chapter thus considers those customary international law principles that might also have some bearing on situations where there is intentional damage to the environment during armed conflict.

As noted, customary international law is one of the traditional sources of international law, as specified in article 38(1)(b) of the Statute of the International Court of Justice.[1] The classic description of what constitutes a rule of customary

1 Opened for signature 26 June 1945, 1 UNTS 16, 59 Stat. 1031 (entered into force 24 October 1945) (ICJ Statute).

international law was given by the International Court of Justice in the *North Sea Continental Shelf Cases*.[2] The Court confirmed that customary international law evolves over time,[3] and is derived from sufficient evidence of both the 'settled practice' of States, as well as *opinio juris* ('recognition as law'), which it described as 'a belief that this practice is rendered obligatory by the existence of a rule of law requiring it'.[4]

As such, the traditional view of customary international law is that it encompasses an objective element, as well as a subjective or 'psychological' element.[5] Much has been written about these two elements and how they should be properly formulated,[6] which is also relevant to the process by which they may be adduced in relation to a specific customary rule. As noted by the International Court of Justice:[7]

> [i]t is of course axiomatic that the material of customary international law is to be looked for primarily in the actual practice and *opinio juris* of States

For the purposes of examining the relevant customary rules, reference to principles of customary international law in this book will relate to this traditional view of how it is to be adduced, despite some disagreement in relation to the use of this methodology.[8] That said, it should be noted in passing that there is no longer uniform agreement among international law commentators

2 See *North Sea Continental Shelf Cases (Federal Republic of Germany v. Denmark and Federal Republic of Germany v. The Netherlands)* (Judgment) [1969] ICJ Rep 3, para 77. There have been prior definitions of international custom, some dating back to early international law treatises: for details, see Christiana Ochoa, 'The Individual and Customary International Law Formation' (2007) 48 *Virginia Journal of International Law* 119, 122. See also Dana Zartner Falstrom, 'Can International Law Survive the 21st Century? Yes, with Patience, Persistence and a Peek at the Past' (2007) 8 *San Diego International Law Journal* 291.

3 Note, however, the observations of Judge Lachs of the International Court of Justice that: '[t]he first instruments that men sent into outer space traversed the air space of States and circled above them in outer space, yet the launching States sought no permission, nor did the other States protest. This is how the freedom of movement into outer space, and in it, came to be *established and recognised as law within a remarkably short period of time*': Dissenting Opinion of Judge Lachs, *North Sea Continental Shelf Cases (Federal Republic of Germany v. Denmark and Federal Republic of Germany v. The Netherlands)* [1969] ICJ Rep 3, 230 (emphasis added).

4 *North Sea Continental Shelf Cases (Federal Republic of Germany v. Denmark and Federal Republic of Germany v. The Netherlands)* (Judgment) [1969] ICJ Rep 3, para 77.

5 Erik V. Koppe, *The Use of Nuclear Weapons and the Protection of the Environment During International Armed Conflict* (2006), 166.

6 See, for example, Heather Cash, 'Security Council Resolution 1593 and Conflicting Principles of International Law: How the Future of the International Criminal Court is at Stake' (2007) 45 *Brandeis Law Journal* 573, 592.

7 *Continental Shelf (Libyan Arab Jamahiriya v. Malta)* (Judgment) [1985] ICJ Rep 13, para 27.

8 See, for example, John O. McGinnis and Ilya Somin, 'Symposium: Global Constitutionalism: Global Influence on U.S. Jurisprudence: Should International Law be Part of our Law?' (2007) 59 *Stanford Law Review* 1175, 1199.

that the two distinct elements specified by the Court in the late 1960s are still 'necessary' to prove the existence of a rule of customary international law.[9] There have, for example, been suggestions that, if one were to apply the classical definition of customary international law, some human rights norms would fail to qualify as custom, due to a lack of consensus as to State practice.[10]

There has therefore developed a growing body of contemporary academic literature – described as 'revisionist accounts of custom formation'[11] – that questions the traditional understanding of what constitutes a rule of customary international law. For example, it has been suggested that the practice of non-State actors is also relevant for the formation of customary international humanitarian law.[12]

What is not disputed is that, unlike conventional principles, which usually bind only those parties to the relevant treaty,[13] rules of customary international law are binding on *all* States, subject to the (possible) exception of the so-called 'persistent objector' rule.[14] Consequently, the customary international law principles applicable to the conduct of armed conflict – and the associated

---

9 See, for example, International Law Association, 'Final Report of the International Committee on the Formation of Customary (General) International Law; Principles Applicable to the Formation of General Customary International Law', in *Report of the Sixty-Ninth Conference* (2000), 712, 720–1.

10 See, for example, McGinnis and Somin, *supra* note 8, 1200; Jochen A. Frowein, Book Review – Theodor Meron, 'The Humanization of International Law' (2007) 101 *American Journal of International Law* 680, 683.

11 See, for example, Iain Scobbie, 'The approach to customary international law in the Study', in Elizabeth Wilmshurst and Susan Breau (eds), *Perspectives on the ICRC Study on Customary International Humanitarian Law* (2007), 15, 24 and the commentaries listed in the corresponding footnote.

12 See Jonathan Somer, 'Jungle justice: passing sentence on the equality of belligerents in non-international armed conflict' (2007) 89:867 *International Review of the Red Cross* 655, 662 and the corresponding footnotes. See also Decision on Defence Motion for Interlocutory Appeal on Jurisdiction, *Prosecutor v Duško Tadić*, Case No. IT-94-1, Appeals Chamber, 2 October 1995, para 108.

13 See Vienna Convention on the Law of Treaties, opened for signature 23 May 1969, 1155 UNTS 331; 8 ILM 679 (entered into force 27 January 1980) (VCLT), article 34. This general rule, reflected in the maxim *pacta tertiis nec nocent nec prosunt*, represents a principle of customary international law: D.J. Harris, *Cases and Materials on International Law* (6th ed, 2004), 847. Articles 35 and 36 of the VCLT respectively specify the circumstances whereby obligations and rights may be created for a third State, based on the consent of that State.

14 See *Fisheries Case* (*United Kingdom v. Norway*) (Judgment) [1951] ICJ Rep 116, page 131. Some commentators have, however, cast doubt as to whether the persistent objector rule exists: see, for example, Jean-Marie Henckaerts, 'Study on customary international humanitarian law: A contribution to the understanding and respect for the rule of law in armed conflict' (2005) 87:857 *International Review of the Red Cross* 175, 181 and corresponding footnote. Nevertheless, it has been pointed out that the authors of the ICRC Study on International Humanitarian Law (discussed below) appear to have accepted that the concept of a persistent objector may have some relevance in relation to the applicability of several (suggested) customary *jus in bello* rules: see Scobbie, *supra* note 11, 34.

obligations to prevent their breach and to punish the perpetrators of international crimes – bind each State, regardless of any specific treaty obligations it may, or may not, have accepted.

For this reason, it is appropriate to investigate what customary international law principles are relevant to the intentional destruction of the environment during warfare. As noted in chapter 2, some of the significant *jus in bello* treaties have not been universally ratified. For example, notwithstanding its wide acceptance, there is a diverse group of States that have not ratified the 1977 Additional Protocol I.[15] This compromises not only the extent to which those *jus in bello* treaty rules can have widespread application, but also how they may apply as comprehensive and binding standards in relation to the conduct of armed conflict. As Michael Bothe has put it, the number of States who have, up until now, withheld ratification of that instrument, highlights its 'non-universality'.[16]

There are also other important reasons why a consideration of the relevant rules of customary international law applying to armed conflict is necessary. Even though, as illustrated in the previous chapter, the development of a significant corpus of conventional *jus in bello* rules has meant that international humanitarian law is 'heavily regulated by treaty',[17] customary international law is still a 'critically important' part of this area of law.[18] In fact, it is true to say that the *jus in bello* rules were founded in customary law,[19] and it is generally agreed

---

15 These include India, Indonesia, Iran, Iraq, Israel, Malaysia, Pakistan, Papua New Guinea, Singapore, Somalia, Sri Lanka, Sudan, Thailand, Turkey and the United States: see International Committee of the Red Cross, 'State Parties to the Following International Humanitarian Law and Other Related Treaties' <www.icrc.org/ihl/%28SPF%29/party_main_treaties/$File/IHL_and_other_related_ Treaties.pdf> at 19 January 2015.

16 Michael Bothe, 'Customary International Humanitarian Law: Some Reflections on the ICRC Study' (2005) 8 *Yearbook of International Humanitarian Law* 143, 146. The author was, in fact, referring to *both* the 1977 Additional Protocol 1 and Protocol II Additional to the Geneva Conventions of August 12, 1949, and relating to the Protection of Victims of Non-International Armed Conflicts, opened for signature 8 June 1977, 1125 UNTS 609, 16 ILM 1442 (entered into force 7 December 1978) (1977 Additional Protocol II).

17 Daniel Bethlehem, 'The methodological framework of the Study', in Elizabeth Wilmshurst and Susan Breau (eds), *Perspectives on the ICRC Study on Customary International Humanitarian Law* (2007), 3, 6.

18 Dennis Mandsager, 'Introductory Note to Response of Jean-Marie Henckaerts to the U.S. Joint Letter from John Bellinger III, Legal Advisor, U.S. Department of State, and William J. Haynes, General Counsel, U.S. Department of Defense to Dr. Jakob Kellenberger, President, International Committee of the Red Cross, Regarding Customary International Law Study' (2007) 46 *International Legal Materials* 957, 957.

19 Oonagh E. Fitzgerald, 'Implementation of International Humanitarian and Related International Law in Canada', in Oonagh E. Fitzgerald (ed), *The Globalized Rule of Law: Relationships between International and Domestic Law* (2006), 625, 627.

that the treaties still do not encompass the full extent of the rules applicable to the conduct of armed conflict. As Judge Koroma has written:[20]

> [t]he relevant treaty law covers a wide variety of aspects of warfare, but treaty law, by its very nature, is unable to provide a complete picture of the state of the law

This reiterates the observations of the Nuremberg Military Tribunal in 1946[21] – admittedly when there were considerably fewer *jus in bello* treaties in force – that:[22]

> [t]he law of war is to be found not only in treaties, but in the customs and practices of states which gradually obtained universal recognition, and from the general principles of justice applied by jurists and practiced by military courts. This law is not static, but by continual adaptation follows the needs of a changing world. Indeed, in many cases treaties do no more than express and define for more accurate reference the principles of law already existing.

Indeed, as noted, the message conveyed by the opening words of the Martens clause – '[u]ntil a more complete code of the laws of war has been issued ...'[23] – has been repeated in various forms in a number of subsequent *jus in bello* treaties,[24] and clearly indicates that the terms of the relevant treaties are themselves not to be regarded as completely exhaustive, since the principles of the *jus in bello* continue to evolve. This is particularly the case in relation

---

20 Abdul G. Koroma, 'Foreword', in Jean-Marie Henckaerts and Louise Doswald-Beck (eds), *Customary International Humanitarian Law* (Volume I: Rules, 2005), xii, xii.

21 International Military Tribunal for the Trial of the German Major War Criminals at Nuremberg (Nuremberg Military Tribunal).

22 *Judgment of the International Military Tribunal, Trial of the Major War Criminals* 1 October 1946, reprinted in (1947) 41 *American Journal of International Law* 172, 219.

23 See, for example, Hague Convention IV Respecting the Laws and Customs of War on Land, opened for signature 18 October 1907, 3 Martens (3rd) 461 (entered into force 26 January 1910) (1907 Hague Convention IV), preamble para 8.

24 See, for example, the 1977 Additional Protocol I, article 1(2) ('[i]n cases not covered by this Protocol or by other international agreements ...'), and the Convention on Prohibitions or Restrictions on the Use of Certain Conventional Weapons Which May be Deemed to be Excessively Injurious or to Have Indiscriminate Effects, opened for signature 10 April 1981, 1342 UNTS 137 (entered into force 2 December 1983) (1981 Certain Conventional Weapons Convention), preamble para 5 ('... in cases not covered by this Convention and its annexed Protocols or by other international agreements ...'). Both of these provisions make express reference to the applicability of the 'principles of international law derived from established custom'. On the other hand, the formulation of the Martens clause in the preamble of the 1977 Additional Protocol II makes no reference to 'established custom' since, at the time it was formulated, it was assumed that the practice of States had *not* yet developed to the point where there was a body of customary rules in relation to conflicts not of an international nature: Michael Bothe, Karl Josef Partsch and Waldemar A. Solf, *New Rules for Victims of Armed Conflicts: Commentary on the Two 1977 Protocols Additional to the Geneva Conventions of 1949* (1982), 620.

to those armed conflicts that are not of an international nature,[25] which now constitute the majority of armed conflicts[26] and which bear witness to many atrocities (even if this has historically not been readily acknowledged).[27] As noted in previous chapters, there are significantly fewer treaty rules regulating these conflicts as compared to international conflicts.[28]

Indeed, in relation to many non-international armed conflicts, it still remains the case that common article 3 of the 1949 Geneva Conventions[29] remains the most significant applicable conventional provision, notwithstanding the subsequent conclusion of the 1977 Additional Protocol II. Although common article 3 is a very important provision, the rules that it encompasses cannot be considered as representing a completely comprehensive framework.[30] As a consequence, it has been suggested that the 'most significant contribution' of customary international law in relation to the conduct of armed conflict relates to the regulation of internal armed conflicts,[31] where it is seen as filling some important gaps over and above the scope of common article 3 and the 1977 Additional Protocol II.[32]

---

25 Mandsager, *supra* note 18, 957.

26 Carl E. Bruch, 'All's Not Fair in (Civil) War: Criminal Liability for Environmental Damage in Internal Armed Conflict' (2001) 25 *Vermont Law Review* 695, 720 and 749.

27 Mark Findlay, *Governing through Globalised Crime: Futures for international criminal justice* (2008), 198. For example, Henry Stimson, who was involved in the establishment of the Nuremberg Military Tribunal, had suggested to President Roosevelt before that Tribunal was set up that: 'I have great difficulty in finding any means whereby military commissions may try and convict those responsible for excesses committed in Germany both before and during the war which have no relation to the conduct of war': quoted in John Hagan, *Justice in the Balkans: Prosecuting War Crimes in The Hague Tribunal* (2003), 25.

28 Henckaerts, *supra* note 14, 178.

29 Geneva Convention for the Amelioration of the Condition of the Wounded and Sick in Armed Forces in the Field, opened for signature 12 August 1949, 75 UNTS 31 (entered into force 21 October 1950) (1949 Geneva Convention I), Geneva Convention for the Amelioration of the Condition of the Wounded, Sick and Shipwrecked Members of Armed Forces at Sea, opened for signature 12 August 1949, 75 UNTS 85 (entered into force 21 October 1950) (1949 Geneva Convention II), Geneva Convention Relative to the Treatment of Prisoners of War, opened for signature 12 August 1949, 75 UNTS 135 (entered into force 21 October 1950) (1949 Geneva Convention III), Geneva Convention Relative to the Protection of Civilian Persons in Time of War, opened for signature 12 August 1949, 75 UNTS 287 (entered into force 21 October 1950) (1949 Geneva Convention IV).

30 Henckaerts, *supra* note 14, 177–8.

31 *Ibid*, 189.

32 See International Committee of the Red Cross, 'Response of Jean-Marie Henckaerts to the Bellinger/Haynes Comments on Customary International Law Study' (2007) 46 *International Legal Materials* 959, 965. In any event, the 1977 Additional Protocol II has been described as 'notoriously difficult to apply', since States involved in internal armed conflicts often deny its relevance to their situation: Mark Freeman and Gibran van Ert, *International Human Rights Law* (2004), 131.

For example, the 1977 Additional Protocol II does not include a provision expressly applying the principle of distinction (discussed below) to non-international armed conflicts. This is to be compared to the 1977 Additional Protocol I, in which the principle is included as a conventional norm.[33] It is, therefore, important to explore any relevant customary international law rule, since that *will* apply to the conduct of States engaged in a non-international armed conflict. In this regard, in 2005, a United Nations Commission of Enquiry on Darfur noted that:[34]

> a body of customary rules regulating internal armed conflicts has thus evolved in the international community ... For example, some States in their military manuals for their armed forces clearly have stated that the bulk of international humanitarian law also applies to internal conflicts. Other States have taken a similar attitude with regard to many rules of international humanitarian law.

The relevance of customary international law to the conduct of armed conflict has been affirmed on a number of occasions by the International Court of Justice. In the *Legality of the Threat or Use of Nuclear Weapons* Advisory Opinion, the Court noted that:[35]

> a great many rules of humanitarian law applicable in armed conflict are so fundamental to the respect of the human person and "elementary considerations of humanity" ... [and] are to be observed by all States whether or not they have ratified the conventions that contain them, because they constitute intransgressible principles of international customary law

Of relevance also are the observations of the Court in the *Legal Consequences of the Construction of a Wall in the Occupied Palestinian Territory* Advisory Opinion, where, citing the passage quoted above, it expressed the view that:[36]

> these rules [of international humanitarian law] incorporate obligations which are essentially of an *erga omnes* character

---

33 See 1977 Additional Protocol I, article 48 (discussed below).

34 International Commission of Inquiry on Darfur, 'Report of the International Commission of Inquiry on Darfur to the United Nations Secretary-General, Pursuant to Security Council Resolution 1564 of 18 September 2004' (25 January 2005), para 159 <www.un.org/News/dh/sudan/com_inq_darfur.pdf> at 20 January 2015.

35 [1996] ICJ Rep 226, para 79.

36 [2004] ICJ Rep 136, para 157. In *Barcelona Traction, Light and Power Company, Limited (Belgium v. Spain)* (Judgment) [1970] ICJ Rep 3, the International Court of Justice (at para 34) cited obligations that derived from the outlawing of aggression and genocide, as well as from the principles and rules 'concerning the basic rights of the human person', as examples of obligations *erga omnes*. It has also applied the concept of obligations (and rights) owed to the international community as a whole in relation to the right of a 'peoples' to self-determination: *East Timor (Portugal v. Australia)* [1995] ICJ Rep 90, para 29.

This also reflects earlier views expressed by the Court regarding the Convention on the Prevention and Punishment of the Crime of Genocide,[37] in relation to which it concluded that 'the rights and obligations enshrined by the Convention are rights and obligations *erga omnes*'.[38] A similar observation was made by Judge Simma in *Armed Activities on the Territory of the Congo*, when he noted that:[39]

> at least the core of the obligations deriving from the rule of international humanitarian and human rights law are valid *erga omnes*

Yet another reason why it is important to consider any relevant rules of customary international law in relation to intentional environmental damage during armed conflict is that customary law may be of assistance in interpreting any applicable treaty provisions. Article 31(3)(c) of the VCLT specifies that, when interpreting a treaty in accordance with the fundamental rule set out in article 31(1),[40] 'any relevant rules of international law applicable in the relationship between the parties' shall, 'together with the context', also be taken into account.[41]

Thus, just as multilateral treaties 'may have an important role to play in recording and defining rules deriving from custom, or indeed developing them',[42] when interpreting a *jus in bello* treaty, the relevant principles of

37 Opened for signature 9 December 1948, 78 UNTS 277 (entered into force 12 January 1951) (Genocide Convention).

38 *Application of the Convention on the Prevention and Punishment of the Crime of Genocide (Bosnia and Herzegovina v. Yugoslavia)* (Preliminary Objections, Judgment) [1996] ICJ Rep 595, para 31.

39 Separate Opinion of Judge Simma, *Armed Activities on the Territory of the Congo (Democratic Republic of the Congo v Uganda)* [2005] ICJ Rep 168, 334, para 39.

40 In Judgment on the Prosecutor's Application for Extraordinary Review of Pre-Trial Chamber I's 31 March 2006 Decision Denying Leave to Appeal, *Situation in the Democratic Republic of the Congo*, Case No. ICC-01/04-168, Pre-Trial Chamber I, 13 July 2006, the Pre-Trial Chamber of the International Criminal Court, when considering the effect of article 31(1) of the VCLT, confirmed (at para 33) that:
'[t]he context of a given legislative provision is defined by the particular subsection of the law read as a whole in conjunction with the section of an enactment in its entirety. Its objects may be gathered from the chapter of the law in which the particular section is included and its purposes from the wider aims of the law as may be gathered from its preamble and general tenor of the treaty'.
This was expressly adopted by the Appeals Chamber of that Court in Judgment on the Appeal Against the Decision on Joinder rendered on 10 March 2008 by the Pre-Trial Chamber in the Germain Katanga and Mathieu Ngudjolo Chui Cases, *The Prosecutor v. Germain Katanga and Mathieu Ngudjolo Chui*, Case No. ICC-01/04–01/07, Appeals Chamber, 9 June 2008, para 5.

41 VCLT, article 31(3)(c).

42 *Continental Shelf (Libyan Arab Jamahiriya v. Malta)* (Judgment) [1985] ICJ Rep 13, para 27.

customary international law must also be taken into consideration.[43] Indeed, article 31(3)(c) of the VCLT appears to encompass the full gamut of applicable customary international law rules (through the use of the word 'any'), not just those that relate solely to the specific content of the relevant treaty being interpreted. However, unless otherwise indicated, the discussion in this chapter of the relevant rules of customary international law will be confined to those specifically in the area of international humanitarian law.

A norm of customary international law may, in particular circumstances, be identical in its terms to a conventional provision.[44] This may well be the case in relation to some of the customary rules that relate to intentional environmental destruction during armed conflict. However, even if this were always the position (which it is not), it is still useful to consider these customary rules separately from the relevant treaty provisions. The applicable treaty and customary rules have a different normative base from each other,[45] and may ultimately apply in varying ways in relation to a particular factual circumstance. In this regard, the International Court of Justice has observed that:[46]

> [t]here are a number of reasons for considering that, even if two norms belonging to two sources of international law appear identical in content, and even if the States in question are bound by these rules both on the level of treaty-law and on that of customary international law, these norms retain a separate existence ... Rules which are identical in treaty law and in customary international law are also distinguishable by reference to the methods of interpretation and application

In attempting to set out the relevant customary international law rules, and notwithstanding the reasons outlined above as to why this is necessary in order to consider the key questions addressed in this book, it must be borne in mind that an elaboration of the exact scope of these rules, and indeed even their very existence as such – as opposed, for example, to only describing non-binding patterns of behaviour or acts done 'merely for reasons of political expediency'[47] – is sometimes a difficult exercise. Apart from universally accepted and clearly established principles of customary international law about which there is

---

43 Anthony E. Cassimatis, 'International Humanitarian Law, International Human Rights Law, and Fragmentation of International Law' (2007) 56 *International and Comparative Law Quarterly* 623, 634.

44 See, for example, *North Sea Continental Shelf Cases (Federal Republic of Germany v. Denmark and Federal Republic of Germany v. The Netherlands)* (Judgment) [1969] ICJ Rep 3, para 71: see also *Military and Paramilitary Activities in and against Nicaragua (Nicaragua v. United States of America)* (Merits) (Judgment) [1986] ICJ Rep 14, para 177–8.

45 Scobbie, *supra* note 11, 47.

46 *Military and Paramilitary Activities in and against Nicaragua (Nicaragua v. United States of America)* (Merits) (Judgment) [1986] ICJ Rep 14, para 178.

47 *Colombian-Peruvian Asylum Case (Colombia v. Peru)* (Judgment) [1950] ICJ Rep 266, 277.

little doubt, it is often not easy to affirm with certainty the precise terms and scope of a particular customary rule, or indeed whether such a rule exists.[48] As Judge Koroma has noted, customary international law is 'notorious for its imprecision'.[49] Erik Koppe has suggested that:[50]

> customary international law is surrounded by mystery and uncertainty and is one of the most difficult and intangible topics of international law

One should therefore be cautious before definitively asserting that a particular rule of customary international law actually exists. Naturally, the terms of specific rules can be more confidently asserted on the basis of conclusions reached by the International Court of Justice, which are generally to be regarded as authoritative, and perhaps also from decisions in other international courts and/or tribunals, that a particular customary principle has been established. It is harder (although not impossible) to definitively substantiate the existence of a customary rule of international law in the absence of such a judicial pronouncement. That said, even where an international (criminal) court *does* conclude that a customary international law rule exists, that may not necessarily be universally accepted and might instead be viewed with some scepticism.[51]

However, drawing on the jurisprudence of, in particular, the International Court of Justice, there are several fundamental *jus in bello* principles that

---

48 For example, as noted, in the decision of the High Court of Australia in *Polyukhovich v Commonwealth and Another* (1991) 172 CLR 501, two of the Judges considered the question of whether the crime of 'crimes against humanity' existed under customary international law prior to being 'introduced' into article 6(c) of the Charter of the Nuremberg International Military Tribunal, annexed to the 1945 London Agreement for the Establishment of an International Military Tribunal (8 August 1945) 82 UNTS 279. The two Judges came to opposite conclusions on this point: see 172 CLR 501, 587–90 (per Justice Brennan) and 664–77 (per Justice Toohey).

49 Koroma, *supra* note 20, xii.

50 Koppe, *supra* note 5, 173.

51 For example, in Interlocutory Decision on the Applicable Law: Terrorism, Conspiracy, Homicide, Perpetration, Cumulative Charging, *Ayyash et al*, Case No. ST-11-01/I, Appeals Chamber, 16 February 2011, the Appeals Chamber of the Special Tribunal for Lebanon handed down an interlocutory decision in which it concluded *inter alia* that a 'customary rule of international law has evolved on terrorism in time of peace', and then proceeded to set out the first international judicial definition of the crime of terrorism. This decision has been met with considerable criticism and has attracted controversy, with various commentators arguing strongly that this conclusion was incorrect and unsubstantiated; see, for example, Elies van Sliedregt and Larissa van den Herik, 'Introduction: The STL Interlocutory Decision on the Definition of Terrorism – Judicial Ingenuity or Radicalism?' (2011) 24:3 *Leiden Journal of International Law* 651; Kai Ambos, 'Judicial Creativity at the Special Tribunal for Lebanon: Is There a Crime of Terrorism under International Law?' (2011) 24:3 *Leiden Journal of International Law* 655; Ben Saul, 'Legislating from a Radical Hague: The United Nations Special Tribunal for Lebanon Invents an International Crime of Transnational Terrorism' (2011) 24:3 *Leiden Journal of International Law* 677.

clearly represent rules of customary international law, notwithstanding that there may still be some discussion as to their precise scope.[52] These rules include what are commonly referred to as the principle of humanity, the principle of military necessity, the principle of distinction and the principle of proportionality, although they are sometimes categorized in different ways.[53] As a general observation, an analysis of these principles indicates that they only serve to protect the natural environment 'indirectly and inadequately'.[54] This is discussed in more detail below.

In the *Legality of the Threat or Use of Nuclear Weapons* Advisory Opinion, the International Court of Justice described the principle of humanity in the following terms:[55]

> it is prohibited to cause unnecessary suffering to combatants: it is accordingly prohibited to use weapons causing them such harm or uselessly aggravating their suffering. In application of [this] principle, States do not have unlimited freedom of choice of means in the weapons they use

This principle has also been discussed by several commentators.[56] Its elements are largely reflected in the terms of the Martens clause and, for the purposes of this book, are assumed to fall within the principles of military necessity and proportionality. The customary international law principles of military necessity, distinction and proportionality require a more detailed analysis, which is found in the following section.

---

52 For example, Kenneth Watkin has noted that '[t]here may be no other term in international humanitarian law ... which evokes such debate or controversy as 'proportionality'': Kenneth Watkin, 'Assessing Proportionality: Moral Complexity and Legal Rules' (2005) 8 *Yearbook of International Humanitarian Law* 3, 4.

53 For example, some descriptions of the principles of international humanitarian law combine issues of distinction and military necessity into a broader principle known as 'discrimination'. This author prefers to differentiate between these two so as to emphasize the need to distinguish between civilians and combatants without reference to sometimes subjective considerations as to what constitutes a military target in the context of military necessity.

54 John Alan Cohan, 'Modes of Warfare and Evolving Standards of Environmental Protection under the International Law of War' (2003) 15 *Florida Journal of International Law* 481, 491.

55 [1996] ICJ Rep 226, para 78.

56 See, for example, Bruch, *supra* note 26, 702; Rymn James Parsons, 'The Fight to Save the Planet: U.S. Armed Forces, "Greenkeeping," and Enforcement of the Law Pertaining to Environmental Protection During Armed Conflict' (1998) 10 *Georgetown International Environmental Law Review* 441, 447; Marcos A. Orellana, 'Criminal Punishment for Environmental Damage: Individual and State Responsibility at a Crossroad' (2005) 17 *Georgetown International Environmental Law Review* 673, 676–7; Andy Rich, 'The Environment: Adequacy of Protection in Times of War' (2004) 12 *Penn State Environmental Law Review* 445, 448.

Apart from these fundamental and widely accepted principles, there has also been much academic discussion regarding the (possible) development of other customary rules within the scope of the *jus in bello*. An important research project was undertaken by the International Committee of the Red Cross (ICRC), culminating in the publication in 2005 of the 'Study on Customary International Humanitarian Law' (ICRC Study).[57]

The ICRC Study listed 161 separate customary rules of international humanitarian law, of which 147 are stated as being applicable in both international and non-international armed conflicts.[58] Among the rules set out in the ICRC Study are a number that relate specifically to 'The Natural Environment'.[59]

3.2. FUNDAMENTAL CUSTOMARY RULES OF THE *JUS IN BELLO*

Bearing in mind the complexities involved in definitively determining the terms of specific customary rules, this chapter will now consider the scope of the widely accepted *jus in bello* rules of customary international law, and their (possible) relevance to the intentional destruction of the environment during armed conflict. These principles are generally founded upon the proposition noted above, that 'the right of belligerents to adopt means of injuring the enemy is not unlimited'.[60]

As mentioned, it is proposed to examine the three major customary international humanitarian law rules; the principles of military necessity, distinction and proportionality.

---

57 The study was published in two Volumes: Jean-Marie Henckaerts and Louise Doswald-Beck (eds), *Customary International Humanitarian Law* (Volume I: Rules, 2005) and Jean-Marie Henckaerts and Louise Doswald-Beck (eds), *Customary International Humanitarian Law* (Volume II: Practice (2 Parts), 2005).

58 International Committee of the Red Cross, 'Custom as a Source of International Humanitarian Law' (14 September 2007) <www.icrc.org/web/eng/siteeng0.nsf/html/customary-law-publication-140907> at 19 January 2015.

59 ICRC Study, Chapter 14. Three specific rules of customary law are identified: ICRC Study, Rules 43–45. These are discussed later in this chapter.

60 Regulations annexed to the 1907 Hague Convention IV, article 22. See Anthony Leibler, 'Deliberate Wartime Environmental Damage: New Challenges for International Law' (1992) 23 *California Western International Law Journal* 67, 97–8.

3.2.1. PRINCIPLE OF MILITARY NECESSITY[61]

Military necessity has long been regarded as a fundamental element of the *jus in bello* rules.[62] It was included in the Lieber Code,[63] the first formal text of rules applicable to the conduct of warfare, in the following terms:[64]

> [m]ilitary necessity, as understood by modern civilized nations, consists in the necessity of those measures which are indispensable for securing the ends of war, and which are lawful according to the modern law and usages of war

The identification of the principle of military necessity has been described as the 'greatest theoretical contribution to the modern law of war' made by the Lieber Code.[65] The Lieber Code also provided that the requirements of military necessity prohibited 'the wanton destruction of a district'.[66] However, as noted, even though the Lieber Code expressly applied to the conduct of the Union soldiers during the American Civil War, there is little evidence to suggest that the 'total warfare' strategies of Union Generals Ulysses S. Grant, William T. Sherman and Phillip Sheridan were ever punished.[67]

The principle of military necessity was also expressly recognized in a codified version of the laws of armed conflict published in 1987 under the auspices of the ICRC.[68] It has traditionally been regarded as the 'dominant' customary international law rule relating to the *jus in bello*.[69] Historically, it has been invoked to justify from a legal perspective – and thus eliminate legal responsibility – actions committed during warfare that might otherwise have been regarded as violating important humanitarian standards.[70]

---

61 Expressions such as 'necessity of military operations', 'military exigencies, motives and reasons', 'reasons of war', and 'security reasons' have been categorized as synonyms of 'military necessity': Pietro Verri, *Dictionary of the International Law of Armed Conflict* (translated from French into English by Edward Markee and Susan Mutti) (1992), 75.

62 Mark A. Drumbl, 'Waging War against the World: The Need to Move from War Crimes to Environmental Crimes', in Jay E. Austin and Carl E. Bruch (eds), *The Environmental Consequences of War: Legal, Economic, and Scientific Perspectives* (2000), 620, 628.

63 Instructions for the Government of Armies of the United States in the Field, promulgated as General Orders No.100, 24 April 1863 (Lieber Code). See also chapter 2.

64 *Ibid*, article 14.

65 Burrus M. Carnahan, 'Lincoln, Lieber and the Laws of War: The Origins and Limits of the Principle of Military Necessity' (1998) 92 *American Journal of International Law* 213, 213.

66 Lieber Code, article 16.

67 See Bruch, *supra* note 26, 695–7.

68 Frederic de Mulinen, *Handbook on the Law of War for the Armed Forces* (1987), 82–3.

69 Orellana, *supra* note 56, 677.

70 Mark A. Drumbl, 'International Human Rights, International Humanitarian Law, and Environmental Security: Can the International Criminal Court Bridge the Gaps?' (2000) 6 *ILSA Journal of International and Comparative Law* 305, 319.

Prior to World War I, for example, Germany had developed a doctrine of warfare known as *Kriegsraison geht vor Kriegsmanier* ('military necessity comes before military conduct'), by which *any* rule of international humanitarian law was 'outweighed' by considerations of military necessity[71] although, even at that time, its legality was questioned.[72] This doctrine was applied by German military forces during the Second World War, but was subsequently rejected by the United States Military Tribunals in Nuremberg in the *Krupp* case, when it concluded that:[73]

> these rules and customs of warfare are designed specifically for all phases of war ... To claim that they can be wantonly – and at the sole discretion of any one belligerent – disregarded when he considers his own situation to be critical, means nothing more or less than to abrogate the laws and customs of war entirely

The foundation of this rule stemmed from the requirement that all acts of force employed in the context of armed conflict must, *as a minimum*, be necessary from a military perspective.[74] Without further qualification, this might suggest that *any* act that carries with it some perception of military advantage might be permitted, regardless of the context and consequences of such action. This conception of military necessity ignores the fact that the principle operates as a 'critical component' *within* the *jus in bello* rules, rather than as an external element.[75]

In light of the disastrous consequences that could possibly flow from adopting too broad a view of what might be permissible, the *jus in bello* rules developed further qualifications as to what was required in order to comply with this principle. It is now the case that a proper application of the concept of military necessity also requires a consideration of other issues, which are then to be weighed against the military advantage. Thus, although military necessity is a crucial element in determining the permissible application of force during armed conflict:[76]

---

71 Craig J.S. Forrest, 'The Doctrine of Military Necessity and the Protection of Cultural Property During Armed Conflicts' (2007) 37 *California Western International Law Journal* 177, 186.

72 See, for example, Jesse S Reeves, 'The Neutralization of Belgium and the Doctrine of Kriegsraison' (1915) 13:3 *Michigan Law Review* 179.

73 *United States. v. Krupp and Others* (Judgment), United States Military Tribunal, Nuremberg, (1949) X War Crimes Reports 69, quoted in Karen Hulme, *War Torn Environment: Interpreting the Legal Threshold* (2004), 130.

74 Leibler, *supra* note 60, 98.

75 Peter J. Richards and Michael N. Schmitt, 'Mars Meets Mother Nature: Protecting the Environment During Armed Conflict' (1999) 28 *Stetson Law Review* 1047, 1077.

76 Michael A. Newton, Symposium, 'Modern Military Necessity: The Role and Relevance of Military Lawyers' (2007) 12 *Roger Williams University Law Review* 877, 880 (emphasis added).

> it cannot [now] concurrently serve as a convenient rationale for *any* level of unrestrained violence in the midst of an operation

This 'tension' between military necessity and restraint on the part of combatants has been described by Theodor Meron as 'the hallmark' of international humanitarian law.[77] The evolution of this principle was described in another case before the United States Military Tribunal in Nuremberg in the following way:[78]

> [m]ilitary necessity permits a belligerent, subject to the laws of war, to apply any amount and kind of force to compel the complete submission of the enemy with the least possible expenditure of time, life and money. In general, it sanctions measures by an occupant necessary to protect the safety of his forces and to facilitate the success of his operations. It permits the destruction of life of armed enemies and other persons whose destruction is incidentally unavoidable by the armed conflicts of the war ... The destruction of property to be lawful must be imperatively demanded by the necessities of war. Destruction as an end in itself is a violation of international law. There must be some reasonable connection between the destruction of property and the overcoming of the enemy forces

As a result, the customary rule that has emerged establishes a link between the overcoming of an enemy by the use of force, and the level of casualties and destruction (which may include environmental destruction) that is a direct consequence of such actions. The two issues must be considered concurrently.[79] A proper application of the principle requires a balance between the need to achieve a military victory, and the broad concepts of humanity and the limits on warfare, and thus it is intended to restrict 'unbridled barbarity'.[80] In this regard, military necessity is no longer considered as an 'authorization' of actions that would otherwise be a violation of the *jus in bello*; rather it is intended to serve from the opposite perspective – as a 'protection' against excessive military force.[81]

Viewed from this perspective, the rule prohibits acts during armed conflict that are unnecessary to achieve a military advantage; however, where the act is militarily necessary, 'collateral' damage is permitted to the extent that

---

77 Theodor Meron, 'The Humanization of Humanitarian Law' (2000) 94 *American Journal of International Law* 239, 243.

78 *Trial of William List and Others* (The Hostages Trial) (Judgment), United States Military Tribunal, Nuremberg, (1949) VIII *Law Reports of Trials of War Criminals*, Case No. 47, 66.

79 Jackson Maogoto and Steven Freeland, 'The Final Frontier: The Laws of Armed Conflict and Space Warfare' (2007) 23:1 *Connecticut Journal of International Law* 165, 176.

80 Forrest, *supra* note 71, 181.

81 Betsy Baker, 'Legal Protections for the Environment in Times of Armed Conflict' (1993) 33 *Virginia Journal of International Law* 351, 360.

the prevailing military circumstances so require.[82] This is also reflected in article 51 of the 1977 Additional Protocol 1, which elaborates on the principle of proportionality discussed below.

In *Strugar*, Trial Chamber II of the ICTY considered that the concept of military necessity:[83]

> may be usefully defined for present purposes with reference to the widely acknowledged definition of military objectives in Article 52 of Additional Protocol I as "those objects which by their nature, location, purpose or use make an effective contribution to military action and whose total or partial destruction, capture or neutralization, in the circumstances ruling at the time, offers a definite military advantage"

In the same paragraph, the Trial Chamber confirmed that, when there is a legitimate military objective, the question of proportionality becomes relevant in determining military necessity.

Despite the qualifications imposed by this customary rule in relation to when and how force may be employed during armed conflict, there is still uncertainty as to the precise level of military advantage necessary to justify (from a legal perspective) such an act. Historically, the balancing test between military advantage and environmental concerns has been 'stacked heavily against the environment'.[84] It has, for example, been argued that the use of atomic devices in Japan during the Second World War was a military necessity – as a means of shortening the war,[85] notwithstanding the massive environmental devastation and loss of life caused[86] – although of course there remains strong disagreement on this issue.[87] In this regard, the principle of military necessity has been

---

82 Tara Weinstein, 'Prosecuting Attacks that Destroy the Environment: Environmental Crimes or Humanitarian Atrocities?' (2005) 17 *Georgetown International Environmental Law Review* 697, 703.

83 Judgement, *Prosecutor v. Pavel Strugar*, Case No. IT-01-42-T, Trial Chamber II, 31 January 2005, para 295.

84 Peter Sharp, 'Prospects for Environmental Liability in the International Criminal Court' (1999) 18 *Virginia Environmental Law Journal* 217, 241.

85 See Leibler, *supra* note 60, 98 and the corresponding footnote.

86 On 6 August 1945, the crew of the *Enola Gay* dropped an atomic device, nick-named 'Little Boy', over the city of Hiroshima. Three days later, *Bock's Car* dropped 'Fat Man' over Nagasaki. The Japanese Armies surrendered on 14 August 1945. It was estimated that, by the end of 1945, approximately 140,000 people died in Hiroshima and a similar number in Nagasaki, as a direct result of the atomic explosions: Koppe, *supra* note 5, 31–2. Many further deaths were reported in the years following the end of the war.

87 Richards and Schmitt, *supra* note 75, 1052. In December 1963, the Tokyo District Court, referring to the Regulations annexed to the Hague Convention IV Respecting the Laws and Customs of War on Land (1907 Hague IV Regulations), and the Geneva Protocol for the Prohibition of the Use in War of Asphyxiating, Poisonous and Other Gases, and of

criticised for prohibiting 'only that which the most powerful do not find militarily useful or expedient'.[88] Richard Falk has made this point in even starker terms:[89]

> what dominant states find militarily useful in war is unlikely to be prohibited, and, if it is, the prohibition is unlikely to be respected in the next war

It is difficult to form a conclusive view as to the accuracy of this assertion. It is, however, probably the case that the 'malleability' of the principle of necessity is, somewhat ironically, an important reason why it has been widely recognised as representing customary international law.[90]

The rule does establish an obligation on a belligerent to specify the imperative military advantage intended to be gained by a particular act.[91] This would normally require the identification of a connection between the use of force and the suppression of the enemy's military capability. However, one's perception of what is 'necessary' in the context of armed conflict will, of course, depend on the particular point of view and values of the person making that judgment. As a general observation, it might be expected that a military commander, whose principal responsibility would typically be to maximise the military advantage to be gained from any specific action, may have a different – perhaps less 'rigorous' – threshold than a civilian as to what acts are necessary in order to achieve that goal. Whilst this might not always be the case, these differing perspectives are illustrated by comments made by a former military lawyer, who has opined that:[92]

Bacteriological Methods of Warfare, opened for signature 17 June 1925, 26 Martens (3rd) 643 (entered into force 8 February 1928), concluded that 'an aerial bombardment with an atomic bomb on both cities of Hiroshima and Nagasaki was an illegal act of hostility as the indiscriminate aerial bombardment on undefended cities': (1964) 8 *Japanese Annual of International Law* 212, quoted in Yves Beigbeder, *Judging War Criminals: The Politics of International Justice* (1999), 69. This represents one of the very few express statements by a court of law that the use of these bombs violated fundamental principles of international law: Nina H.B. Jørgensen, *The Responsibility of States for International Crimes* (2000), 229. An action brought by Japanese citizens alleging that the use of atomic weapons against Japan constituted a war crime was dismissed by the Supreme Court of Japan on technical jurisdictional grounds: see M. Cherif Bassiouni, 'Introduction to the Symposium: The Normative Framework of International Humanitarian Law: Overlaps, Gaps and Ambiguities' (1998) 8 *Transnational Law and Contemporary Problems* 199, footnotes 136 and 137.

88 Parsons, *supra* note 56, 447.

89 Richard Falk, 'The Environmental Law of War: An Introduction', in Glen Plant (ed), *Environmental Protection and the Law of War: A "Fifth Geneva" Convention on the Environment in Time of Armed Conflict* (1992), 78, 80.

90 Parsons, *supra* note 56, 448.

91 See Verri, *supra* note 61, 75.

92 Colonel James P. Terry, 'The Environment and the Laws of War: The Impact of Desert Storm' (1992) 45 *Naval War College Review* 61, 102, quoting (and endorsing) the comments made by another military commentator, Colonel Harry Summers.

> [j]ust as the military needs to be aware of political, economic and social issues, so our civilian leadership must be aware of the imperatives of military operations

Moreover, the conduct of warfare is often undertaken without full knowledge of the capacities of the enemy. In addition, the perspective of the combatant undertaking the relevant action as to what is necessary will often differ from that of the opposing party.[93] The United States Military Tribunal in Nuremberg has observed that:[94]

> [t]he course of a military operation by the enemy is loaded with uncertainties, such as the numerical strength of the enemy, the quality of his equipment, his fighting spirit, the efficiency and daring of his commanders, and the uncertainty of his intentions

This is relevant to the application of the customary rule, since a determination of whether an action is militarily necessary is to be made on a 'subjective' basis – that is, based on the circumstances as determined by the relevant military commander at the time – even where those acts may, in hindsight, be shown to have been unnecessary. A 'reasonableness' standard is to be applied; however, this generally relates to whether the actions of the relevant military commander were reasonable, not to a more standardised conception of what might be reasonable.

These issues were highlighted in the *Rendulic* case,[95] described as 'a classic example of the operation of military necessity as a defense to environmental crimes'.[96] The United States Military Tribunal hearing the case concluded that '[t]he destructions of public and private property by retreating military forces which would give aid and comfort to the enemy'[97] may be acceptable under the 'necessities of war' exception in article 23(g) of the 1907 Hague Convention IV. This was so, notwithstanding that the Tribunal had concluded that, in all likelihood, there was no military necessity for the acts of the defendant, which had resulted in widespread environmental destruction. Rendulic was acquitted on the specific charges relating to the environmental destruction,[98] principally

93 Yoram Dinstein, *The Conduct of Hostilities under the Law of International Armed Conflict* (2004), 86.

94 *Trial of William List and Others* (The Hostages Trial) (Judgment), United States Military Tribunal, Nuremberg, (1949) VIII *Law Reports of Trials of War Criminals*, Case No. 47, 69.

95 *Id.*

96 Drumbl, *supra* note 70, footnote 25.

97 *Trial of William List and Others* (The Hostages Trial) (Judgment), United States Military Tribunal, Nuremberg, (1949) VIII *Law Reports of Trials of War Criminals*, Case No. 47, 69.

98 He was, however, found guilty of a number of other serious crimes not related to his acts of environmental destruction and sentenced to 20 years imprisonment, which was subsequently reduced to 10 years. He was released from prison in February 1951 and died in Austria in 1971.

because his actions were 'justified' under the circumstances at the time they were undertaken.[99]

This appears to also reflect the terms of the customary rule of military necessity today. Questions of military necessity (and proportionality) are to be considered in light of a reasonable interpretation of prevailing circumstances at the time.[100] As a consequence, if the principle is applied to intentional environmental damage during armed conflict, these subjective factors would, in most cases outweigh concerns about such damage,[101] and thus render lawful what might otherwise be regarded as unacceptable destruction.

Such actions may therefore be lawful within the scope of the rule if it was reasonable for the military commander to believe that they were necessary.[102] In this regard, it has been suggested that the 'justification' of military necessity has been 'overused' as a (legal) reason to excuse environmental damage,[103] thus providing belligerents with a 'means to suspend the laws of armed conflict'.[104]

On the other hand, some commentators believe that the development of broad principles of international environmental law has 'significantly qualified' the application of the customary international law rules of military necessity and proportionality.[105] While, as discussed in chapter 2, the general effectiveness of international environmental law (treaties) during periods of armed conflict is questionable, and it is therefore difficult to specify the precise extent of this development in practice, the International Court of Justice has, as noted in

99 The United States Military Tribunal in Nuremberg concluded:
'[t]here is evidence in the record that there was no military necessity for this destruction and devastation. An examination of the facts in retrospect can well sustain this conclusion. But we are obliged to judge *the situation as it appeared to the defendant at the time*. If the facts were such as would justify the action by the exercise of judgment, after giving consideration to all the factors and existing possibilities, even though the conclusion reached may have been faulty, it cannot be said to be criminal. After giving careful consideration to all the evidence on the subject, we are convinced that the defendant cannot be held criminally responsible although when viewed in retrospect, the danger did not actually exist': *Trial of William List and Others* (The Hostages Trial) (Judgment), United States Military Tribunal, Nuremberg, (1949) VIII *Law Reports of Trials of War Criminals*, Case No. 47, 68–9 (emphasis added).

100 Terry, *supra* note 92, 91.

101 See Baker, *supra* note 81, 361 and the corresponding footnote.

102 See Weinstein, *supra* note 82, 703 and the references in the corresponding footnote.

103 Parsons, *supra* note 56, 448.

104 Ensign Florencio J. Yuzon, 'Deliberate Environmental Modification Through the Use of Chemical and Biological Weapons: "Greening" the Environmental Laws of Armed Conflict to Establish an Environmentally Protective Regime' (1996) 11 *American University Journal of International Law and Policy* 793, 812.

105 See, for example, Orellana, *supra* note 56, 677.

chapter 1, indicated that environmental considerations must be taken into account when assessing issues of necessity and proportionality.[106]

As noted, there has been a growing tendency for provisions relating to the environment to be included in the military manuals of a number of countries,[107] thus reflecting an increasing awareness of the need to take account of the possibility of environmental destruction during armed conflict. Enhanced training programs and more sophisticated education of military units, such in the United States, arc now being undertaken in relation to the environmental effects of preparing for war.[108] As Betsy Baker has (perhaps optimistically) noted, 'the mere fact that these restrictions exist is evidence that the definition of military necessity is changing'.[109]

Nevertheless, as noted, the concept of military necessity – and the underlying assumption that military considerations should, in general, be a predominant consideration – remains an important cornerstone of the laws of armed conflict. Various significant *jus in bello* treaty provisions make specific reference to military necessity, including the principles codified in the 1907 Hague Convention IV – which the United States Military Tribunal described as 'mandatory provisions of International Law'[110] – and also in others such as the 1949 Geneva Convention IV,[111] the 1954 Hague Convention,[112] the 1977 Additional Protocol I,[113] and Protocol III to the 1981 Certain Conventional Weapons Convention.[114]

---

106 *Legality of the Threat or Use of Nuclear Weapons* (Advisory Opinion) [1996] ICJ Rep 226, para 30.

107 This has been particularly the case in Europe, the United States, Australia, and for NATO forces generally: Carl Bruch, 'Closing Remarks', Symposium, 'The International Responses to the Environmental Impacts of War' (2005) 17 *Georgetown International Environmental Law Review* 565, 644.

108 Nancye L. Bethurem, 'Environmental Destruction in the Name of National Security: Will the Old Paradigm Return in the Wake of September 11?' (2002) 8 *Hastings West-Northwest Journal of Environmental Law and Policy* 109, 112–7.

109 Baker, *supra* note 81, 362.

110 *Trial of William List and Others* (The Hostages Trial) (Judgment), United States Military Tribunal, Nuremberg, (1949) VIII *Law Reports of Trials of War Criminals*, Case No. 47, 69.

111 1949 Geneva Convention IV, article 53.

112 Hague Convention for the Protection of Cultural Property in the Event of Armed Conflict, opened for signature 14 May 1954, 249 UNTS 240 (entered into force 7 August 1956) (1954 Hague Convention), article 2.

113 1977 Additional Protocol I, article 54.

114 Protocol III on Prohibitions or Restrictions on the Use of Incendiary Weapons to the Convention on Prohibitions or Restrictions on the Use of Certain Conventional Weapons Which May be Deemed to be Excessively Injurious or to Have Indiscriminate Effects, opened for signature 10 April 1981, 1342 UNTS 137 (entered into force 2 December 1983) (Protocol III to the 1981 Certain Conventional Weapons Convention), article 2(4).

This serves to reinforce the importance of military necessity as a relevant consideration as to what may be lawful in the context of armed conflict. Whilst this might seem logical to some, it has even been suggested that, in certain circumstances, military necessity may be used as a justification for acts that would otherwise violate a *jus in bello* treaty, even where this possibility is *not* mentioned in the text itself,[115] although others have dismissed this assertion.[116]

In relation to the issue of environmental damage during armed conflict, it thus remains the case that military necessity is a highly relevant factor. Drawing upon 'existing international legal obligations and from State practice concerning the protection of the environment against the effects of armed conflict',[117] the ICRC draft 'Guidelines for Military Manuals and Instructions on the Protection of the Environment in Times of Armed Conflict', which provides a model for national military manuals,[118] provides that:[119]

> (8) Destruction of the environment *not justified by military necessity* violates international humanitarian law ...
> (9) The general prohibition to destroy civilian objects, *unless such destruction is justified by military necessity*, also protects the environment

This also reflects the view expressed in 1992 by the United Nations General Assembly, when it stressed that:[120]

> destruction of the environment, not justified by military necessity and carried out wantonly, is clearly contrary to existing international law

The 1990s proposal to establish a 'Fifth Geneva' Convention on the Protection of the Environment in Time of Armed Conflict encompassed the suggestion that, in certain circumstances, military necessity was not to be regarded as a

---

115 Forrest, *supra* note 71, 192–3.

116 Dinstein, *supra* note 93, 18–9.

117 International Committee of the Red Cross, 'Guidelines for Military Manuals and Instructions on the Protection of the Environment in Times of Armed Conflict' (ICRC Environment Guidelines), article I(1), (1995) <www.icrc.org/eng/resources/documents/misc/57jn38.htm> at 19 January 2015. See also Hans-Peter Gasser, 'Guidelines for Military Manuals and Instructions on the Protection of the Environment in Times of Armed Conflict' (1996) 311 *International Review of the Red Cross* 230, appendix.

118 Aurelie Lopez, 'Criminal Liability for Environmental Damage Occurring in Times of Non-International Armed Conflict: Rights and Remedies' (2007) 18 *Fordham Environmental Law Review* 231, 251.

119 ICRC Environmental Guidelines, articles III(8) and III(9) (emphasis added).

120 United Nations General Assembly Resolution 47/37 (25 November 1992) UN Doc A/RES/47/37 on the Protection of the Environment in Times of Armed Conflict, preamble para 5.

justification for the destruction of the environment.[121] As noted, this proposal did not receive widespread support at the time.

The relevance of military necessity within the *jus in bello* rules involving the destruction of property is also reflected in some of the comparable international criminal law treaty provisions. These instruments embody several principles of international humanitarian law,[122] and represent important enforcement mechanisms.[123] This has seen a 'shift [of] some state-to-state aspects of international humanitarian law to individual criminal responsibility'.[124] In addition, the customary *jus in bello* principles also represent a 'source' of law for international criminal law, in particular as it is to be interpreted by the ICC. The Rome Statute provides that the Court shall apply:[125]

> ... where appropriate, applicable treaties and the principles and rules of international law, including the established principles of the international law of armed conflict

Article 8(2)(a) of the Rome Statute incorporates '[g]rave breaches' of the 1949 Geneva Conventions within the definition of 'war crimes'. Mirroring the wording from the 1949 Geneva Conventions,[126] article 8(2)(a)(iv) of the Rome Statute defines one war crime as:[127]

> [e]xtensive destruction and appropriation of property, *not justified by military necessity* and carried out unlawfully and wantonly

In similar vein, article 3(b) of the ICTY Statute provides that a violation of the laws or customs of war includes 'wanton destruction of cities, towns or villages, or devastation *not justified by military necessity*'.[128]

---

121 Nicholas G. Alexander, 'Notes and Comments: Airstrikes and Environmental Damage: Can the United States be Held Liable for Operation Allied Force?' (2000) 11 *Colorado Journal of International Environmental Law and Policy* 471, 491–2.

122 Elaina I. Kalivretakis, 'Are Nuclear Weapons Above the Law? A Look at the International Criminal Court and the Prohibited Weapons Category' (2001) 15 *Emory International Law Review* 683, 688.

123 Karima Bennoune, 'Do We Need International Law to Protect Women in Armed Conflict?' (2006–2007) 38 *Case Western Reserve Journal of International Law* 363, 364.

124 Meron, *supra* note 77, 243.

125 Rome Statute of the International Criminal Court, opened for signature 17 July 1998, 2187 UNTS 3; 37 ILM 999 (entered into force 1 July 2002) (Rome Statute), article 21(1)(b).

126 See 1949 Geneva Convention I, article 50; 1949 Geneva Convention II, article 51; 1949 Geneva Convention IV, article 147.

127 Rome Statute, article 8(2)(a)(iv) (emphasis added).

128 Statute of the International Tribunal for the Prosecution of Persons Responsible for Serious Violations of International Humanitarian Law Committed in the Territory of the Former Yugoslavia since 1991, 32 ILM 1159 (entered into force 25 May 1993) (ICTY Statute), article 3(b) (emphasis added).

As noted in previous chapters, article 8(2)(b)(iv) of the Rome Statute defines a war crime arising from particular actions that cause 'widespread, long-term and severe damage to the natural environment'. The definition expressly provides that the 'direct overall military advantage anticipated' by the attack is a relevant consideration in determining whether the provision has been violated.[129] It is unclear as to how 'direct' the military advantage must be in order to be 'necessary'.[130] There is also disagreement as to how widely the notion of 'military advantage' might extend. Some commentators suggest that it is very broad and is 'not necessarily restricted to tactical gains'.[131] Others have suggested that the concept of military advantage may not as expansive as military necessity.[132]

Irrespective of (or perhaps due to) the uncertainties that the concept of military advantage introduces, its inclusion in article 8(2)(b)(iv) of the Rome Statute gives rise to a number of similar concerns as those raised by notions of military necessity, and reflects the fact that, as indicated in the preparatory works of the Rome Statute, the protection of the environment was regarded as a 'consideration secondary to military objectives'.[133]

3.2.2. PRINCIPLE OF DISTINCTION[134]

In the *Legality of the Threat or Use of Nuclear Weapons* Advisory Opinion, the International Court of Justice stated that one of the 'cardinal principles' of the *jus in bello* rules involves:[135]

129 Rome Statute, article 8(2)(b)(iv). The provision also introduces a 'proportionality' test, by requiring that the environmental damage be 'clearly excessive' in relation to the military advantage: Drumbl, *supra* note 62, 629. The principle of proportionality is discussed later in this chapter and article 8(2)(b)(iv) is analyzed in detail in chapter 4.

130 Richards and Schmitt, *supra* note 75, 1076.

131 See, for example, Dinstein, *supra* note 93, 86.

132 See, for example, Drumbl, *supra* note 62, 629.

133 Orellana, *supra* note 56, 694.

134 As with military necessity, some commentators have used different names to describe the principle of distinction. Michael Schmitt, for example, describes the principle as 'discrimination', which he suggests includes three components: distinction, minimizing collateral damage and incidental injury, and proportionality: Michael N. Schmitt, 'Humanitarian Law and the Environment' (2000) 28 *Denver Journal of International Law and Policy* 265, 311–2. Aurelie Lopez appears to use the word 'discrimination' as a synonym of 'distinction', as it is referred to in this chapter: Lopez, *supra* note 118, 247–8. The customary international law principles as described in this chapter encompass all of these concepts and, particularly because there is also considerable overlap between the underlying rationale for each, it is felt that they are all adequately covered within the framework used here, notwithstanding that there is some variation as to how these concepts may be classified.

135 [1996] ICJ Rep 226, para 78.

> the protection of the civilian population and civilian objects and establishes the distinction between combatants and non-combatants; States must never make civilians the object of attack and must consequently never use weapons that are incapable of distinguishing between civilian and military targets

This statement reflects in general terms the customary international law principle of distinction, which is another fundamental norm of international humanitarian law. Its importance, particular in relation to the protection of civilians and civilian objects, is underscored by the assertion of Kristin Dorman that:[136]

> [i]n combat situations, the entire body of international humanitarian law can be reduced to the obligation to observe the principle of distinction

While this probably underestimates the importance of the other fundamental customary principles of international humanitarian law discussed in this chapter, it does highlight the essential need for all belligerents to ensure that only non-civilian targets (and persons) are attacked. In 1969, the United Nations General Assembly affirmed that:[137]

> (a) ... the right of the parties to a conflict to adopt means of injuring the enemy is not unlimited;
> (b) ... it is prohibited to launch attacks against the civilian populations as such;
> (c) ... distinction must be made at all times between persons taking part in the hostilities and members of the civilian population to the effect that the latter be spared as much as possible

These principles were recognised at the time as reflecting customary international law.[138] In *Galic*, the Trial Chamber of the ICTY, when referring to article 52(2) of the 1977 Additional Protocol I (see below), and citing the dicta of the International Court of Justice quoted above, described the customary principle of distinction in the following terms:[139]

> the provision in question explicitly confirms the customary rule that civilians must enjoy general protection against the danger arising from hostilities. The prohibition against attacking civilians stems from a fundamental principle of international

136 Kristin Dorman, 'Proportionality and Distinction in the International Criminal Tribunal for the Former Yugoslavia' (2005) 12 *Australian International Law Journal* 83, 84.

137 United Nations General Assembly Resolution 2444 (XXIII) (19 December 1968) UN Doc A/RES/2444(XXIII) on the Respect for Human Rights in Armed Conflicts, para 1.

138 Royal Swedish Ministry for Foreign Affairs, 'Conventional Weapons, their Deployment and Effects from a Humanitarian Aspect: Recommendations for the Modernization of International Law' (1973), 18.

139 Judgement and Opinion, *Prosecutor v. Stanislav Galic*, Case No. IT-98-29-T, Trial Chamber I, 5 December 2003, para 45 (emphasis in original).

> humanitarian law, the principle of distinction, which obliges warring parties to distinguish *at all times* between the civilian population and combatants and between civilian objects and military objectives and accordingly to direct their operations only against military objectives

The principle of distinction interacts with the principle of military necessity, which prescribes those measures that are appropriate to overpower the enemy. Attackers must identify the prospective target in advance of the attack as one that is militarily legitimate, since attacks not directed at a legitimate military target are prohibited. In undertaking this assessment, attackers must distinguish between civilians and civilian objects on the one hand, and military objectives (combatants or objects) on the other, and also only use weapons capable of distinguishing between them.

In relation to weapons that are incapable of discriminating between legitimate military targets and civilians and civilian objects, which would include certain methods and means of armed conflict that are capable of significant environmental damage (such as biological weapons, chemical weapons and the use of environmental modification), some specific prohibitions exist under the relevant *jus in bello* treaties discussed in chapter 2.[140]

With respect to the protection of civilians and civilian objects from attack, the principle of distinction requires that attacks can only be directed towards military objectives. The term 'military objectives' was first introduced in the non-binding 1923 Hague Rules of Air Warfare, which specified that:[141]

> [a]n air bombardment is legitimate only when is directed against a military objective, i.e. an objective whereof the total or partial destruction would constitute an obvious military advantage for the belligerent

The expression has subsequently been defined in article 52(2) of the 1977 Additional Protocol I in the following terms:[142]

---

140 See, for example, Convention on the Prohibition of the Development, Production and Stockpiling of Bacteriological (Biological) and Toxic Weapons and on their Destruction, opened for signature 10 April 1972, 1015 UNTS 163 (entered into force 26 March 1975) (1972 Biological Weapons Convention); Convention on Prohibitions or Restrictions on the Use of Certain Conventional Weapons Which May be Deemed to be Excessively Injurious or to Have Indiscriminate Effects, opened for signature 10 April 1981, 1342 UNTS 137 (entered into force 2 December 1983); Convention on the Prohibition of the Development, Production, Stockpiling and Use of Chemical Weapons and on Their Destruction, opened for signature 13 January 1993, 32 ILM 800 (entered into force 29 April 1997) (1993 Chemical Weapons Convention).

141 Rules concerning the Control of Wireless Telegraphy in Time of War and Air Warfare, drafted by a Commission of Jurists at The Hague, December 1922 – February 1923, article 24(1).

142 1977 Additional Protocol I, article 52(2).

> [a]ttacks shall be limited strictly to military objectives. In so far as objects are concerned, military objectives are limited to those objects which by their nature, location, purpose or use make an effective contribution to military action and whose total or partial destruction, capture or neutralization, in the circumstances ruling at the time, offers a definite military of advantage

This definition of the principle of distinction has been repeated in several subsequent *jus in bello* treaties,[143] and is regarded by many commentators as representing customary international law.[144] Civilian objects are regarded as all objects that are not military objectives.[145] What is unusual about this is that the definition of civilians and civilian objects follows a 'negative approach',[146] in that it is expressed in terms of what they are *not*. This formula has been criticized as 'vague, nebulous, and highly subjective'.[147] Indeed, it has been suggested that, as a result of the approach taken in article 52(1) of the 1977 Additional Protocol I, restrictions imposed by the *jus in bello* in relation to incidental loss of civilian life are 'probably [made] derogable in the case of military necessity'.[148]

On the other hand, some commentators contend that this negative approach has the advantage of ensuring that there is 'no undistributed middle between the categories of combatants (or military objectives) and civilians (or civilian objects)'.[149]

---

143 These include Protocol II on Prohibitions or Restrictions on the Use of Mines, Booby-Traps and Other Devices to the Convention on Prohibitions or Restrictions on the Use of Certain Conventional Weapons Which May be Deemed to be Excessively Injurious or to Have Indiscriminate Effects, opened for signature 10 April 1981, 1342 UNTS 137 (entered into force 2 December 1983) (Protocol II to the 1981 Certain Conventional Weapons Convention), article 2(4); Protocol III to the 1981 Certain Conventional Weapons Convention, article 1(3); Second Protocol to the Hague Convention of 1954 for the Protection of Cultural Property in the Event of Armed Conflict, opened for signature 26 March 1999, 38 ILM 769 (entered into force 9 March 2004), article 1(f).

144 See, for example, Dinstein, *supra* note 93, 83 and the corresponding footnote.

145 1977 Additional Protocol I, article 52(1); Protocol II to the 1981 Certain Conventional Weapons Convention, article 2(5).

146 Dinstein, *supra* note 93, 114.

147 See Dorman, *supra* note 136, 88 and the corresponding footnotes.

148 Timothy J. Heverin, 'Case Comment: Legality of the Threat or Use of Nuclear Weapons: Environmental and Humanitarian Limits on Self-Defense' (1997) 72 *Notre Dame Law Review* 1277, 1300. That author does, however, indicate that this is not always the case, referring to the terms of article 57(5) of the 1977 Additional Protocol I, which deals with 'Precautions in Attack' and provides that '[n]o provision of this Article may be construed as authorizing any attacks against the civilian population, civilians or civilian objects'.

149 See, for example, Dinstein, *supra* note 93, 114 and the corresponding footnote. However, at the same time, that author (at 84) agrees that the 'abstract character' of the definition of military objectives in article 52(2) of the 1977 Additional Protocol I gives rise to 'divergent interpretations'.

Despite ongoing improvement in general weapons systems accuracy, compliance with the principle of distinction has become increasingly complicated. The changing nature of armed conflict means that the lines between lawful targets and protected objects may become increasingly blurred, as might the distinction between civilians and combatants, due primarily to the growing involvement of civilians and civilian activities in armed conflicts.[150] Since civilians can sometimes play a direct part in hostilities, it is perhaps now more accurate to state that the customary principle of distinction requires a differentiation to be made between 'combatants, legal or otherwise,[151] and those civilians who do not take a direct part in hostilities'.[152]

The principle of distinction was initially articulated in the St. Petersburg Declaration of 1868 in the following terms:[153]

> the only legitimate object which States should endeavour to accomplish during war is to weaken the military forces of the enemy

It was subsequently incorporated into a number of the *jus in bello* instruments,[154] as well as United Nations General Assembly Resolutions.[155] More recently, the principle was affirmed in the 1977 Additional Protocol as follows:[156]

> [i]n order to ensure respect for and protection of the civilian population and between civilian objects, the Parties to the conflict shall at all times distinguish between the civilian population and combatants and civilian objects and military objectives and accordingly shall direct their operations only against military objectives

---

150 Maogoto and Freeland, *supra* note 79, 178.

151 Unlawful combatants may not be entitled to all of the protections that apply to lawful combatants under the relevant *jus in bello* treaties: see *Ex Parte Quirin*, United States Supreme Court, (1942) 317 US 1.

152 Kenneth Watkin, 'Controlling the Use of Force: A Role for Human Rights Norms in Contemporary Armed Conflict' (2004) 98 *American Journal of International Law* 1, 17.

153 Declaration Renouncing the Use, in Time of War, of Explosive Projectiles under 400 Grammes Weight, opened for signature 29 November 1868, 18 Martens (1st) 474 (entered into force 11 December 1868), para 3.

154 See, for example, the 1907 Hague IV Regulations. In *Legality of the Threat or Use of Nuclear Weapons* (Advisory Opinion) [1996] ICJ Rep 226, the International Court of Justice noted (at para 80) that, in 1946, the Nuremberg Military Tribunal had held that, already by 1939, the 1907 Hague IV Regulations 'were recognized by all civilized nations and were regarded as being declaratory of the laws and customs of war'.

155 See, for example, United Nations General Assembly Resolution 2444 (XXIII) (19 December 1968) UN Doc A/RES/2444(XXIII) on the Respect for Human Rights in Armed Conflicts, para 1.

156 1977 Additional Protocol I, article 48.

It is generally regarded that the rule specified in article 48 is 'identical or similar to' the customary international law principle of distinction,[157] although some doubts have been expressed on this point.[158] A violation of the principles of distinction constitutes a grave breach of the 1977 Additional Protocol I,[159] although it is unfortunately the case that, as expressed by the then Foreign Minister of France in 2009, 'there are few conflicts in which [the principle of distinction] is fully respected'.[160] The obligation to distinguish between civilian and military targets under the 1977 Additional Protocol I is reinforced by article 57(1), which provides *inter alia* that:

> [i]n the conduct of military operations, constant care shall be taken to spare the civilian population, civilians and civilian objects

Moreover, various precautions are to be taken by those who plan or decide upon an attack, including:[161]

> [d]oing everything feasible to verify that the objectives to be attacked are neither civilians nor civilian objects and are not subject to special protection but are military objectives within the meaning of [article 52(2)] and that it is not prohibited by the provisions of [the 1977 Additional Protocol I] to attack them

As noted, the 1977 Additional Protocol I relates only to international armed conflicts. There are no comparable provisions in the 1977 Additional Protocol II, which regulates non-international armed conflicts, although that instrument does contain a more general prohibition of attacks against 'the civilian population as such, as well as individual civilians'.[162] However, it is now widely agreed that the customary principle of distinction applies to non-international armed conflicts, a conclusion that was affirmed by the Appeals Chamber of the ICTY in 1995,[163] both in relation to the requirement to distinguish between combatants and non-combatants, as well as to the deployment of weapons. In relation to the latter, the Appeals Chamber stated that:[164]

> elementary considerations of humanity and common sense make it preposterous that the use by States of weapons prohibited in armed conflicts between themselves be

157 See, for example, Henckaerts, *supra* note 14, 187, where the author summarizes some of the conclusions of the ICRC Study; Watkin, *supra* note 152, 15; Dinstein, *supra* note 93, 82.

158 See, for example, Dorman, *supra* note 136, 86.

159 See 1977 Additional Protocol I, article 85(3).

160 Bernard Kouchner, 'The savaging of humanitarian law', *The International Herald Tribune* (Paris), 29 January 2009, 8.

161 1977 Additional Protocol I, article 57(2)(a)(i).

162 1977 Additional Protocol II, article 13(2).

163 Decision on Defence Motion for Interlocutory Appeal on Jurisdiction, *Prosecutor v Duško Tadić*, Case No. IT-94-1, Appeals Chamber, 2 October 1995, para 100–18.

164 *Ibid*, para 119.

> allowed when States try to put down rebellion by their own nationals on their own territory. What is inhumane, and consequently proscribed, in international wars, cannot but be inhumane and inadmissible in civil strife…

It also reiterated the same point a little later by stating that:[165]

> [p]rinciples and rules of humanitarian law reflect "elementary considerations of humanity" widely recognized as the mandatory minimum for conduct in armed conflicts of any kind. No one can doubt the gravity of the acts at issue, nor the interest of the international community in their prohibition

The principle of distinction has also been included in the relevant Statutes of the various international criminal justice mechanisms, including the ICC. For example, the Rome Statute regards 'other serious violations [apart from grave breaches of the Geneva Conventions] of the laws and customs applicable in international armed conflict' as war crimes. These include:[166]

> (i) Intentionally directing attacks against the civilian population as such or against individual civilians not taking direct part in hostilities;
> (ii) Intentionally directing attacks against civilian objects, that is, objects which are not military objectives

In terms of the protection of the environment, the ICRC Environment Guidelines confirm that the principles of distinction and proportionality are relevant, but only in the following very general terms:[167]

> [i]n addition to the specific rules set out below [including those set out in articles III(8) and III(9) quoted above], the general principles of international law applicable in armed conflict – such as the principle of distinction and the principle of proportionality – provide protection to the environment. In particular, only military objectives may be attacked and no methods or means of warfare which cause excessive damage shall be employed. Precautions shall be taken in military operations as required by international law

The ICRC Environment Guidelines do, however, make specific reference to particular rules contained in various of the *jus in bello* treaties that restrict attacks by incendiary weapons on 'forests or other kinds of plant cover', but they confirm that these are subject to a 'military objectives' exception, mirroring the terms of the Protocol III of the 1981 Certain Conventional Weapons Convention, discussed in chapter 2.[168]

---

165 *Ibid*, para 129.
166 Rome Statute, articles 8(2)(b)(i) and 8(2)(b)(ii).
167 ICRC Environment Guidelines, article II(4).
168 Protocol III to the 1981 Certain Conventional Weapons Convention, article 2(4).

Overall, there are some fundamental practical difficulties in attempting to apply the principle of distinction, which is based on a subjective assessment of what is a civilian object (as opposed to a military objective). The relevant *jus in bello* treaties do not provide clear guidance as to what objects constitute military objectives, and thus what would be a civilian object. There is some conjecture, for example, as to whether the concept of military objectives is restricted to inanimate objects, or also extends to military personnel.[169] The lack of a sufficiently clear definition on this point has led to various criticisms, including that it has actually 'contributed to the commission of serious human rights abuses'.[170]

In applying the principle of distinction, specific objects such as hospitals, schools and churches would normally be designated as civilian objects, meaning that 'military operations [should] simply remove them from their pool of potential targets.'[171] This remains, however, subject to overarching military necessity in the specific circumstances prevailing at the time, so that, for example, 'if a church steeple is used by snipers, it [may become] a military objective'.[172]

The difficulties associated with lawful target selection in relation to the principle of distinction are also complicated by the utilization of specific objects for *both* civilian and military purposes. These so-called 'dual use objects'[173] give rise to complex legal questions, principally as to when a civilian object may cease to have the protection afforded to such an object and instead 'becomes' a legitimate military objective, based upon the specific uses of that object.[174]

The practical difficulties associated with the principle of distinction are exacerbated even further when one attempts to apply it as a mechanism that is applicable to the environment during armed conflict. Armed conflict will always take place within the 'environment' and, by its inherent nature, will invariably have a destructive environmental effect, sometimes to a very significant level.

169 See Dinstein, *supra* note 93, 84–5 and the corresponding footnotes.

170 See Dorman, *supra* note 136, 89 and the corresponding footnote.

171 Richards and Schmitt, *supra* note 75, 1078.

172 Dinstein, *supra* note 93, 86.

173 See, for example, Meron, *supra* note 77, 275; Dinstein, *supra* note 93, 96–7.

174 The difficulties associated with this question were highlighted in the case of the NATO bombing campaign in Serbia and Kosovo in 1999, where one of the targets was the Serbian Television and Radio Station in Belgrade, which was at the same time being used for both military and non-military purposes. For a discussion, see Steven Freeland, 'The Bombing of Kosovo and the Milosevic Trial: Reflections on Some Legal Issues' [2002] *Australian International Law Journal* 150. The same problems associated with the application of the *jus in bello* principles to objects serving both civilian and military purposes also extend to so-called 'dual use satellites' in outer space: see, for example, Steven Freeland, 'The Applicability of the *Jus in Bello* Rules of International Humanitarian Law to the Use of Outer Space' (2006) 49 *Proceedings of the Colloquium on the Law of Outer Space* 338.

Whilst it may be the case that the requirement to target only military objectives provides 'a measure of indirect environmental protection',[175] such prohibitions can have a tangible and direct effect in this way only to the extent that the environment is to be accorded civilian 'status'.[176]

Yet, it is unclear precisely how, and indeed whether, the 'natural environment' can for practical (and legal) purposes be designated as a civilian 'object'. It has been suggested that the 'nebulous character' of the natural environment means that it is difficult to consider it as an object *per se*, and to thus accord it with a civilian status for the purposes of applying the existing customary and conventional *jus in bello* rules.[177] On the other hand, various commentators, as well as the ICRC Study, have concluded that the natural environment *is* to be regarded as a civilian object, and thus subject to the same protections under international humanitarian law as other civilian objects.[178]

The provisions of the 1977 Additional Protocol I do not provide any significant clarification in this regard. Article 55 is included in the chapter of that instrument that deals with 'Civilian Objects', but no definition of the 'natural environment' is included. This is, in part, symptomatic of the broader difficulties associated with reaching a widespread consensus as to what the definition of the environment should be for the purposes of both the customary rules of international humanitarian law,[179] as well as the wide range of international environmental law treaties.

As a result, until it has been widely accepted – and, ideally, codified – that the environment is a civilian objective for the purposes of the *jus in bello* and thus not subject to attack, the application of the principle of distinction in order to protect the environment in a significant way will remain problematic. It also indicates that, under international criminal law, the various (war) crimes associated with protecting civilian objects are not particularly well-suited to addressing the issue of intentional environmental destruction during warfare. Thus, as discussed in detail in chapter 5, a 'stand-alone' crime is required that is specifically directed towards such acts in clear terms.

---

175 Hulme, *supra* note 73, 120.

176 Heverin, *supra* note 148, 1294.

177 Baker, *supra* note 81, 364.

178 Henckaerts, *supra* note 14, 191. The relevant (suggested) rules concerning the natural environment in the ICRC Study are discussed later in this chapter. See also Hulme, *supra* note 73, 126.

179 See Baker, *supra* note 81, 364 and the corresponding footnotes.

3.2.3. PRINCIPLE OF PROPORTIONALITY

Even when a belligerent attacks a legitimate military objective, the extent of military force used, and any injury and damage inflicted upon civilians and civilian property, should not be disproportionate to any expected military advantage. Thus, an attack may still be unlawful if the civilian losses and/or damage are unacceptably high in the circumstances, taking into account the military contingencies. In these circumstances, the military objective does not cease to be a military objective; rather this principle of proportionality represents 'a further restriction' on attacks on such objectives.[180]

The application of this principle requires that those who plan or decide upon an attack assess beforehand any potential 'collateral' or 'incidental' damage that may reasonably be expected to result from military action. They are then required to:[181]

> [t]ake all *feasible* precautions in the choice of means and methods of attack with a view to avoiding, and in any event to minimizing, incidental loss of civilian life, injury to civilians and damage to civilian objects

Having undertaken such an assessment, the attack must be cancelled or suspended:[182]

> if it becomes apparent that … the attack may be expected to cause incidental loss of civilian life, injury to civilians, damage to civilian objects, or a combination thereof, which would be excessive in relation to the concrete and direct military advantage anticipated

As with the principle of military necessity, it is often difficult to apply the proportionality principle in practice, given that this requires a subjective assessment of the relative 'values' to be ascribed to military advantage *vis-à-vis* civilian injury and damage. In this regard, considerations of military necessity may again be relevant. The 1977 Additional Protocol I does not define what is 'feasible' in terms of the precautions to be taken; however, that term has been defined in Protocol II to the 1981 Certain Conventional Weapons Convention, as follows:[183]

---

180 See Dinstein, *supra* note 93, 120 and the corresponding footnote.

181 1977 Additional Protocol I, article 57(2)(a)(ii) (emphasis added).

182 *Ibid*, article 57(2)(b).

183 Article 3(4) (emphasis added). The relevant 'circumstances' are elaborated upon in a subsequent amendment to Protocol II to the 1981 Certain Conventional Weapons Convention, adopted in 1996.

> practicable or practically possible taking into account *all circumstances* ruling at the time, including humanitarian and *military considerations*

This definition has also been taken up by a number of States. For example, when ratifying the Optional Protocol to the Convention on the Rights of the Child on the Involvement of Children in Armed Conflict,[184] the United States included in its interpretive declaration an identical definition of the term 'feasible measures', as it is used in article 1 of that instrument. In addition, several countries have included a similar definition in interpretive declarations made at the time of their respective ratifications of the 1977 Additional Protocol 1.[185] To the extent that this limited definition of what is feasible itself represents a customary norm of international humanitarian law, it will have a diluting effect on the principles to which it applies. This would inevitably limit the extent to which the principle of proportionality offers tangible protection to the environment in the course of armed conflict.

When the issue of incidental damage was first addressed, it was considered to be the case that, if an attack was directed towards a clear military objective, then *unavoidable* injury to civilians and civilian objects was to be regarded as 'collateral damage' and, as such, (legally) acceptable.[186] Thus, the Lieber Code stated that:[187]

> [m]ilitary necessity admits of all direct destruction of life or limb of armed enemies, and other persons whose destruction is *incidentally unavoidable* in the armed contests of war

This is no longer the position. Over time, the principle of proportionality has developed into a fundamental customary rule of international humanitarian law,[188] involving a 'balancing' test between the military advantage to be gained and the risk of injury to civilians and/or damage to civilian objects. In *Kupreskic et al,* the ICTY Trial Chamber described articles 57 and 58 of the 1977 Additional Protocol I as:[189]

---

184 Opened for signature 25 May 2000, 2173 UNTS 222 (entered into force 12 February 2002) (Children in Armed Conflict Protocol), article 1. See also Steven Freeland, 'Mere Children or Weapons of War – Child Soldiers and International Law' (2008) 29 *University of La Verne Law Review* 19, 37.

185 See Michael J. Dennis, 'Current Development: Newly Adopted Protocols to the Convention on the Rights of the Child' (2000) 96 *American Journal of International Law* 789, 791.

186 Dinstein, *supra* note 93, 119.

187 Lieber Code, article 15 (emphasis added).

188 Dinstein, *supra* note 93, 120.

189 Judgement, *Prosecutor v. Zoran Kupreskic, Mirjan Kupreskic, Vlatko Kupreskic, Drago Josipovic, Dragan Papic and Vladimir Santic*, Case No. IT-95-16-T, Trial Chamber, 14 January 2000, para 524.

> now [being] part of customary international law, not only because they specify and flesh out general pre-existing norms, but also because they do not appear to be contested by any State, including those which have not ratified the Protocol

In addition, in her dissenting opinion in the *Legality of the Threat or Use of Nuclear Weapons* Advisory Opinion, Judge Higgins noted that:[190]

> [t]he principle of proportionality, even if finding no specific mention, is reflected in many provisions of Additional Protocol 1 to the Geneva Conventions of 1949. Thus even a legitimate target may not be attacked if the collateral civilian casualties would be disproportionate to the specific military gain from the attack

As indicated by the Judge, the term 'proportionality' was not expressly referred to in the 1977 Additional Protocol I. One author suggests that this was due to objections raised by States, apparently because of what they perceived to be the difficulties in 'trying to compare things that were not viewed as comparable'.[191] The thrust of the principle as described by Judge Higgins is reflected in articles 51(4) and 51(5)(b) of the 1977 Additional Protocol I, the combined effect of which is to prohibit 'indiscriminate attacks', including:[192]

> [a]n attack which may be expected to cause incidental loss of civilian life, injury to civilians, damage to civilian objects, or a combination thereof, which would be excessive in relation to the concrete and direct military advantage anticipated

Stated in this way, the rule incorporates a requirement that the incidental loss not be excessive in relation to the 'concrete and direct' military advantage. These qualifying words are also not defined in the 1977 Additional Protocol I, and have been described as 'vague'.[193] Yet, even though their precise meaning is unclear, their inclusion in the treaty does appear to further restrict the number of lawful targets that may be subject to an attack.[194]

The principle may also apply to *unintended* incidental damage,[195] and thus might be regarded as imposing a 'reasonableness'[196] test in terms of 'what may be expected', although this will be based on the actual circumstances as perceived

---

190 [1996] ICJ Rep 226, 587 (para 20). The Judge's comments on this point did not form part of her reasons for dissenting from the conclusion of the majority of the Court.

191 Hulme, *supra* note 73, 127.

192 1977 Additional Protocol I, article 51(5)(b).

193 Drumbl, *supra* note 62, 629.

194 Hulme, *supra* note 73, 126.

195 See, for example, Judgement, *Prosecutor v. Zoran Kupreskic, Mirjan Kupreskic, Vlatko Kupreskic, Drago Josipovic, Dragan Papic and Vladimir Santic*, Case No. IT-95-16-T, Trial Chamber, 14 January 2000, para 524.

196 W. Jason Fisher, 'Targeted Killing, Norms, and International Law' (2007) 45 *Columbia Journal of Transnational Law* 711, 728 and 743.

by the person(s) deciding to undertake the attack. In this regard, the Trial Chamber of the ICTY has stated that:[197]

> [i]n determining whether an attack was proportionate it is necessary to examine whether a reasonably well-informed person in the circumstances of the actual perpetrator, making reasonable use of the information available to him or her, could have expected excessive civilian casualties to result from the attack

It seems likely that damage to specific environmental components[198] might also constitute incidental damage for the purposes of determining the applicability of the principle of proportionality. In this sense, Timothy Heverin has concluded that the requirements of proportionality 'should prohibit environmental damage which exceeds military objectives'.[199] However, even this rather optimistic viewpoint may not be sufficient – or sufficiently clear – to comprehensively protect the environment. For example, it has been argued that 'a very substantial military advantage' may render as legitimate:[200]

> attacks against military targets which are known or can reasonably be assumed to cause *grave environmental harm*. At a minimum, actions resulting in *massive environmental destruction*, especially where they do not serve a clear and important military purpose, would be questionable.

Indeed, the Committee Report immediately added that the NATO attacks on the Serbian petro-chemical industries in 1999, which did in fact lead to significant environmental damage, 'may well have served a clear and important military purpose'.[201]

It should be noted that, although there are some differences (for example, the addition of the word 'clearly' excessive, and the requirement to consider the 'overall' military advantage),[202] article 8(2)(b)(iv) of the Rome Statute, which is discussed in detail in chapter 4, is in similar terms to article 52(5)(b) of the 1977

197 Judgement and Opinion, *Prosecutor v. Stanislav Galic*, Case No. IT-98-29-T, Trial Chamber I, 5 December 2003, para 58.

198 These may include, for example, endangered species, fragile habitats and natural heritage sites: Hulme, *supra* note 73, 127.

199 Heverin, *supra* note 148, 1294.

200 Final Report to the Prosecutor of the International Criminal Tribunal for the former Yugoslavia by the Committee Established to Review the NATO Bombing Campaign Against the Federal Republic of Yugoslavia, 13 June 2000, 39 ILM 1257 (Committee Report), para 22 (emphasis added).

201 *Id.*

202 It has been suggested that the inclusion of the word 'overall' in article 8(2)(b)(iv) of the Rome Statute 'somewhat broadens the scope of military advantages which may be taken into account': Ruth G. Wedgwood, 'Proportionality, Cyberwar, and the Law of War', in Michael N. Schmitt and Brian T. O'Donnell (eds), *Computer Network Attack and International Law* (2002), 219, 225.

Additional Protocol I. In addition to the specific reference to environmental damage, article 8(2)(b)(iv) designates the following, if done within the context of an international armed conflict, as a war crime:[203]

> [i]ntentionally launching an attack in the knowledge that such attack will cause incidental loss of life or injury to civilians or damage to civilian objects ... which would be clearly excessive in relation to the concrete and direct overall military advantage anticipated

The 1977 Additional Protocol I does not clearly specify a 'maximum' amount of collateral damage beyond which any attack would always be regarded as violating the principle of proportionality. In terms of environmental damage, the treaty does provide for a 'widespread, long-term and severe' threshold of destruction[204] (as does article 8(2)(b)(iv) of the Rome Statute). In this regard, acts that damage the environment significantly may be inconsistent with the principle of proportionality.[205] However, as discussed in chapter 2, the precise limits of this damage threshold are difficult to determine and, in any event, may in practical terms be so high as to have no real effect in all but the most extreme of circumstances.[206] Moreover, as noted, there is much uncertainty as to what constitutes 'excessive' incidental damage in any specific circumstance.[207]

Indeed, as previously noted, one need only recall the (majority) conclusion of the International Court of Justice in the *Legality of the Threat or Use of Nuclear Weapons* Advisory Opinion.[208] The Court had been asked by the General Assembly for an Advisory Opinion on the question: 'Is the threat or use of nuclear weapons in any circumstance permitted under international law?'[209] The Resolution referring this question to the Court was highly controversial, having been adopted by less than a two-thirds majority of the General Assembly and having been opposed by France, Russia, the United Kingdom and the United States.[210]

---

203 Rome Statute, article 8(2)(b)(iv).

204 1977 Additional Protocol I, articles 35(3) and 55(1).

205 See Baker, *supra* note 81, 367 and the corresponding footnote.

206 William Fenrick, when discussing the inclusion of this threshold in article 8(2)(b)(iv) of the Rome Statute, suggests that it 'probably exclude[s] the sort of damage caused by heavy shelling during World War I battles on the Western Front': William J. Fenrick, 'Article 8: War Crimes – article 8(2)(b)(iv)', in Otto Triffterer (ed), *Commentary on the Rome Statute of the International Criminal Court: Observers' Notes, Article by Article* (1999), 197.

207 See, for example, Dinstein, *supra* note 93, 122 and the corresponding footnotes.

208 [1996] ICJ Rep 226.

209 United Nations General Assembly Resolution 49/75K (15 December 1994) UN Doc A/RES/49/75 on Request for an Advisory Opinion from the International Court of Justice on the Legality of the Threat or Use of Nuclear Weapons, para 11.

210 The vote was 78 in favour, 43 against and 38 abstentions.

This division was also reflected by the divided opinion rendered by the Court.[211] As noted, the Court did confirm that environmental considerations should be taken into account in the 'calculus' of proportionality under the *jus in bello*.[212] However, while confirming that the threat or use of a nuclear weapon – which will inevitably cause very substantial environmental damage – should comply with the requirements of international law relating to armed conflict, in particular the principles of international humanitarian law, the Court could not say categorically that it would in every circumstance constitute a violation of international law. As Christyne Vachon has stated, the Court 'failed to take the strongest step of all, [by] declaring [nuclear] weapons illegal'.[213]

Another complexity relating to the principle of proportionality is the suggestion that it may actually serve to *prolong* a war, and perhaps therefore lead to greater loss of life in the long term. The Lieber Code asserts that '[t]he more vigorously wars are pursued, the better it is for humanity. Sharp wars are brief'.[214] As noted, this 'logic' was applied by the United States to justify its use of two atomic devices against Japan in 1945.[215]

This is a difficult legal (and moral) argument to substantiate, and will almost always involve speculation and hypothetical scenarios. In any event, there is no express requirement to take such a suggestion into account in the balancing exercise to be applied although, of course, it might be a factor considered by those planning an attack, under the broad rubric of 'military advantage anticipated'.

The ICTY has developed a 'cumulative effect' test in determining what may be disproportionate incidental damage arising from an attack. Drawing upon the terms of the Martens clause, the Trial Chamber has suggested that, even though a single attack may in and of itself not give rise to unacceptable civilian casualties and/or damage:[216]

> in case of repeated attacks, all or most of them falling within the grey area between indisputable legality and unlawfulness, it might be warranted to conclude that

---

211 Beigbeder, *supra* note 87, 69–70.

212 Meron, *supra* note 77, 241.

213 Christyne J. Vachon, 'Sovereignty Versus Globalization: The International Court of Justice's Advisory Opinion on the Threat or Use of Nuclear Weapons' (1998) 26 *Denver Journal of International Law and Policy* 691, 724.

214 Lieber Code, article 29, quoted in Meron, *supra* note 77, 241.

215 United States President Harry S. Truman, when justifying the use of the atomic bombs against Japan, was reported to have said that it was necessary 'in order to shorten the agony of war, in order to save the lives of thousands and thousands of young Americans': quoted in Beigbeder, *supra* note 87, 68.

216 Judgement, *Prosecutor v. Zoran Kupreskic, Mirjan Kupreskic, Vlatko Kupreskic, Drago Josipovic, Dragan Papic and Vladimir Santic*, Case No. IT-95-16-T, Trial Chamber, 14 January 2000, para 526.

the cumulative effect of such acts entails that they may not be in keeping with international law

This approach was designed to broaden the effectiveness of the principle of proportionality as a way of limiting attacks, by suggesting that a series of 'legal' attacks may, once accumulated, be regarded as disproportionate in the circumstances, and thus illegal. There is nothing to suggest that this approach could not be applied to a series of attacks that may, in sum, cause significant environmental damage. However, this 'cumulative effect' test was subsequently misinterpreted in the Committee Report, highlighting further the difficulties that often arise when applying the principle of proportionality in practice. While describing the approach outlined in *Kupreskic et al* as 'a progressive statement of the applicable law with regard to the obligation to protect civilians', the Committee Report suggests that it was 'somewhat ambiguous and its application far from clear'.[217]

In fact, the Committee Report applies its own version of the cumulative test – which was very different from that outlined by the ICTY Trial Chamber – in determining whether the NATO bombing campaign was within the acceptable limits of the principle of proportionality. It looked at the absolute numbers in relation to Operation Allied Force (38,400 sorties, 10,484 strike sorties and 23,614 bombs released) and apparently concluded that, when considered in this context, the number of civilian deaths (approximately 500) was not sufficiently high so as to violate the principle.[218]

By adopting an approach based on absolute numbers, and thus testing the proportionality issue within the context of the totality of the campaign, the Committee Report gave no guidance as to what level of civilian deaths *would* have been disproportionate.[219] This is essentially the reverse position from what appears to have been contemplated by the ICTY Trial Chamber in *Kupreskic et al.* Ironically, the approach set out in the Committee Report would mean that

217 Committee Report, para 52.

218 This conclusion is even more startling when one considers that, in undertaking its campaign, NATO forces undertook a strategy of high altitude bombing, which ensured that it suffered no losses – an example of what might be regarded as 'zero casualty warfare', except that there were, clearly, casualties on the other side. It has been suggested that this in itself might have been inconsistent with a good faith adherence to the principle of proportionality: Mary Ellen O'Connell, Book Review – Judith Gardam, 'Necessity, Proportionality and the Use of Force by States' (2006) 100 *American Journal of International Law* 973, 977 (commenting on the 'chief conclusion' of the reviewed book in relation to the NATO bombing campaign).

219 Steven Freeland, 'Commentary on the Findings of the Prosecutor of the International Criminal Tribunal for the Former Yugoslavia into the actions of NATO in Serbia and Kosovo during 1999', in André Klip and Göran Sluiter (eds), *Annotated Leading Cases of International Criminal Tribunals, Volume V, The International Criminal Tribunal for the Former Yugoslavia, 1999–2000* (2003), 38.

single attacks that appear to have violated the principle of proportionality, and were thus (*prima facie*) illegal, have suddenly 'become' legal in the context of the whole attack. As Andreas Laursen has put it, this involves 'dilut[ing] the significance of a single attack by pouring it into a sea of integrated attacks'.[220] To undertake such an approach effectively denies the restraining power of the principle of proportionality and would appear to be 'irreconcilable with humanitarian considerations',[221] as well as any environmental concerns.

Overall, despite the rationale underlying the principle of proportionality – that those undertaking attacks act in a way that is least likely to result in civilian injury and/or damage – there is no clear and generally accepted 'formula' as to how to properly balance this requirement with the military advantages associated with a legitimate military objective.[222] The ICTY Trial Chamber in *Kupreskic et al*, when it was considering articles 57 and 58 of the 1977 Additional Protocol I (which also reflect customary international law), noted that:[223]

> these two provisions leave a *wide margin of discretion* to belligerents by using language that might be regarded as leaving the last word to the attacking party

Thus, as with the principles of military necessity and distinction, the application of proportionality involves subjective assessments by military personnel, whose main goal will, typically, be to maximise the military success to be gained by the actions undertaken by themselves or those within their command or control. Even though the principle of proportionality continues to evolve, and there appears to be increasing awareness among military leaders about the effects of armed conflict on the environment,[224] environmental considerations may, in practice, often still be far removed from such aims. In addition, the balancing of interests required between military necessity and proportionality will invariably involve subjective value judgements.[225] This necessitates the attribution of a 'value' to the competing military and humanitarian interests[226] and, as noted,

220 Andreas Laursen, 'NATO, the War over Kosovo, and the ICTY Investigation' (2002) 17 *American University International Law Review* 765, 793.

221 *Ibid*, 796 and 813.

222 Rich, *supra* note 56, 450.

223 Judgement, *Prosecutor v. Zoran Kupreskic, Mirjan Kupreskic, Vlatko Kupreskic, Drago Josipovic, Dragan Papic and Vladimir Santic*, Case No. IT-95-16-T, Trial Chamber, 14 January 2000, para 524 (emphasis added).

224 See Richards and Schmitt, *supra* note 75, 1047, who explain (at 1083) that, compared with the increased concern for the environment shown today among both military and political leaders, '[w]ar's devastating consequences on the environment – no new phenomenon – resulted in barely a murmur fifty years ago'.

225 See, for example, Jonathan Pearlman, 'Dead baby and teenager ruled to be 'collateral casualties'', *The Sydney Morning Herald* (Sydney) 13 May 2008 <www.smh.com.au/news/national/dead-children-ruled-collateral-casualties/2008/05/12/1210444338885.html> at 29 March 2013.

226 Richards and Schmitt, *supra* note 75, 1082.

different parties will have varying conceptions of what this relative value should be, depending on their perspective, aims and involvement in the activities.

Furthermore, the question (and relevance) of damage or injury caused indirectly by military action – described by Peter Richards and Michael Schmitt as the 'reverberating effects' of an attack[227] – makes a determination of what might be legally acceptable even more difficult. This is also relevant to environmental damage during armed conflict. Just as, for example, shutting off access to power following an attack on an electricity grid might cause severe civilian injury and damage, so might an attack on a dam, resulting in water not being made available for irrigation purposes, lead to potentially significant environmental damage. Yet, some might argue that, in these circumstances, it was not the attack that caused the damage, and that these 'indirect' effects should not be included in an assessment of the overall proportionality of the military actions undertaken.

While every circumstance is to be considered on a case-by-case basis, this illustrates further the complexities involved in seeking to apply the principle of proportionality. These difficulties led the ICRC Study to conclude that 'the exact scope and application of the principle of proportionality in attack' was 'still open to question'.[228] In the end, the balancing test required when assessing whether a particular action satisfies the principle of proportionality is, in practice, only likely to be possible in cases where there is an 'obvious' imbalance.[229]

The difficulties in undertaking this assessment are even greater in a case necessitating a weighing of military advantage against any incidental damage to the environment.[230] The sometimes intangible nature of the environment, coupled with the time that some forms of environmental damage may take to manifest (unlike damage to civilians and civilian objects), mean that, in most cases, ascribing the appropriate 'weight' to the environment will be very difficult. In these circumstances, therefore, it is more likely that environmental concerns will simply be ascribed with minimal weight when compared to the more immediate military advantages. This is notwithstanding the fact that the environmental damage may ultimately prove to be very significant. Thus, the effectiveness of the principle of proportionality in restraining actions that may, or are intended to cause significant environmental damage during armed conflict may be significantly limited in practice.

---

227 *Ibid*, 1081.
228 Henckaerts, *supra* note 14, 191 and 197.
229 Dorman, *supra* note 136, 98.
230 Hulme, *supra* note 73, 127.

3.2.4. THE FUNDAMENTAL CUSTOMARY *JUS IN BELLO* RULES ARE IMPORTANT BUT INSUFFICIENT

The discussion in this chapter has thus far focussed on the fundamental principles of the *jus in bello* and their possible application in relation to situations where acts are done with intent to cause significant damage to the environment during armed conflict. In summary, while the customary principles of military necessity, distinction and proportionality do represent very important rules with respect to the conduct of armed conflict, there is still some uncertainty as to the precise standards with which combatants must, in practice, comply so as to not be in violation of the principles.

If anything, these uncertainties are further exacerbated when one considers the extent to which these principles might be applied in a practical way so as to address intentional destruction of the environment during hostilities. Whilst, in general terms, the principles do restrict some activities that may otherwise have significant destructive consequences for the environment, it is difficult to be much more precise than that. It is unlikely that these principles would adequately address such acts, and thus provide for criminal accountability in appropriate circumstances.

This is due to the uncertainties both as to the exact scope of each principle, as well as the relative 'value' to be ascribed to the environment when undertaking the relevant balancing tests. It may be that a significant military advantage – or at least military considerations – could outweigh even extreme damage to the environment in certain circumstances.[231]

Indeed, it is not entirely clear as to whether, and if so, how, some of these rules apply to the environment at all. For example, as noted, an argument could be made that there is still an open question as to whether the natural environment can properly be considered as a civilian 'object' for the purposes of the principle of distinction, although the ICRC Study seems quite clear on this point (see below). Whilst the principle of proportionality is relevant, it has been suggested that, when applied to the environment, the principle would 'prohibit very little'.[232]

In this context, the ICRC Study lists a number of (suggested) rules of customary international law that relate directly to the issue of environmental damage caused during armed conflict. For the reasons discussed below, one cannot – unlike in relation to the principles of military necessity, distinction and proportionality –

231 Karen Hulme, 'Natural Environment', in Elizabeth Wilmshurst and Susan Breau (eds), *Perspectives on the ICRC Study on Customary International Humanitarian Law* (2007), 204, 216.

232 *Ibid*, 215.

assert unequivocally that these represent rules of customary international law in every respect. However, it is still appropriate to consider them, particularly in relation to the question of how, if at all, they may expand the scope of the fundamental rules discussed above, so as to apply specifically to such acts.

3.3. ADDITIONAL PROTECTION? ICRC STUDY – RULES RELATING TO 'THE NATURAL ENVIRONMENT'[233]

In December 1995, the 26th International Conference of the Red Cross and Red Crescent endorsed a recommendation by the Intergovernmental Group of Experts for the Protection of War Victims, which invited the ICRC:[234]

> to prepare, with the assistance of experts in IHL [international humanitarian law] representing various geographical regions and different legal systems, and in consultation with experts from governments and international organizations, a report on customary rules of IHL applicable in international and non-international armed conflicts, and to circulate the report to States and competent international bodies

This report, subsequently known as the 'Study on Customary International Humanitarian Law', was published nearly 10 years later 'after extensive research and widespread consultation with [over 100] experts'.[235] The aim of the ICRC Study was not to determine the customary nature of each provision of the *jus in bello* treaties, but rather to analyse various issues regulated by those treaties that had not been universally ratified,[236] in order to establish 'what rules of customary international law can be found inductively on the basis of State practice in relation to these issues'.[237] This exercise was undertaken partly to overcome what were considered to be 'problems related to the application of humanitarian treaty law' – specifically the fact that the conventional *jus in bello rules* were *per se* only binding on the States Parties to each instrument – as well as the lack of treaty provisions directly applicable to non-international armed conflicts.[238]

As an initial comment, and without in any way wishing to detract from the significance and quality of the research involved, the ICRC Study does not have the authoritative status of a decision of the International Court of Justice; nor

233 For a detailed discussion of the findings of the ICRC Study in relation to these rules, and the State practice upon which its conclusions are based, see Henckaerts and Doswald-Beck, *supra* note 57, Volume I, chapter 14 and Volume II, Part 1, chapter 14, sections A, B and C.

234 Meeting of the Intergovernmental Group of Experts for the Protection of War Victims, Geneva, 23–27 January 1995, Recommendation II.

235 Henckaerts, *supra* note 14, 177.

236 *Ibid*, 187.

237 Henckaerts and Doswald-Beck, *supra* note 57, Volume I, xxx.

238 Henckaerts, *supra* note 14, 177.

in some cases does it express definitive conclusions. Whilst the ICRC Study is, obviously, an important document and has already been referred to in international criminal law[239] and domestic jurisprudence,[240] as well as having an impact upon the practice of States and on international relations,[241] from the perspective of determining its legal significance, it is probably best regarded as 'teachings of the most highly qualified publicists', which is expressed in the ICJ Statute as a 'subsidiary' source of law.[242]

Moreover, there have been suggestions that the ICRC Study does 'not properly evaluate its primary sources',[243] and that the methodology employed, and some of the conclusions reached, should thus not be regarded as 'the last word on the subject' but rather:[244]

> the appropriate *starting point* in a review of State practice and *opinio juris* relevant to the crystallisation of custom

Indeed, in relation to the three rules included in the ICRC Study that relate specifically to the natural environment, Karen Hulme has opined that:[245]

> [w]hile many of the authors' conclusions will be welcomed by those familiar with environmental law ... it is suggested that they are not yet all a true reflection of the status of customary law

239 For example, the Appeals Chamber of the ICTY included references to a specific rule (Rule 129) contained in the ICRC Study, when engaging in a survey of customary international law in relation to the scope of the crime of deportation as a crime against humanity under article 5(d) of the ICTY Statute: Judgement, *Prosecutor v. Milomir Stakic*, Case No. IT-97-24-A, Appeals Chamber, 22 March 2006, para 296–7, 305 and 307. In Decision on Joint Defence Interlocutory Appeal of Trial Chamber Decision on Rule 98 *bis* Motions for Acquittal, *Prosecutor v. Enver Hadzihasanovic and Amir Kubura*, Case No. IT-01-47-AR73.3, Appeals Chamber, 11 March 2005, the Appeals Chamber of the ICTY referred to the contents of military manuals as set out in the ICRC Study in relation to various aspects of international humanitarian law, including attacks on civilian objects in non-international armed conflicts (para 30), the prohibition against pillage (para 38), and the general protection of cultural property (para 45).

240 See, for example, *Hamdan v. Rumsfeld, Secretary of Defense, et al* (29 June 2006), United States Supreme Court, 548 U.S. 557 (2006).

241 Bothe, *supra* note 16, 176.

242 See ICI Statute, article 38(1)(d). Of course, decisions of the International Court of Justice (and other international courts and tribunals) also fall within the scope of this provision. However, a conclusion by the International Court of Justice that a particular (alleged) rule of customary international law does (or does not) exist is regarded as authoritative.

243 Scobbie, *supra* note 11, 40.

244 Bethlehem, *supra* note 17, 14 (emphasis added).

245 Hulme, *supra* note 231, 237. See also Joshua H. Joseph, 'Rethinking Yamashita: Holding Military Leaders Accountable for Wartime Rape' (2007) 28 *Women's Rights Law Reporter* 107, 117 and the corresponding footnote.

Even one of its editors has conceded that the ICRC Study is not a definitive reflection of the rules of customary international humanitarian law, albeit for different reasons:[246]

> [t]his study should not be seen ... as the final word on custom; it is not exhaustive because the formation of customary law is an ongoing process

The ICRC Study lists 161 separate rules that were considered by its authors to be principles of customary international law, based on an extensive examination of relevant State practice. These rules were formulated in six parts:

(i) Principle of Distinction (encompassing ICRC Study Rules 1–24);
(ii) Specifically Protected Persons and Objects (Rules 25–45) – which includes the specific provisions that relate to the natural environment;
(iii) Specific Methods of Warfare (Rules 46–69);
(iv) Weapons (Rules 70–86);
(v) Treatment of Civilians and Persons *Hors de Combat* (Rules 87–138); and
(vi) Implementation (Rules 139–161).

As can be seen from these general headings, the structure of the ICRC Study reflects some of the important areas of the *jus in bello* principles, although the authors admit that it is not a comprehensive analysis of all relevant rules. For example, the authors argue that the 'great majority' of the 1949 Geneva Conventions, including common article 3, which have been almost universally ratified, along with the 1907 Hague IV Regulations, are already 'considered to be part of customary international law'.[247] As a result, the customary nature of the provisions of those instruments was not the subject of the ICRC Study.

While several of the general rules specified in the ICRC Study might have indirect relevance to intentional destruction of the environment during armed conflict[248] – in similar vein to some of the *jus in bello* treaty provisions[249] – these do not appear to add any greater direct protection to the environment, or possibilities for direct accountability for such acts, than already exists within the broad scope of those instruments. In the context of this book, therefore, the most

---

246 Jean-Marie Henckaerts, 'Assessing the Laws and Customs of War: The Publication of Customary International Humanitarian Law' (2006) 13 *Human Rights Brief* 8, 8.

247 Henckaerts and Doswald-Beck, *supra* note 57, Volume I, xxx.

248 See, for example, Rule 71 – 'The use of weapons which are by nature indiscriminate is prohibited'; Rule 72 – 'The use of poison or poisoned weapons is prohibited'; Rule 73 – 'The use of biological weapons is prohibited'; Rule 74 – 'The use of chemical weapons is prohibited'; Rule 76(e) – 'The use of herbicides as a method of warfare is prohibited if they ... would cause widespread, long-term and severe damage to the natural environment'.

249 See, for example, 1972 Biological Weapons Convention; 1981 Certain Conventional Weapons Convention; 1993 Chemical Weapons Convention.

directly relevant rules specified in the ICRC Study are those that specifically apply to the 'Natural Environment' (Rules 43–45).

3.3.1. ICRC STUDY – RULE 43

Rule 43 of the ICRC Study provides as follows:

> The general principles on the conduct of hostilities apply to the natural environment:
> A. No part of the natural environment may be attacked, unless it is a military objective.
> B. Destruction of any part of the natural environment is prohibited, unless required by imperative military necessity.
> C. Launching an attack against a military objective which may be expected to cause incidental damage to the environment which would be excessive in relation to the concrete and direct military advantage anticipated is prohibited.

The rule is expressed to apply in relation to both international and non-international armed conflicts.[250] As noted, the ICRC Study concluded that the environment is to be considered as a civilian object and, as such, is protected by the same fundamental rules that protect other civilian objects, notably the principles of distinction and proportionality.[251] This is reflected in the terms of Rule 43. The rule also seeks to confirm that the general principles of international humanitarian law discussed in this chapter are also applicable to the environment, although Parts A and C appear to extend this to any 'part' of the environment. This is an unclear term – how small is a part of the environment? In any event, this could be construed as setting too low a threshold and, thus, may not in fact constitute a rule of customary international law.[252]

Even assuming that the environment is, in fact, a civilian object[253] – Michael Bothe suggests that the rule operates to protect the natural environment as 'an element of the civilian world, to the extent that it really is'[254] – Parts A and C of this rule are largely concerned with the protections afforded by the existing fundamental principles of distinction and proportionality. Thus, they do not add any further level of regulation or restriction on acts conducted during armed

250 Henckaerts and Doswald-Beck, *supra* note 57, Volume I, 143.

251 Henckaerts, *supra* note 14, 191.

252 Hulme, *supra* note 231, 214.

253 See, for example, *ibid*, where that author asserts (at 209) that this is 'universally accepted although a little clumsy'. See also Michael N. Schmitt, Charles H.B. Garraway and Yoram Dinstein, *The Manual on the Law of Non-International Armed Conflict* (2006), where the drafting committee (at para 4.2.4) state that 'the natural environment is a civilian object'.

254 Bothe, *supra* note 16, 167.

conflict. It is also noted that Part C omits the word 'natural', but it has been suggested that this is a simple oversight.[255]

Part B appears to be slightly broader in scope, particularly due to the inclusion of the word 'imperative'. This word is also included in Rule 50 of the ICRC Study, which prohibits *inter alia* the destruction of property, 'unless required by imperative military necessity'.[256] While it might be argued that the word in reality adds nothing – a 'necessity' *is* an 'imperative' – the inclusion of this qualifying word in Rule 43 probably does set a higher threshold standard as to what level of military necessity must be present before an attack on the environment may be legitimate, although Karen Hulme suggests that, '[i]f this were truly the law, it is being ignored on a daily basis.[257]

Once again, however, the presence of military 'necessity' is to be determined from the perspective of the military commander at the time. Moreover, the problems associated with a determination of what constitutes military necessity still remain, as does the issue of exactly what is an imperative (as opposed to a 'non-imperative') necessity. These factors combine to make it difficult to determine precisely when the rule might be violated in all but the most obvious situations – in which case the word 'imperative' would then seem to be unnecessary.

3.3.2. ICRC STUDY – RULE 44

Rule 44 of the ICRC Study provides as follows:

> Methods and means of warfare must be employed with due regard to the protection and preservation of the natural environment. In the conduct of military operations, all feasible precautions must be taken to avoid, and in any event to minimize, incidental damage to the environment. Lack of scientific certainty as to the effects on the environment of certain military operations does not absolve a party to the conflict from taking such precautions.

This rule is stated to apply in relation to international armed conflicts and 'arguably' to non-international armed conflicts.[258] The relevant State practice upon which the ICRC Study relies to assert the existence of this rule, particularly

255 Hulme, *supra* note 231, 208.
256 ICRC Study, Rule 50.
257 Hulme, *supra* note 231, 211.
258 Henckaerts and Doswald-Beck, *supra* note 57, Volume I, 147.

in relation to the application of a form of the 'precautionary principle',[259] which has been incorporated into the second sentence, has been described as 'somewhat scarce'.[260]

In the *Gabcíkovo-Nagymaros Project Case*, the International Court of Justice referred to the fact that the parties in that dispute had agreed on the 'need to take environmental concerns seriously and to take the required precautionary measures'.[261]

However, the Court made no reference to the precautionary principle, and this reference to the agreement of two States to a dispute (Hungary and Slovakia) does not in itself provide any significant evidence of State practice regarding the status of that principle. In fact, there is a divergence of opinion as to the precise scope of the precautionary principle, with some commentators citing a lack of clarity and the 'potential for over-regulation' if it is applied too broadly.[262] In any event, although it has been recognized by some as having crystallized into a general principle of international *environmental* law, there is not yet consensus that the precautionary principle itself represents a customary rule in international law,[263]

---

259 The precautionary principle is based on a premise that the lack of scientific certainty as to the negative environmental consequences of a particular action should not be used as a justification to carry out that action. In practical terms, this is often thought to reverse the burden of proof as to the consequences of an action, placing it on those who claim that an activity is *not* damaging; Elli Louka, *International Environmental Law: Fairness, Effectiveness, and World Order* (2006), 50. See also Principle 15 of the Declaration of the United Nations Conference on Environment and Development, (14 June 1992) UN Doc A/CONF.151/26 (Volume 1), 31 ILM 874 (Rio Declaration); article 2(2)(a) of the Convention for the Protection of the Marine Environment of the North East Atlantic, opened for signature 22 September 1992, 32 ILM 1069, (entered into force 25 March 1998). For a detailed discussion of the precautionary principle see Patricia Birnie and Alan Boyle, *International Law and the Environment* (2nd ed, 2002); Philippe Sands, *Principles of International Environmental Law* (2nd ed, 2005), 266–79; James Cameroon and Juli Abouchar, 'The Precautionary Principle: A Fundamental Principle of Law and Policy for the Protection of the Global Environment' (1991) 14:1 *Boston College International and Comparative Law Review* 1; Lothar Gündling, 'The Status in International Law of the Principle of Precautionary Action' (1990) 5 *International Journal of Estuarine and Coastal Law* 23.

260 Bothe, *supra* note 16, 167.

261 [1997] ICJ Rep 7, para 113.

262 The Asia Pacific Forum of National Human Rights Institutions and the Advisory Council of Jurists, 'Human Rights and the Environment: Final Report and Recommendations', (September 2007), 22 footnote 69, <www.ohchr.org/Documents/Issues/ClimateChange/Submissions/Asia_Pacific_Forum_of_NHRIs_1_HR_and_Environment_ACJ_Report_Recommendations.pdf> at 21 January 2015.

263 See, for example, *Beef Hormones, EC Measures Concerning Meat and Meat Products* (Decision), Panel of the World Trade Organisation, WT/DS26/AB/R, WT/DS48/AB/R, 16 January 1998, para 123 ('[w]hether it [the precautionary principle] has been widely accepted by Members as a principle of *general* or *customary international law* appears less than clear') (emphasis in original); *EC – Measures Affecting Biotech Products* (Decision), Panel of the World Trade Organisation, DS291, 292, 293, 26 September 2006, para 7.88 ('the legal status of the precautionary principle remains unsettled').

although the concept of precautionary measures, or a precautionary approach,[264] has been incorporated into several major international environmental treaties and various 'soft law' instruments.[265] This, however, gives rise to further uncertainty since, as discussed in chapter 2, there is no clear consensus as to whether, and to what extent, such treaties remain applicable during times of armed conflict, in the absence of an express provision to that effect.

In addition, there is no significant State practice referred to in the military manuals identified in the ICRC Study regarding the applicability of the precautionary principle.[266] In the end, the authors of the ICRC Study do not state that the precautionary principle is customary law, but rather that it 'has been gaining increasing recognition'.[267] It is therefore doubtful as to whether this international environmental concept can automatically be applied in the context of, and to the rules of armed conflict; still less can it be claimed to represent a customary rule of international humanitarian law.

In any event, it will often (although not always) be the case that the destructive effects on the environment that would result from military action will be predictable, in which case the precautionary principle, based as it is on a lack of scientific certainty, would not be relevant. The commentary to Rule 44 in the ICRC Study refers to the fact that, in the *Gabcíkovo-Nagymaros Project Case*, the International Court of Justice readily accepted that:[268]

---

264 The International Law Commission (ILC) has suggested that '[t]here are differing views whether or not the "precautionary principle" has been established as customary international law' and that a 'less disputed expression of "precautionary approach" could satisfy the basic necessity to introduce the special consideration of scientific uncertainties ...': International Law Commission, Draft Principles on the Allocation of Loss in the Case of Transboundary Harm Arising out of Hazardous Activities: General Commentary, contained in Report of the International Law Commission to the General Assembly on its work of the fifty-eighth session, UN Doc A/61/10 (2006), 222–3, para 6.

265 See Vienna Convention for the Protection of the Ozone Layer, opened for signature 22 March 1985, 26 ILM 1529 (entered into force 22 September 1988), which refers in paragraph 5 of its preamble to 'precautionary measures'. See also, for example, Convention for the Protection of the Marine Environment of the North East Atlantic, opened for signature 22 September 1992, 32 ILM 1069, (entered into force 25 March 1998), article 2(2)(a) ('precautionary principle'); United Nations Framework Convention on Climate Change, opened for signature 4 June 1992, 31 ILM 849 (entered into force 21 March 1994), article 3(3) ('precautionary measures'); Declaration of the United Nations Conference on Environment and Development, (14 June 1992) UN Doc A/CONF.151/26 (Volume 1), 31 ILM 874, Principle 15 ('precautionary approach'). While there have been some suggestions that these various expressions have different meanings, the prevailing interpretation seems to be that the 'difference in terminology is not significant': see Lotta Viikari, *The Environmental Element in Space Law: Assessing the Present and Charting the Future* (2008), 160–1 and the corresponding footnotes.

266 Hulme, *supra* note 231, 221.

267 Henckaerts and Doswald-Beck, *supra* note 57, Volume I, 150.

268 *Gabcíkovo-Nagymaros Project (Hungary v. Slovakia)* (Judgment) [1997] ICJ Rep 7, para 53. Article 33 of the document referred to dealt with the international law concept of 'necessity'

> the concerns expressed by Hungary for its natural environment in the region affected by the Gabcíkovo Nagymaros Project related to an "essential interest" of that State, within the meaning given to that expression in Article 33 of the Draft of the International Law Commission

It should perhaps be noted, however, that it might, in many cases, be difficult for the International Court of Justice to 'overrule' a claim of an essential interest by a State, unless it is 'completely disingenuous'.[269]

The Court went on to note that the ILC had expressed the view that a state of necessity may arise where there was 'a grave danger to ... the ecological preservation of all or some of [the] territory' of a State,[270] although at the same time it referred to difficulties associated with the[271]

> verification of the existence ... of the "peril" invoked by Hungary, of its "grave and imminent" nature, as well as of the absence of any "means" to respond to it

Without providing an explanation as to how this might point directly towards the formation of Rule 44 as a customary norm, the ICRC Study uses the comments of the Court to highlight the fact that there has been a movement over the past few decades towards the protection of the environment from degradation 'caused by humans'. This is undoubtedly the case, but it does not necessarily support the assertion that the rule as stated has acquired the status of a customary principle of international law.

Indeed, Michael Bothe, while regarding the arguments presented in the ICRC Study in relation to this rule as 'convincing', makes the point that it 'is not really based on actual practice'.[272] Karen Hulme questions the inclusion of the term 'preservation'.[273] While these concerns may not detract from the desirability of

---

as a 'circumstance precluding wrongfulness'. Its terms were superseded by article 25 of the 'Responsibility of States for internationally wrongful acts', which was noted by the United Nations General Assembly in United Nations General Assembly Resolution 56/83 'Responsibility of States for internationally wrongful acts' (12 December 2001) UN Doc A/RES/56/83 on the Responsibility of States for Internationally Wrongful Acts. Article 25 does differ from the terms of article 33(1) of the earlier draft, specifically in relation to the reference to the interest of 'the international community as a whole': Andreas Laursen, *Changing International Law to Meet New Challenges: Interpretation, Modification and the Use of Force* (2006), 243.

269 *Ibid*, 245.

270 [1997] ICJ Rep 7, para 53. See also *LG&E Energy Corp., LG&E Capital Corp., LG&E International Inc. v. Argentine Republic* (Decision on Liability), International Centre for Settlement of Investment Disputes (ICSID), Case No. ARB/02/1, 3 October 2006, para 251.

271 [1997] ICJ Rep 7, para 54.

272 Bothe, *supra* note 16, 167.

273 Hulme, *supra* note 231, 219.

the rule as stated, they again raise doubts as to whether it represents customary international law.

3.3.3. ICRC STUDY – RULE 45

Rule 45 of the ICRC Study provides as follows:

> The use of methods or means of warfare that are intended, or may be expected, to cause widespread, long-term and severe damage to the natural environment is prohibited. Destruction of the natural environment may not be used as a weapon.

Like Rule 44, this rule is stated to apply in relation to international armed conflicts and 'arguably' to non-international armed conflicts.[274] The ICRC Study concludes that the 'new' rule introduced into the 1977 Additional Protocol I in relation to the natural environment, and reflected in articles 35(3) and 55(1) (which at the time had no basis in existing customary international law),[275] has subsequently:[276]

> received such extensive support in State practice that it has crystallized into customary law, even though some States have persistently maintained that the rule does not apply to nuclear weapons and that they may, therefore, not be bound by it in respect of nuclear weapons

In its commentary to Rule 45, the ICRC Study suggests that:

> [i]t appears that the United States is a 'persistent objector' to the first part of this rule. In addition, France, the United Kingdom and the United States are persistent objectors with regard to the application of the first part of this rule to the use of nuclear weapons

In fact, one of the editors of the ICRC Study has raised the possibility that the United States may even be a persistent objector:[277]

> with respect to Rule 45 *in general, including for conventional weapons*, but this is a case the US would have to make

All of these qualifications cast doubt as to the strength of the claim that Rule 45 does reflect a rule of customary international law. It has been pointed out,

---

274 Henckaerts and Doswald-Beck, *supra* note 57, Volume I, 151.
275 Hulme, *supra* note 231, 204.
276 Henckaerts, *supra* note 14, 191.
277 International Committee of the Red Cross, *supra* note 32, 963 (emphasis added).

for example, that the three countries referred to should be regarded as 'specially affected'[278] States, and that their contrary practice therefore precludes the formation of a rule of custom.[279] In any event, these three States are significant nuclear powers,[280] giving rise to doubt as to whether there is sufficient State practice to lead to the establishment of the customary rule, at least in relation to nuclear weapons.

Michael Matheson has suggested that the significant reliance by nuclear weapon States on a policy of deterrence – sometimes referred to as the 'MAD' principle (Mutually Assured Destruction) during the Cold War period – means that there is insufficient *opinio juris* that the use of nuclear weapons is completely prohibited.[281] In this regard, the International Court of Justice concluded in the *Legality of the Threat or Use of Nuclear Weapons* Advisory Opinion that:[282]

> the members of the international community are profoundly divided on the matter of whether non-recourse to nuclear weapons over the past 50 years constitutes the expression of an *opinio juris*. Under these circumstances the Court does not consider itself able to find that there is such an *opinio juris*

Other commentators have suggested that it is very difficult to judge the current status of the principles incorporated in articles 35(3) and 55(1) of the 1977 Additional Protocol I.[283] However, even if one accepts that Rule 45 does represent a rule of custom, it does not appear to add significantly to the existing prohibitions in articles 35(3) and 55(1), in relation to which, as discussed in chapter 2, there are a number of significant practical limitations.

It should also be noted that the first sentence of Rule 45 repeats the threshold requirement of 'widespread, long-term and severe' damage found in articles 35(3) and 55(1) of the 1977 Additional Protocol I. The difficulties associated with ascertaining a precise meaning for this threshold, and the practical problems of actually proving that it has been reached in all but the most extreme cases, are also discussed in chapter 2. On this issue, one of the editors of the ICRC Study

---

278 See *North Sea Continental Shelf Cases (Federal Republic of Germany v. Denmark and Federal Republic of Germany v. The Netherlands)* (Judgment) [1969] ICJ Rep 3, para 73.

279 Scobbie, *supra* note 11, 36; Hulme, *supra* note 231, 233.

280 It is generally acknowledged that there are currently nine or ten countries with existing nuclear weapons capability. Besides the United States, France and the United Kingdom, these comprise Russia, China, India, Pakistan, Israel and North Korea, and (possibly) Iran: see, for example, Bothe, *supra* note 16, who, in 2005, suggested (at 168) that there were nine nuclear powers at that time.

281 Michael J. Matheson, 'The Opinion of the International Court of Justice on the Threat or Use of Nuclear Weapons' (1997) 91 *American Journal of International Law* 417, 425.

282 [1996] ICJ Rep 226, para 67.

283 See, for example, Hulme, *supra* note 73, 108.

has acknowledged that the fact that Rule 45 has come into existence, at least with regard to conventional weapons:[284]

> may not actually have much meaning, as the threshold of the cumulative conditions of long-term, widespread and severe damage is very high

It is not clear whether the second sentence of Rule 45 adds anything further as regards the protection of the environment over and above the fundamental principles of distinction and proportionality, which themselves also incorporate general rules regarding precautions to be taken when undertaking attacks, and the prohibition of wanton destruction of property.[285] However, to the extent that it reflects a customary rule, that sentence is significant, in that it plainly does focus upon the prohibition of deliberate damage of the environment during warfare.

It should be noted that Rule 45 refers to the destruction of the environment as a 'weapon'. This reinforces the fact that, as noted in chapter 1, in certain circumstances of intentional damage, the environment can be thought of not only as a victim, but in terms of being utilized as a weapon, as for example, through the employment of the specific techniques of environmental modification proscribed by the ENMOD Convention.[286]

Overall, the three specific rules in the ICRC Study that directly relate to the natural environment largely reflect the terms of the relevant *jus in bello* provisions, although in some respects they purport to extend the protections further. However, although the ICRC Study will undoubtedly have a practical impact and contribute to the wider dissemination of international humanitarian law, these rules should not themselves be regarded as 'additional' customary principles over and above the fundamental and generally accepted customary international humanitarian law principles of military necessity, distinction and proportionality. As such, they do not impose any additional specific *binding* (customary) obligations with respect to the conduct of armed conflict, including in relation to acts perpetrated with intent to cause significant damage to the environment.

---

284 International Committee of the Red Cross, *supra* note 32, 963.

285 Hulme, *supra* note 231, 234.

286 Convention on the Prohibition of Military or Any Other Hostile Use of Environmental Modification Techniques, opened for signature 10 December 1976, 1108 UNTS 151; 16 ILM 88 (entered into force 5 October 1978).

3.4. A GENERAL CUSTOMARY OBLIGATION TO TAKE ACCOUNT OF THE ENVIRONMENT DURING ARMED CONFLICT?

As has been noted, in the *Legality of the Threat or Use of Nuclear Weapons* Advisory Opinion, the International Court of Justice held that environmental considerations must be taken into account in determining the legality of acts taken during warfare.[287] In support of its conclusions, the Court quoted Principle 24 of the Rio Declaration, which provides that:[288]

> Warfare is inherently destructive of sustainable development. States shall therefore respect international law providing protection for the environment in times of armed conflict and cooperate in its further development, as necessary

The Court also concluded that articles 35(3) and 55 of the 1977 Additional Protocol I:[289]

> embody a general obligation to protect the natural environment against widespread, long-term and severe environmental damage

Taking all of these factors into account, the Court concluded that:[290]

> while the existing international law relating to the protection and safeguarding of the environment does not specifically prohibit the use of nuclear weapons, it indicates important environmental factors that are properly to be taken into account in the context of the implementation of the principles and rules of the law applicable in armed conflict

The significance of the natural environment is also underscored by the conclusion of the International Court of Justice in the *Gabcíkovo-Nagymaros Project Case*, noted above, that the protection of the environment may represent an 'essential interest' of a State, in the sense that this is relevant in determining the applicability of an alleged state of necessity as a circumstance precluding wrongfulness. In that case, the Court concluded that:[291]

> the awareness of the vulnerability of the environment and the recognition that environmental risks have to be assessed on a continuous basis have become much stronger [since 1977]

---

287 [1996] ICJ Rep 226, para 30.
288 Rio Declaration, Principle 24.
289 [1996] ICJ Rep 226, para 31.
290 *Ibid*, para 33.
291 [1997] ICJ Rep 7, para 112.

All of these observations may suggest the existence of a general rule of customary international law – requiring States to undertake a 'due regard' (due diligence?) assessment of the environment in the calculation of their military actions during armed conflict.[292] This standard is included in a document prepared in 1994 by the International Institute of Humanitarian Law which, when drafting principles related to naval warfare, concluded that:[293]

> [m]ethods and means of warfare should be employed with *due regard* for the natural environment taking into account the relevant rules of international law

This is also reflected in the first sentence of Rule 44 of the ICRC Study. Various aspects of the fundamental principles of customary humanitarian law discussed in this chapter, as well as the relevant provisions of the *jus in bello* treaties, reinforce this general obligation. This can be seen, for example, in the application of the principle of distinction, to the extent that the environment is to be regarded as a civilian object, and taking into account the conclusion of the International Court of Justice that environmental concerns form part of the calculation of the principles of necessity and proportionality.

In addition, articles 35(3) and 55(1) of the 1977 Additional Protocol I require a combatant to refrain from certain 'methods or means of warfare which are intended, or may be expected' to cause a specified level of environmental damage. In other words, in each of these examples, consideration of the environmental consequences of a proposed action must be taken *before* that action is to be implemented.

Karen Hulme goes further, by suggesting that military decision makers must now recognise 'the basic customary obligation to have regard to protection of the environment'.[294] However, this appears to over-state the position. On the one hand, it is widely acknowledged that States are responsible under customary international law:[295]

---

292 Hulme, *supra* note 73, 108.

293 International Institute of Humanitarian Law, 'San Remo Manual on International Law Applicable to Armed Conflicts at Sea' (1995) 309 *International Review of the Red Cross* 595, para 44 (emphasis added).

294 Hulme, *supra* note 73, 108.

295 Declaration of the United Nations Conference on the Human Environment (Stockholm Declaration) (16 June 1972) UN Doc A/CONF.48/14/Rev.1, Principle 21. Principle 21 was subsequently declared by the United Nations General Assembly as 'lay[ing] down the basic rules' governing the international responsibility of States in regard to the environment: United Nations General Assembly Resolution 2996 (XXVII) (15 December 1972) UN Doc A/RES/2996(XXVII) on the International Responsibility of States in Regard to the Environment.

> to ensure that activities within their jurisdiction and control do not cause damage to the environment of other states or of areas beyond the limits of national jurisdiction

This represents an extension of a more general principle of 'good neighbourliness', which can be traced back to a well-established Roman Law principle *sic utere tuo, ut alienum non laedas* ('use your own property in such a manner as to not injure that of another').[296] The obligation was specifically applied to the environment, in more restrictive terms, by the International Court of Justice in the *Legality of the Threat or Use of Nuclear Weapons* Advisory Opinion, when it confirmed that:[297]

> [t]he existence of the general obligation of States to ensure that activities within their jurisdiction and control *respect the environment* of other States or of areas beyond national control is now part of the corpus of international law relating to the environment

However, the obligation not to cause environmental damage does not expressly extend to situations of armed conflict. For one thing, this would be a *non sequitor*, given that warfare is inherently destructive of the environment, so that most decisions made in relation to military activities would be in breach of any such obligation. Moreover, various States, including the United States, take the view that Principle 21 is not applicable to environmental damage caused during armed conflict.[298] The statements of the International Court of Justice referred to above emphasise the relevance of environmental concerns in the decision-making process, but do not appear to impose a positive obligation upon the military to *protect* the environment during armed conflict.

This is also implied from a report produced by the International Institute of Humanitarian Law, which, when dealing with non-international conflicts, deliberately avoids listing a general obligation to protect the environment as a rule of customary international law.[299] Rather, it appears, at least at the current time, that the strongest that any such obligation can be put is that military leaders must consider the possibility of damaging environmental consequences of any action they are planning, and that such considerations must form part of the decision-making process as to whether, and how, to implement that action.

However, on a practical level, it is not clear whether this advances the protection of the environment during armed conflict in a significant way. As with the

296 Viikari, *supra* note 265, 150.

297 [1996] ICJ Rep 226, para 29 (emphasis added).

298 See Aaron Schwabach, 'Ecocide and Genocide in Iraq: International Law, the Marsh Arabs, and Environmental Damage in Non-International Conflicts' (2004) 15 *Colorado Journal of International Law and Policy* 1, 16–7 and the corresponding footnote.

299 Hulme, *supra* note 231, 221.

principles underpinning the Martens clause, the International Court of Justice has indicated that environmental concerns are already to be regarded as relevant when assessing the fundamental principles of necessity and proportionality. Environmental considerations do differ somewhat from elements of humanity, in that they do not (necessarily) have an anthropocentric focus.

However, as has been discussed, when applying fundamental principles of customary international humanitarian law, a balancing test must be undertaken, which will invariably include considerations of military advantage. Not only is this a difficult exercise, but it is characterised by subjectivity and uncertainty, and the threshold level of environmental damage generally stipulated must be so high as to be unattainable in all but the most extreme circumstances. Moreover, the military advantage of any proposed action would normally be uppermost in the minds of military leaders, making it questionable as to how extensively they would view the environmental consequences of such actions in most situations.

In addition, the different obligations that appear to apply with respect to the environment, depending upon whether an armed conflict is of an international or non-international nature,[300] make it difficult to reach definitive conclusions as to the precise extent of the protections that customary international law currently provides. Nevertheless, it is clear that environmental concerns now do form part of a military assessment of any proposed action, and that regard for the environment might possibly represent a separate rule of customary international law, as opposed to just being a part of the rules of necessity and proportionality.

Whilst such a conclusion may indicate a certain momentum within international law towards greater protection of the environment during armed conflict, it remains to be seen whether, in practice, such a rule (if it does exist) would serve to alter the behaviour of those engaged in armed conflict. In any event, it seems prudent to bear in mind Judge Koroma's words quoted earlier, regarding the 'notorious ... imprecision' of customary international law. Even if a relatively new customary principle can be identified, its scope beyond existing principles may be difficult to ascertain. Moreover, it would not necessarily provide for an adequate or effective mechanism for enforcement of any such obligation. The clear 'codification' of obligations relating the environment during armed conflict is therefore preferable, particularly in the context of specifying and clarifying the relevant international criminal law rules.

---

300 See the commentary to the ICRC Study in relation to Rules 44 and 45. See also Schmitt, Garraway and Dinstein, *supra* note 253, where the drafting committee (at para 4.2.4) lists only one specific rule in relation to the environment in the context of non-international armed conflicts, as follows:
'Damage to the natural environment during military operations must not be excessive in relation to the military advantage anticipated from those operations'.

3.5. CONCLUSION

It has been noted by Rymn Parsons that the law protecting the environment during armed conflict 'must be dynamic and progressive'.[301] Other commentators have suggested that, to be of use to the international community, law must lend itself to a 'consistent, uniform and practical application'.[302]

As discussed in this chapter, there are a number of important principles of customary international law that underpin the conduct of armed conflict. These principles apply to all States, operating independently of, as well as coincidentally with, the provisions of all relevant *jus in bello* treaties to which combatants may be bound. However, in terms of their applicability to intentional destruction of the environment during hostilities, there are significant shortcomings in relation to these customary rules. In general terms, the rules that purport to protect the environment are characterized by vagueness and uncertainty. This stems from several factors, including the difficulties associated with specifying the precise scope of the rules, the traditional and overarching importance accorded to considerations of military advantage, the imprecise 'status' and value to be accorded to the environment, the inherently destructive nature of warfare, and the overall complexity of the issues involved.

As with the *jus in bello* treaties, compliance with the customary rules of international humanitarian law has often been inconsistent and demonstrative of a lack of political will. Despite a strong imperative for effective enforcement, this is 'difficult to translate into effective policies'.[303] The also makes effective implementation and enforcement harder to achieve. It is in this regard that documents such as the ICRC Study may have the greatest impact, because they highlight the existence of a separate body of international humanitarian law over and above the *jus in bello* treaties, even though the ICRC Study itself may not in all respects accurately describe the precise scope of those customary principles.

However, the *realpolitik* means that, to the extent that there is any accountability for violations of customary principles (and perhaps also of the *jus in bello* treaties), these are likely to take place against the backdrop of the concept of *vie victis* ('woe to the vanquished'), with only the actions of the losing combatants likely to fall under judicial scrutiny,[304] although it is true that this was

---

301 Parsons, *supra* note 56, 483.

302 Richards and Schmitt, *supra* note 75, 1086.

303 Adam Roberts, 'The Laws of War: Problems of Implementation in Contemporary Conflicts' (1995) 6 *Duke Journal of Comparative and International Law* 11, 70.

304 Forrest, *supra* note 71, 196.

(potentially) not the case in relation to the actions of NATO in Operation Allied Force.

Even though it is clear that the environment has become a more relevant consideration in the assessment of all military actions, the reluctance of States to adequately address the protection of the environment in times of war by way of binding international legal instruments also impacts upon, and reflects, the lack of an effective complementary regime of customary rules. Thus, as with the *jus in bello* treaties, there is no comprehensive regime under customary law that adequately addresses intentional damage to the environment during armed conflict. As a consequence, it is submitted that, within the treaty and customary principles of the *jus in bello* and international environmental law, there is currently no effective system for imposing a rigorous level of responsibility for such acts.

At the same time, however, there have been significant advances in international *criminal* law and jurisprudence over the past two decades, and there is now widespread acceptance of the imperative to establish effective mechanisms and legal regimes to prosecute specific international crimes. It is therefore necessary to now turn to an examination of the existing applicable international criminal law provisions to determine whether they provide a sufficient degree of accountability for such acts. The next chapter thus examines the existing crimes articulated in the Rome Statute in order to determine their applicability (or otherwise) to such acts. This will include, in particular, a detailed analysis of the war crime specified in article 8(2)(b)(iv) of the Rome Statute.

CHAPTER FOUR

REGULATION OF THE INTENTIONAL DESTRUCTION OF THE ENVIRONMENT DURING WARFARE UNDER THE EXISTING ROME STATUTE REGIME

4.1. INTRODUCTION – THE CORE INTERNATIONAL CRIMES IN THE ROME STATUTE

The previous two chapters have examined the extent to which the intentional destruction of the environment during armed conflict has been addressed under the *jus in bello* (and international environmental) treaties, and any applicable customary international law. Where appropriate, reference has also been made (as will also be the case in this chapter) to relevant jurisprudence of international courts and tribunals. However, as noted, there has been little judicial precedent directly relating to the prosecution of persons who have been specifically charged with intentionally causing such destruction.[1]

Based on the analysis undertaken in the previous chapters, the conclusion thus far is that, from the perspective of individual criminal responsibility for such acts, the various 'sources' of international law that apply in those two regimes of law are not sufficient, particularly when viewed in the light of the (potentially) adverse consequences that could flow from damage of this type, and the idea that a person would have the specific *intention* to significantly damage the environment. This analysis has in fact reinforced the clear imperatives for the sanctioning of such behaviour within the scope of international *criminal* law that were raised in chapter 1.

1 Eric Talbot Jensen and James J. Teixeira, Jr., 'Prosecuting Members of the U.S. Military for Wartime Environmental Crimes' (2005) 17 *Georgetown International Environmental Law Review* 651, 655.

This has been the approach taken by the international community. Whilst the Rome Statute of the International Criminal Court[2] does not contain a specific crime of 'crimes against the environment' within the crimes in relation to which the Court has jurisdiction,[3] it does, as noted, set out the circumstances in which a particular level of damage to the 'natural environment' may constitute a war crime. This represents the first time that the environment has been expressly mentioned in the constitutive documents of any of the international criminal courts and tribunals that have been established, commencing from (and including) the Nuremberg Military Tribunal.[4]

As a general observation, the inclusion of such a provision within the statute of what is intended to be a *permanent* international criminal court might, potentially, represent an important preliminary step towards individual criminal responsibility for intentional environmental destruction during armed conflict. However, of course, the effectiveness of any proscriptive provision largely depends upon its express terms. It is therefore necessary to carefully analyse the precise scope and effect of this provision – and any other crime set out within the Rome Statute that might be relevant – to understand whether, in fact, they serve to expand the reach of the mechanisms of international justice that are designed to 'end impunity' for those who commit international crimes. The reality of the severe damage that has been, and inevitably will in the future be, inflicted on the environment during the course of armed conflict, necessitates an understanding of whether the existing obligations are adequate, or whether, in fact, they may need further development and amendment.

The discussion below therefore examines whether, and to what extent (if at all), the position may have changed with the coming into force on 1 July 2002 of the Rome Statute, and the establishment of the International Criminal Court (ICC or the Court). In accordance with the provisions of the Rome Statute, the ICC has jurisdiction in relation to the specific crimes of genocide (as defined in article 6), crimes against humanity (article 7), war crimes (article 8), as well as the more recently defined crime of aggression.[5] These crimes, particularly

2 Opened for signature 17 July 1998, 2187 UNTS 3; 37 ILM 999 (entered into force 1 July 2002) (Rome Statute).

3 This was partly due to the fact that such an idea 'regrettably' failed to gain support in the deliberations leading to the finalization of the Rome Statute; Alfred Rest, 'Enhanced Implementation of International Environmental Treaties by Judiciary – Access to Justice in International Environmental Law for Individuals and NGOs: Efficacious Enforcement by the Permanent Court of Arbitration' (2004) 1 *Macquarie Journal of International and Comparative Environmental Law* 1, 18.

4 International Military Tribunal for the Trial of the German Major War Criminals at Nuremberg (Nuremberg Military Tribunal).

5 The Rome Statute originally provided that the Court would exercise jurisdiction over the crime of aggression once a provision defining the crime and setting out the conditions under

those of genocide, crimes against humanity and war crimes, are broadly regarded as reflecting universal international criminal norms.[6] Shortly after the conclusion of the Rome Statute, a Trial Chamber of the International Criminal Tribunal for the Former Yugoslavia[7] described the legal effect of the provisions of the Rome Statute in the following way:[8]

> [d]epending on the matter at issue, the Rome Statute may be taken to restate, reflect or clarify customary rules or crystallise them, whereas in some areas it creates new law or modifies existing law. At any event, the Rome Statute by and large may be taken as constituting an authoritative expression of the legal views of a great number of States.

Since that time, it has been asserted that the Rome Statute 'purports to contain a comprehensive stocktaking of the current status of customary international law'.[9] This may be over-stating the case, since, as noted, particular aspects of the definitions of crimes in the Rome Statute, in particular, that of crimes against humanity and war crimes, and the crime of aggression as a whole, most likely also include(d) elements of 'progressive development', rather than already well-established principles of law. Hence, arguments persist as to whether every aspect of the definitions of the crimes in the Rome Statute may (or may not) be identical to current customary international law.[10] Nevertheless, it is generally

---

which such jurisdiction would be exercised was adopted in accordance with the procedures set out in that instrument; Rome Statute, article 5(2). This was a 'last-minute addition' to the Rome Statute: Danesh Sarooshi, 'The Statute of the International Criminal Court' (1999) 48 *International and Comparative Law Quarterly* 387, 400. A definition of the crime was agreed at the first Review Conference of the Rome Statute, held in Kampala, Uganda from 31 May to 11 June 2010: see Resolution RC/Res.6, adopted at the Review Conference on 11 June 2010 <www.icc-cpi.int/iccdocs/asp_docs/Resolutions/RC-Res.6-ENG.pdf> at 23 January 2015. See Rome Statute, articles 8 *bis*, 15 *bis* and 15 *ter*.

6 For the purposes of the discussion that follows, these three crimes are referred to in this and the following chapter as the 'core' international crimes.

7 International Tribunal for the Prosecution of Persons Responsible for Serious Violations of International Humanitarian Law Committed in the Territory of the Former Yugoslavia since 1991 (ICTY).

8 Judgement, *Prosecutor v. Furundzjia*, Case No. IT-95-17/1, Trial Chamber, 10 December 1998, para 227.

9 Elaina I. Kalivretakis, 'Are Nuclear Weapons Above the Law? A Look at the International Criminal Court and the Prohibited Weapons Category' (2001) 15 *Emory International Law Review* 683, 684.

10 See, for example, Antonio Cassese, *International Criminal Law* (2003), 91–4. See also Judgement, *Prosecutor v. Zoran Kupreskic, Mirjan Kupreskic, Vlatko Kupreskic, Drago Josipovic, Dragan Papic and Vladimir Santic*, Case No. IT-95-16-T, Trial Chamber, 14 January 2000, where the Trial Chamber (at para 545) noted that '[b]y requiring that crimes against humanity be committed in either internal or international armed conflict, the [United Nations] Security Council, in establishing the [ICTY], may have defined the crime in Article 5 [of the Statute of the ICTY] more narrowly than is necessary under customary international law'. It may be pertinent, however, to note that Antonio Cassese was the Presiding Judge in that specific case.

acknowledged that the parameters of the core international crimes (that is, other than the crime of aggression) represent acts that are a gross affront to fundamental norms.

This chapter examines these three core international crimes, as they currently appear in the Rome Statute, with a view to determining whether, individually or *in toto*, they are sufficient to address the need for individual criminal responsibility under international criminal law relating to the intentional destruction of the environment during armed conflict. In addition, therefore, to a detailed analysis of the only provision of the Rome Statute that makes express reference to the environment – the war crime defined in article 8(2)(b)(iv) – it is also necessary to undertake a (briefer) examination of other core international crimes to determine whether such acts might also (or instead) fall within their scope, even though they are not specifically directed towards environmental damage.

Following this analysis, it will be possible to form a conclusion as to whether article 8(2)(b)(iv), together with, or independently of the other core international crimes within the Rome Statute, comprehensively meet what is argued in this book to be an impending need.[11] If it is concluded that the current regime within the Rome Statute does not fulfil the requirements to address such acts, a consideration of alternate 'solutions' is necessary. This would inevitably involve amendment of the Rome Statute.

This analysis will serve to test the underlying hypothesis that has motivated this book – that there is a need to define a new *sui generis* crime to cover such behaviour, since it is not adequately addressed in the existing Rome Statute regime and, as has been demonstrated in previous chapters, neither is it properly dealt with under the *jus in bello* and international environmental law treaties, nor under customary international law. Before undertaking this analysis, however, it is necessary to briefly set out the background to the establishment of the Court, as well as its general jurisdictional mandate.

4.2. FINALIZATION OF THE ROME STATUTE AND ESTABLISHMENT OF THE ICC

In December 1989, in response to a request by Trinidad and Tobago, the United Nations General Assembly asked the International Law Commission (ILC) to resume its work on the drafting of a constitutive instrument for an international

11 Refer in particular to chapter 1.

criminal court,[12] with a jurisdiction that would specifically include drug trafficking.[13] The ILC subsequently completed a draft statute (ILC Draft Statute) and submitted it to the United Nations General Assembly in 1994.[14] Even though the final version of the Rome Statute differs quite significantly from the ILC Draft Statute,[15] in the words of the chairperson of the ILC Working Group that produced the draft, it 'got the diplomatic ball rolling again'.[16]

Following receipt of the ILC Draft Statute, the General Assembly established the *Ad Hoc* Committee on the Establishment of an International Criminal Court (*Ad Hoc* Committee), which met twice in 1995 and then reported back to the General Assembly. Having considered the *Ad Hoc* Committee's report, the General Assembly subsequently created the Preparatory Committee on the Establishment of an International Criminal Court (PrepCom), to prepare a consolidated draft text for submission to a diplomatic conference. The PrepCom, which began meeting in 1996, completed its work at its final session in March-April of 1998.

---

12 The ILC had in 1948 already been invited by the United Nations General Assembly to 'study the desirability and possibility of establishing an international judicial organ for the trial of persons charged with genocide and other crimes': United Nations General Assembly Resolution 260 (III) B (9 December 1948) UN Doc A/RES/260(III)B on the Prevention and Punishment of the Crime of Genocide: Study by the International Law Commission of the Question of an International Criminal Jurisdiction. However, the political realities associated with the Cold War meant that, in 1954, the United Nations General Assembly halted work on the drafting of any statute for such a proposed court: see United Nations General Assembly Resolution 897 (IX) (4 December 1954) UN Doc A/RES/897(IX) on the Draft Code of Offences Against the Peace and Security of Mankind, and United Nations General Assembly Resolution 898 (IX) (14 December 1954) UN Doc A/RES/898(IX) on the International Criminal Jurisdiction. For a description of the evolution of the system of international criminal justice leading up to the establishment of the ICC, see Steven Freeland, 'International Criminal Justice and the Death Penalty', in Jon Yorke (ed), *The Right to Life and the Value of Life: Orientations in Law, Politics and Ethics* (2010), 193, 195–210.

13 United Nations General Assembly Resolution 44/39 (4 December 1989) UN Doc A/RES/44/39 on the International Criminal Responsibility of Individuals and Entities Engaged in Illicit Trafficking in Narcotic Drugs across National Frontiers and Other Transnational Criminal Activities, para 1.

14 Draft Statute for an International Criminal Court, Report of the International Law Commission on Its Forty-sixth session, United Nations General Assembly Official Records 49th Sess., Supp. No. 10, UN Doc A/49/10 (1994) (ILC Draft Statute).

15 It has been suggested by Gilbert Bitti that the ILC Draft Statute was 'a common law orientated' document, and consequently required considerable amendment in order to accommodate the concern among civil lawyers that the Court would become an 'International Criminal *Common* Law Court': Gilbert Bitti, 'Two Bones of Contention Between Civil and Common Law: The Record of the Proceedings and the Treatment of a Concursus Delictorum', in Horst Fischer, Claus Kress and Sascha Rolf Lüder (eds), *International and National Prosecution of Crimes Under International Law: Current Developments* (2001), 273, 273–4 (emphasis added).

16 James Crawford, 'The drafting of the Rome Statute', in Philippe Sands (ed), *From Nuremberg to The Hague: The Future of International Criminal Justice* (2003), 109, 110.

The General Assembly convened the United Nations Diplomatic Conference of Plenipotentiaries on the Establishment of an International Criminal Court (1998 Rome Conference), which was held from 15 June to 17 July 1998, 'with a view to finalizing and adopting a convention on the establishment of an international criminal court'.[17] The 1998 Rome Conference was attended by delegations from over 160 States, 30 Intergovernmental Organizations and 230 Non-Governmental Organizations. During the Rome Conference, several complex questions remained to be agreed and, at times, the discussions were tortured and acrimonious, not the least because many of the States had not participated in the *Ad Hoc* Committee or PrepCom discussions, and were thus confronted with a detailed draft perhaps even for the first time.[18] Right up until the final vote, there remained a number of contentious issues – particularly in relation to the extent of the Court's proposed jurisdiction, as well as its relationship with the United Nations Security Council. In the end, however, the delegates adopted the final terms of the Rome Statute, albeit not unanimously.[19]

As events at the 1998 Rome Conference and subsequently have illustrated, the establishment of the Court was as political an event as it was significant in the evolution of international criminal justice. The negotiating States, as well as the other stakeholders at the 1998 Rome Conference, represented a multitude of differing views as to how the Court should be structured. The final terms of the Rome Statute were, in many respects, the result of an 'enduring tension inherent in multilateral negotiations between sovereignty and universality',[20] which by necessity required a 'solution' based upon political compromise, even including the question as to what crimes should be included in the Rome Statute,[21] in order that the (draft) instrument could be presented as 'marketable' to the delegates.[22]

---

17 United Nations General Assembly Resolution 52/160 (15 December 1997) UN Doc A/RES/52/160 on the Establishment of an International Criminal Court, para 3.

18 Kalivretakis, *supra* note 9, 697–8.

19 Of those represented at the 1998 Rome Conference, 120 States voted to adopt the Rome Statute. There were 21 abstentions and seven States (China, Iraq, Israel, Libya, Qatar, Yemen and the United States of America) voted against the resolution.

20 Tim McCormack and Sue Robertson, 'Jurisdictional Aspects of the Rome Statute for the New International Criminal Court' (1999) 23 *Melbourne University Law Review* 635, 636.

21 For example, a number of States had argued that the definition of war crimes should include a provision prohibiting the use of nuclear weapons. As the 1998 Rome Conference was drawing to a close, these States largely agreed to compromise on this point – with the result that such a provision was not included in the Rome Statute – since they were prepared to 'put the larger goal of achieving an international criminal court first': see Kalivretakis, *supra* note 9, 702.

22 Sharon A. Williams, 'The Rome Statute on the International Criminal Court – Universal Jurisdiction or State Consent – To Make or Break the Package Deal', in Michael J.N. Schmitt (ed), *International Law Studies Volume 75: International Law Across the Spectrum of Conflict* (2000), 539, 546.

Of course, compromise may also produce an instrument reflecting an agreement of the 'lowest common denominator', with the result that important issues may be consciously omitted from the final version, and the language utilized may give rise to more than one possible meaning, uncertainties as to interpretation, and/or weaknesses that may dilute its effectiveness in addressing the object(s) of the instrument.

The precise scope of those crimes included in the final version of Rome Statute thus represented a *realpolitik* compromise in the circumstances – a careful balance between the important goals of the Court on the one hand,[23] and a politically acceptable series of definitions of proscribed acts to be regarded as international crimes within the jurisdiction of the Court on the other. In such a context, a more expansive approach towards prohibiting intentional environmental damage during armed conflict beyond the scope of what was agreed under article 8(2)(b)(iv) was not politically feasible.

Certainly, also, the specific proscription of intentional environmental destruction was not among the principal concerns of the delegates to the 1998 Rome Conference. In the end, the conclusion of the Rome Statute, and the establishment four years later of the ICC, were considered by the international community to be the most important outcome, notwithstanding the impossibility of satisfying the concerns of all stakeholders.

If anything, the importance of the principles represented by the aims of the Court has grown further since 1998. The finalization of the Rome Statute and its subsequent ratification (thus far) by 123 States (as at January 2015),[24] demonstrates a broader acceptance of the principle underlying the establishment of the ICC – the need for more effective enforcement of the universal norms of international criminal law. Given that the Rome Statute was finalized 17 years ago, and the ICC has been operating for more than a decade, it is argued that the imperatives for individual criminal responsibility have expanded in the intervening years, and it is now appropriate to (re)consider the issue of intentional environmental destruction during armed conflict in greater detail. This is the main focus of this and the following chapter.

The Court has been given the mandate to play a role when certain international crimes have (allegedly) been committed. As a permanent court, it differs from the *ad hoc* international criminal Tribunals. The ICTY and ICTR were set up

---

23 See Rome Statute, preamble.

24 Palestine acceded to the Rome Statute on 2 January 2015. The Rome Statute will enter into force for Palestine on 1 April 2015: see ICC Website, 'Palestine' <www.icc-cpi.int/en_menus/asp/states%20parties/asian%20states/Pages/Palestine.aspx> at 23 January 2015.

as 'UN subsidiary organs'[25] *in response to* specific events, and were always intended to have a limited life-span, as is indicated by the formulation of the Completion Strategy for those Tribunals.[26]

By contrast, the ICC is 'a permanent institution',[27] established under a treaty and, as such, is formally independent of the United Nations, although there is clearly an ongoing relationship between the two institutions on several key issues.[28] The Court has the power to exercise its jurisdiction with respect to circumstances that may occur *in the future* – that is, at any time after the Rome Statute came into force (1 July 2002).[29] Unless the Assembly of States Parties to the Rome Statute decides to completely alter the nature or focus of the Court, the ICC will remain in place for the long-term, and will have the potential to play a role in circumstances where 'the most serious crimes of concern to the international community as a whole' have been committed.[30]

In this sense, the Court represents an important mechanism of international criminal justice in relation to values and norms that are accepted universally among the international community now and into the future. These values and norms will continue to evolve further in the years and decades following the establishment of the ICC, in order to address relevant concerns regarding

25 Sarooshi, *supra* note 5, 389.

26 See *inter alia* United Nations Security Council Resolution 1503 (28 August 2003) UN Doc S/Res/1503 on the International Criminal Tribunal for the former Yugoslavia and International Criminal Tribunal for Rwanda; United Nations Security Council Resolution 1534 (26 March 2004) UN Doc S/Res/1534 on the International Criminal Tribunal for the former Yugoslavia and International Criminal Tribunal for Rwanda; United Nations Security Council Resolution 1966 (22 December 2010) UN Doc S/Res/1966/2010 on the International Tribunal for the Prosecution of Persons Responsible for Serious Violations of International Humanitarian Law Committed in the Territory of the Former Yugoslavia since 1991and the International Criminal Tribunal for the Prosecution of Persons Responsible for Genocide and Other Serious Violations of International Humanitarian Law Committed in the Territory of Rwanda and Rwandan Citizens Responsible for Genocide and Other Such Violations Committed in the Territory of Neighbouring States between 1 January 1994 and 31 December 1994.

27 Rome Statute, article 1.

28 There is a Negotiated Relationship Agreement between the International Criminal Court and the United Nations (4 October 2004) ICC Doc ICC-ASP/3/Res.1, whose purpose is to 'define (…) the terms on which the United Nations and the Court shall be brought into relationship' (article 1(1)). In addition, there are a number of provisions in the Rome Statute that formalize various aspects of the specific relationship between the Court and the United Nations Security Council: see, for example, Rome Statute, articles 13(b), 16, 53(2)(c), 53(3)(a), 87(5), 87(7) and 115(b).

29 It should be noted, however, that if a State becomes a State Party to the Rome Statute after 1 July 2002, the Court may exercise its jurisdiction only with respect to crimes committed after the entry into force of the Rome Statute for that State, unless that State has made a declaration as a non-State Party under article 12(3) of the Rome Statute: see Rome Statute, article 11(2).

30 Rome Statute, preamble para 4.

actions taken within the context of armed conflict that may, for example, threaten the future of humanity. It is contended in this book that the integrity of the environment has become such a concern not just in respect of its own sake, but also from a human rights perspective,[31] and that it is important to adequately address the intentional destruction of the environment during armed conflict from an international criminal law perspective, utilizing the ICC acting in accordance with the Rome Statute.

As United Nations Secretary-General, Kofi Annan, put it following the adoption of the Rome Statute at the conclusion of the 1998 Rome Conference, the Court's creation is:[32]

> a gift of hope to future generations, and a giant step forward in the march towards universal human rights and the rule of law

Not only is this important in and of itself; it also reaffirms the interrelationship noted in chapter 1 between the maintenance of international peace and security – one of the principal purposes of the United Nations[33] – and respect for fundamental human rights. The United Nations Security Council has acknowledged that:[34]

> peace and security, development and human rights are the pillars of the United Nations system and the foundations for collective security and well-being, and … in this regard … development, peace and security and human rights are interlinked and mutually reinforcing

4.3. JURISDICTION OF THE COURT – A BRIEF OVERVIEW

Obviously, the Court can legitimately operate only within the limits of its jurisdiction. A specific crime will only fall within its jurisdiction where express provision has been made within the Rome Statute. There is, however, nothing that restricts the inclusion in the future of further crimes within the Rome

31 See, for example, Laura Horn and Steven Freeland, 'More than Hot Air: Reflections on the Relationship between Climate Change and Human Rights' (2009) 13 *University of Western Sydney Law Review* 101.

32 Statement by the United Nations Secretary-General Kofi Annan at the Ceremony Celebrating the Adoption of the Statute of the International Criminal Court (18 July 1998) <http://untreaty.un.org/cod/icc/rome/proceedings/contents.htm> at 29 March 2013.

33 See United Nations Charter, opened for signature 26 June 1945, 1 UNTS 16; 59 Stat. 1031 (entered into force 24 October 1945) (United Nations Charter), article 1(1).

34 United Nations Security Council Resolution 1674 (28 April 2006) UN Doc S/RES/1674/2006 on the Protection of Civilians in Armed Conflict, preamble para 3.

Statute, as long as the necessary procedural requirements are complied with. As noted, the Rome Statute currently provides that the following crimes, when committed after 1 July 2002,[35] fall within the jurisdiction of the ICC:

(a) The crime of Genocide – when committed 'with intent to destroy, in whole or in part, a national, ethnical, racial or religious group, as such';[36]
(b) Crimes against Humanity – when committed 'as part of a widespread or systematic attack directed against any civilian population, with knowledge of the attack';[37]
(c) War Crimes – which usually involves a breach of 1949 Geneva Conventions[38] and/or the laws and customs of armed conflict, 'in particular when committed as part of a plan or policy or as part of a large-scale commission of such crimes';[39]
(d) The Crime of Aggression as recently defined.[40]

The specific requirements of the crimes of genocide, crimes against humanity and war crimes, particularly as they might possibly be applied to intentional destruction of the environment during armed conflict, are discussed in 4.4 below. In general terms, and as is the case with serious crimes under the national law of most countries, these crimes require the presence of both a 'physical' element (*actus reus)* and a 'mental' element (*mens rea*).[41] In terms of the mental element, the Rome Statute sets out a 'default' standard of *mens rea*, which specifies that, for individual criminal responsibility to apply, the material (physical) elements of the crime must be 'committed with intent and knowledge', as those terms are defined in the provision.[42] As discussed in chapter 5, this standard is expressed to apply '[u]nless otherwise provided', so that specific crimes may have *mens rea* requirements that differ from the article 30 standard.

---

35 This is, as noted, subject to the terms of article 11(2) of the Rome Statute.
36 See Rome Statute, chapeau article 6.
37 *Ibid*, article 7(1).
38 Geneva Convention for the Amelioration of the Condition of the Wounded and Sick in Armed Forces in the Field, opened for signature 12 August 1949, 75 UNTS 31 (entered into force 21 October 1950) (1949 Geneva Convention I), Geneva Convention for the Amelioration of the Condition of the Wounded, Sick and Shipwrecked Members of Armed Forces at Sea, opened for signature 12 August 1949, 75 UNTS 85 (entered into force 21 October 1950) (1949 Geneva Convention II), Geneva Convention Relative to the Treatment of Prisoners of War, opened for signature 12 August 1949, 75 UNTS 135 (entered into force 21 October 1950) (1949 Geneva Convention III), Geneva Convention Relative to the Protection of Civilian Persons in Time of War, opened for signature 12 August 1949, 75 UNTS 287 (entered into force 21 October 1950) (1949 Geneva Convention IV).
39 See Rome Statute, article 8(1).
40 See *ibid*, article 8 *bis*.
41 See, for example, William A. Schabas, *Genocide in International Law* (2000), chapters 4 and 5.
42 Rome Statute, article 30(1). Article 30 is discussed in more detail in chapter 5.

Articles 12 and 13 of the Rome Statute specify the '[p]reconditions to the exercise of jurisdiction' and 'exercise of jurisdiction' by the Court respectively.[43] In summary, the Court can exercise its jurisdiction in relation to these crimes in the following circumstances:

(a) where an (alleged) crime has been committed on the territory of a State Party to the Rome Statute;[44]
(b) where a national(s) of a State Party to the Rome Statute is alleged to have committed a crime;[45]
(c) where a situation in which a crime(s) 'appears to have been committed' has been referred to the Prosecutor of the ICC by the United Nations Security Council acting under Chapter VII of the United Nations Charter;[46] or
(d) where a non-State Party to the Rome Statute lodges a declaration with the Registrar of the ICC, accepting the jurisdiction of the Court with respect to the 'crime in question'.[47]

As also noted, the ICC is also subject to a specific *ratione temporis*, as set out in article 11 of the Rome Statute.

The jurisdiction of the Court is subject to the principle of 'complementarity' as established under the Rome Statute. In essence, this means that primary responsibility for the prosecution of these crimes lies with States, and that the ICC therefore operates as a 'court of last resort'. This in itself demonstrates a shift in emphasis from the culture of impunity that had existed before the 1990s, during which time States had been reluctant to try their own nationals for war crimes, 'and even more [so] where crimes against humanity or genocide [were] concerned'.[48]

---

43 There are specific provisions relating to the exercise by the Court of its jurisdiction over the crime of aggression, which have the effect that such jurisdiction will not exist until 1 January 2017 at the earliest, subject to various requirements: see *ibid*, article 15 *bis* and 15 *ter*.

44 *Ibid*, article 12(2)(a).

45 *Ibid*, article 12(2)(b).

46 *Ibid*, article 13(b).

47 *Ibid*, article 12(3). For a discussion of the declaration process by non-State Parties under article 12(3) of the Rome Statute, see Steven Freeland, 'How Open Should the Door Be? – Declarations by non-States Parties under Article 12(3) of the Rome Statute of the International Criminal Court' (2006) 75(2) *Nordic Journal of International Law* 211; Carsten Stahn, 'Why some Doors may be Closed Already: Second Thoughts on a 'Case-by-Case' Treatment of Article 12(3) Declarations' (2006) 75 *Nordic Journal of International Law* 243.

48 Philippe Sands, 'After Pinochet: the role of national courts', in Philippe Sands (ed), *From Nuremberg to The Hague: The Future of International Criminal Justice* (2003), 68, 72.

Article 17 of the Rome Statute applies the complementarity principle in terms of the 'admissibility' of a case. A case is determined by the Court as being inadmissible *inter alia* where:[49]

> (a) [t]he case is being investigated or prosecuted by a State which has jurisdiction over it, unless the State is unwilling or unable genuinely to carry out the investigation or prosecution;
> (b) [t]he case has been investigated by a State which has jurisdiction over it and the State has decided not to prosecute the person concerned, unless the decision resulted from the unwillingness or inability of the State genuinely to prosecute;
> (...)
> (d) [t]he case is not of sufficient gravity to justify further action by the Court.

This creates what has been described as a 'presumption in favour of prosecution in domestic courts',[50] given that it accords a priority to national jurisdiction. This is to be contrasted with the 'primacy' principle under which the *ad hoc* Tribunals operate.[51]

---

49 Rome Statute, article 17(1). The Appeals Chamber of the ICC considered the meaning of the words 'is being investigated' in article 17(1)(a) of the Rome Statute in Judgment on the appeal of the Republic of Kenya against the decision of Pre-Trial Chamber II of 30 May 2011 entitled "Decision on the Application by the Government of Kenya Challenging the Admissibility of the Case Pursuant to Article 19(2)(b) of the Statute", *The Prosecutor v. William Samoei Ruto, Henry Kiprono Kosgey and Joshua Arap Sang*, Case No. ICC-01/09-01/11 OA, Appeals Chamber, 30 August 2011, and Judgment on the appeal of the Republic of Kenya against the decision of Pre-Trial Chamber II of 30 May 2011 entitled "Decision on the Application by the Government of Kenya Challenging the Admissibility of the Case Pursuant to Article 19(2)(b) of the Statute", *The Prosecutor v. Francis Kirimi Muthaura, Uhuru Muigai Kenyatta and Mohammed Hussein Ali*, Case No. ICC-01/09-02/11 OA, Appeals Chamber, 30 August 2011.

50 Sarooshi, *supra* note 5, 395.

51 See articles 9(1) and 9(2) of the Statute of the International Tribunal for the Prosecution of Persons Responsible for Serious Violations of International Humanitarian Law Committed in the Territory of the Former Yugoslavia since 1991, 32 ILM 1159 (entered into force 25 May 1993) (ICTY Statute). The Appeals Chamber of the ICTY confirmed the legitimacy of its primacy in Decision on Defence Motion for Interlocutory Appeal on Jurisdiction, *Prosecutor v Duško Tadić*, Case No. IT-94-1, Appeals Chamber, 2 October 1995, para 49–64. See also articles 8(1) and 8(2) of the Statute of the International Criminal Tribunal for the Prosecution of Persons Responsible for Genocide and Other Serious Violations of International Humanitarian Law Committed in the Territory of Rwanda and Rwandan Citizens Responsible for Genocide and Other Such Violations Committed in the Territory of Neighbouring States, between 1 January 1994 and 31 December 1994, 33 ILM 1598 (entered into force 8 November 1994) (ICTR Statute), which is in slightly wider terms. It has been suggested that the wording in the ICTR Statute, which was drafted after the ICTY Statute, reflected a broader consensus at that time among the United Nations Security Council as to the concept of primacy: Bartram S. Brown, 'Primacy or Complementarity: Reconciling the Jurisdiction of National Courts and International Criminal Tribunals' (1998) 23 *The Yale Journal of International Law* 391, 402.

Article 17(2) specifies those circumstances in which the ICC may determine the 'unwillingness' of a State in a particular case. This may arise in the following situations:[52]

> (a) [t]he proceedings were or are being undertaken or the national decision was made for the purpose of shielding the person concerned from criminal responsibility (...);
> (b) ... an unjustified delay in the proceedings ... inconsistent with an intent to bring the person concerned to justice;
> (c) [t]he proceedings were or are not being conducted independently or impartially, and ... are being conducted in a manner ... inconsistent with an intent to bring the person concerned to justice.

In assessing these circumstances, the Court is to have regard to 'the principles of due process recognized by international law'.[53] In determining a State's (in)ability in a particular case, the Court has to consider whether:[54]

> due to a total or substantial collapse or unavailability of its national judicial system, the State is unable to obtain the accused or ... evidence or otherwise [is] unable to carry out its proceedings

4.4. APPLICABILITY OF THE ROME STATUTE TO THE INTENTIONAL DESTRUCTION OF THE ENVIRONMENT DURING WARFARE

The remainder of this chapter will examine the extent to which the intentional destruction of the environment during armed conflict is addressed by the existing core international crimes defined in the Rome Statute. As mentioned, article 8(2)(b)(iv) expressly refers to, and 'criminalizes' damage to the natural environment in certain specific circumstances. The terms of this provision will be discussed in 4.4.3.1.

Before turning to that specific provision, however, one should first consider whether it may be possible that other provisions of the Rome Statute might also be relevant to such acts; specifically, whether the crimes of genocide, crimes against humanity and other war crimes (apart from article 8(2)(b)(iv)) also address the issue to any significant degree. In this way, one can gain a more comprehensive understanding of the applicability of the Rome Statute to acts that might constitute a crime against the environment, as well as the significance

52 Rome Statute, article 17(2).
53 *Ibid.*
54 *Ibid*, article 17(3).

of article 8(2)(b)(iv) itself within the broader schema of crimes within the instrument. In undertaking this exercise, however, it is important to bear in mind that, as provided in the Rome Statute itself:[55]

> [t]he definition of a crime shall be strictly construed and shall not be extended by analogy. In case of ambiguity, the definition shall be interpreted in favour of the person being investigated, prosecuted or convicted.

4.4.1. INTENTIONAL DESTRUCTION OF THE ENVIRONMENT AS GENOCIDE?

The word 'genocide' was first coined in 1944 by a Polish-Jewish lawyer, Raphael Lemkin. It was derived from a combination of other words – the first part derived from the Greek root *génos* (meaning 'race or tribe'), and the second part from the Latin *caedere* (or *cide*) ('killing').[56] In a book published that year, Lemkin restated the idea he first suggested in the 1930s, that acts that amount to the crime of 'barbarity' constituted a crime against international law.[57] In his book, Lemkin presented a definition of genocide,[58] and conceived of the crime as one directed against specific 'groups', involving:[59]

> the criminal intent to destroy or to cripple permanently a human group. The acts are directed against groups, as such, and individuals are selected for destruction only because they belong to these groups.

---

55 *Ibid*, article 22(2). Reference should also be made to the rules relating to the interpretation of treaties discussed in chapter 2. In May 2008, the ICC Appeals Chamber endorsed its earlier finding when considering its methodology for the interpretation of the Rome Statute, where it had stated:
'The rule governing the interpretation of a section of the law is its wording read in context and in light of its object and purpose. The context of a given legislative provision is defined by the particular sub-section of the law read as a whole in conjunction with the section of an enactment in its entirety. Its objects may be gathered from the chapter of the law in which the particular section is included and its purposes from the wider aims of the law as may be gathered from its preamble and general tenor of the treaty': Judgment on the appeal of Mr. Germain Katanga against the decision of Pre-Trial Chamber 1 entitled "Decision on the Defence Request Concerning Languages", *The Prosecutor v. Germain Katanga*, Case No. ICC-01/04-01/07 (OA3), Appeals Chamber, 27 May 2008, para 39.

56 Raphael Lemkin, *Axis Rule in Occupied Europe, Laws of Occupation, Analysis of Government, Proposals for Redress* (1944), 79. See also Decision on the Prosecution's Application for a Warrant of Arrest against Omar Hassan Ahmad Al Bashir, *The Prosecutor v. Omar Hassan Ahmad Al Bashir ('Omar Al Bashir')*, Case No. ICC-02/05-01/09, Pre-Trial Chamber I, 4 March 2009, para 114.

57 In 1933, Lemkin proposed at the fifth International Conference for the Unification of Criminal Law that the destruction of racial, religious or social groups constituted a crime under international law. He advocated for the creation of both the crime of barbarism and the crime of vandalism: Nina H.B. Jørgensen, *The Responsibility of States for International Crimes* (2000), 32.

58 Lemkin, *supra* note 56, 79.

59 Raphael Lemkin, 'Genocide as a Crime Under International Law' (1947) 44 *American Journal of International Law* 145, 146.

Lemkin argued that persons who committed genocide should be subject to trial in an international court, since 'a country which makes a policy of genocide cannot be trusted to try its own offenders'.[60] He lobbied to have the issue of genocide as an international crime discussed at the United Nations General Assembly and, with the support of the United States this item was eventually placed on the agenda of that body. The concept was soon taken up widely by the international community, which agreed with the assertion that genocide was indeed a crime under international law.[61] There then followed a number of General Assembly Resolutions,[62] culminating in the conclusion and unanimous adoption by the General Assembly of the 1948 Genocide Convention.[63] This landmark instrument set out an agreed definition of genocide that has since remained largely unchanged.

The 1948 Genocide Convention was the first major treaty dealing with human rights issues to be concluded under the auspices of the United Nations. The International Court of Justice has said of the 1948 Genocide Convention that:[64]

> [t]he origins of the Convention show that it was the intention of the United Nations to condemn and punish genocide as "a crime under international law" involving a denial of the right of existence of entire human groups, a denial which shocks the conscience of mankind and results in great losses to humanity, and which is contrary to moral law and to the spirit and aims of the United Nations

However, despite the significance of the 1948 Genocide Convention, and the recognition of genocide as an international crime, the precise 'meaning' of the crime, as it had been defined, was not judicially considered for many years. The

60 Raphael Lemkin, 'Genocide: A Modern Crime' (1945) 4 *Free World* 39, reproduced in Samuel Totten and Paul R. Bartrop (eds), *The Genocide Studies Reader* (2009), 6, 10.

61 Some commentators consider that *acts* of genocide had in fact been included within the concept of 'crimes against humanity' as applied in the indictments brought under the Charter of the Nuremberg International Military Tribunal, annexed to the 1945 London Agreement for the Establishment of an International Military Tribunal (8 August 1945) 82 UNTS 279 (Nuremberg Charter): see, for example, Restatement of the Law: Third Restatement of U.S. Foreign Relations Law, Volume 2 (1987), 165, ss 702, Reporters' comments, para 3. However, it is now accepted that, even though they have some common elements, what distinguishes genocide from crimes against humanity is that the crime of genocide is a 'crime of intent' in which a specific 'group' is targeted, and not merely specific individuals within that group; or put another way, 'the ultimate victim of genocide is the group': Judgement on Defence Motions to Acquit, *Prosecutor v Dusko Sikirica, Damir Dosen and Dragan Kolundzjia*, Case No. IT-95-8-T, Trial Chamber, 3 September 2001, para 89.

62 See, for example, United Nations General Assembly Resolution 96(I) (11 December 1946) UN Doc A/RES/96/(I) on the Crime of Genocide.

63 Convention on the Prevention and Punishment of the Crime of Genocide, opened for signature 9 December 1948, 78 UNTS 277 (entered into force 12 January 1951) (Genocide Convention).

64 *Reservations to the Convention on the Prevention and Punishment of the Crime of Genocide* (Advisory Opinion) [1951] ICJ Rep 16, 23.

lack of political will, coupled with the prevailing geopolitical circumstances in the period between the end of the 1940s and the beginning of the 1990s, meant that the prosecution of such a crime at the international level (or indeed at the national level), was neither a high priority nor, in many situations, a practical possibility. It was not until after the establishment of the *ad hoc* Tribunals dealing with the conflicts in the former Yugoslavia and in Rwanda that an international criminal court began to consider the precise meaning of the definition in any detail.

The crime of genocide as defined in the Rome Statute is drawn from the definition in the 1948 Genocide Convention, which was incorporated as well into the ICTY and ICTR Statutes.[65] The preamble of the 1948 Genocide Convention notes that 'at all periods of history genocide has inflicted great losses on humanity'.[66] Genocide has been referred to as the 'crime of crimes'[67] and is often regarded as the most heinous of all violations of human rights.[68] The particular distinguishing characteristic of the crime of genocide is, as noted, its focus on 'groups', specifically the intended destruction of entire human groups.[69] It therefore incorporates a very specific *dolus specialis*, being the intent 'to destroy, in whole or in part, a national, ethnical, racial or religious group, as such'.[70]

From the perspective of any prosecution, this specific intent is far more complex in terms of proof than a *general* intent, which might instead be inferred on the basis of a 'reasonable person' test,[71] and/or the default standard of intent set out in article 30 of the Rome Statute referred to above. For this reason, genocide is a

---

65 There are, however, some differences between the ICTY and ICTR Statutes on the one hand and the Rome Statute on the other. For example, article 4(3) of the ICTY Statute and article 2(3) of the ICTR Statute respectively specify that the following acts are punishable:
'(a) Genocide;
(b) Conspiracy to commit genocide;
(c) Direct and public incitement to commit genocide;
(d) Attempt to commit genocide;
(e) Complicity in genocide'.
By contrast, these acts are not included in article 6 of the Rome Statute, but are instead incorporated into article 25(3), which deals with individual criminal responsibility and applies to each of the crimes within the jurisdiction of the Court. The act of 'directly and publicly incit[ing] others to commit genocide' is, however, specifically referred to in article 25(3)(e) of the Rome Statute.

66 Genocide Convention, preamble para 2.

67 Judgement, *Prosecutor v. Jean-Paul Akayesu*, Case No. ICTR-96-4-T, Trial Chamber, 2 September 1998, para 16.

68 Daniela de Vito, Aisha Gill and Damien Short, 'Rape characterized as genocide' (2009) 10 *SUR International Journal on Human Rights* 29, 37.

69 *Ibid*, 36.

70 1948 Genocide Convention, article II. See also, for example, Rome Statute, chapeau article 6.

71 M. Cherif Bassiouni, 'The Normative Framework of International Humanitarian Law: Overlaps, Gaps, and Ambiguities', in Michael J.N. Schmitt (ed), *International Law Studies Volume 75: International Law Across the Spectrum of Conflict* (2000), 1, 9.

very difficult crime to prove in the absence of a clear 'paper trail' – which would not normally exist[72] – although, as noted, there have since 1998 been several successful prosecutions of genocide before the *ad hoc* Tribunals, and counts of genocide have also been alleged against an accused before the ICC.[73]

As for the 'physical' elements, the crime comprises a number of acts of genocide relating to four specific types of group. The concept of genocide based on an intention to destroy a 'political' group was not included in the final definition, despite having been part of both Lemkin's original draft definition, as well as the more general description adopted by the United Nations General Assembly two years prior to the conclusion of the 1948 Genocide Convention.[74] Attempts to include political and social groups into the Rome Statute definition were also not accepted at the 1998 Rome Conference, as the majority of States present did not want to alter a definition that was clearly recognized under customary international law.[75]

Similarly, the definition of genocide does not appear to include actions intended to destroy (in whole or in part) a group based on their *culture*. Thus, from a legal perspective, there exists no formal concept under international criminal law of 'cultural genocide'.[76]

These observations raise a more general point; that the existing definitions of a number of the core international crimes in the relevant treaties, including that of genocide, are now too narrow and require revision or 'upgrading',[77] in order to make them accord more closely with the both contemporary values and reality.[78] This has, for example, been the case, as has been noted, with respect to the sexual violence provisions within the definition of crimes

72 See, for example, *id*, where Cherif Bassiouni suggests that the 1948 Genocide Convention was drafted with the Nazi Germany experience in mind, which left behind a very detailed paper trail, but that this is a 'situation [that] never has been repeated'.

73 See Second Warrant of Arrest for Omar Hassan Ahmad Al Bashir, *The Prosecutor v. Omar Hassan Ahmad Al Bashir ('Omar Al Bashir')*, Case No. ICC-02/05-01/09, Pre-Trial Chamber I, 12 July 2010. This accused was also currently charged with five counts of crimes against humanity and two counts of war crimes. However, as noted, the investigation by the ICC Prosecutor in the case concerning President Bashir is currently suspended.

74 See United Nations General Assembly Resolution 96(I) (11 December 1946) UN Doc A/RES/96/(I) on the Crime of Genocide, para 4.

75 See Gerhard Werle, *Principles of International Criminal Law* (2005), 191.

76 See, however, *id*, where the author has suggests that article 6(e) of the Rome Statute ('Forcibly transferring children of the group to another group') 'defines a form of cultural genocide'.

77 Steven Freeland, 'Towards Universal Justice – Why Countries in the Asia / Pacific Region should Embrace the International Criminal Court' (2007) 5 *New Zealand Journal of Public and International Law* 49, 63.

78 See, for example, Jørgensen, *supra* note 57, 41, where that author, when discussing the notion of State responsibility for genocide, argues that '[t]he time now seems ripe to bring the [1948] Genocide Convention up to date'.

against humanity,[79] and is a topic that is the subject of ongoing debate among academic commentators.[80] Yet, history has thus far indicated that there is a strong reluctance among States to amend the definition of the crime of genocide, particularly given its 'crime of crime' connotations.

Putting this question aside, acts designed to destroy a group (in whole or in part) could possibly involve the intentional destruction of the environment during armed conflict, as a way of attempting to render impossible the group's ability to continue to exist. The Rome Statute specifies that '[d]eliberately inflicting on the group conditions of life calculated to bring about its physical destruction' would fall within the type of acts that constitute genocide.[81] Under this provision, it is not necessary for the intended 'results' – the *actual* destruction of the group – to occur. A 'result requirement' had been proposed by the United States in the drafting process of the Elements of Crimes under the Rome Statute (Elements of Crimes),[82] but was ultimately not included in the final version.[83]

Although the footnote to article 6(c) in the Elements of Crimes also envisages a broader range of circumstances,[84] acts such as the draining of the marshes in southern Iraq, or the poisoning of water wells or destruction of forests upon which local indigenous groups depend, could arguably fall within this description, although it is still necessary to demonstrate that the special intention

79 Compare article 5(g) of both the ICTY Statute and the ICTR Statute, which each refer to 'rape', with article 7(1)(g) of the Rome Statute, which covers 'Rape, sexual slavery, enforced prostitution, forced pregnancy, enforced sterilization, or any other form of sexual violence of comparable gravity'.

80 See, for example, Cassese, *supra* note 10, 118, who points to the conflict between the generally agreed customary law definition of the crime of torture and the narrower treaty definition, which is limited to certain acts 'inflicted by or at the instigation of or with the consent or acquiescence of a public official or other person acting in an official capacity': Convention against Torture and Other Cruel, Inhuman or Degrading Treatment or Punishment, opened for signature 10 December 1984, 23 ILM 1027 (entered into force 26 June 1987), article 1(1).

81 Rome Statute, article 6(c).

82 Article 9(1) of the Rome Statute provides for the adoption of the 'Elements of Crimes' by a two-thirds majority of the Assembly of States Parties to the Rome Statute, and specifies that their function is to 'assist the Court in the interpretation and application of articles 6, 7 and 8' of the Rome Statute. See also Elements of Crimes (9 September 2002) ICC Doc ICC-ASP/1/3 (part II-B), article 6(c).

83 Wiebke Rückert and Georg Witschel, 'Genocide and Crimes Against Humanity in the Elements of Crimes', in Horst Fischer, Claus Kress and Sascha Rolf Lüder (eds), *International and National Prosecution of Crimes Under International Law: Current Developments* (2001), 59, 68.

84 This footnote provides that:
'[t]he 'conditions of life' may include, but is not necessarily restricted to, deliberate deprivation of resources indispensable for survival, such as food or medical services, or systematic expulsion from homes'.
See also Judgement, *Prosecutor v. Jean-Paul Akayesu*, Case No. ICTR-96-4-T, Trial Chamber, 2 September 1998, para 506.

to 'destroy' is directed towards the *physical* destruction (in whole or in part) of the relevant group,[85] rather than, for example, their 'way of life'. As noted, proving the existence of this *dolus specialis* is a difficult exercise, and reliance on this provision to address the intentional destruction of the environment will therefore rarely be applicable.

In addition, it may be that the targeted group that is the subject of the environmental destruction does not constitute one of the established groupings within the definition of genocide. As noted, these groups were 'chosen' very carefully by the international community when agreeing the definition of the crime, and various additional groups suggested by Lemkin and others were deliberately not included. That said, the categorization into (one or more of) the four specified groups in the definition of the crime of genocide may not be quite as clear-cut as might first appear. Indeed, William Schabas has suggested that the listed groups 'resist efforts at precise definition'.[86]

In *Akayesu*,[87] Trial Chamber I of the ICTR, when considering the appropriate classification of the Tutsi, concluded the grouping of 'ethnic[al]' came closest, but that to use this categorization was 'troublesome' for the Trial Chamber, since the Tutsi could not meaningfully be distinguished from the Hutu on the basis of language and culture.[88] In the end the Trial Chamber preferred its own interpretation[89] of what constituted a group for the purposes of genocide.[90]

---

85 Judgement, *Prosecutor v Goran Jelesić*, Case No. IT-95-10-I, Trial Chamber, 14 December 1999, para 78–83. It has been held by the ICTY Appeals Chamber that the 'in part' requirement refers to a 'substantial part of that group': see Judgement, *Prosecutor v Radislav Krstić*, Case No. IT-98-33-A, Appeals Chamber, 19 April 2004, para 8 and the various references made in para 8–13.

86 Schabas, *supra* note 41, 109.

87 Judgement, *Prosecutor v. Jean-Paul Akayesu*, Case No. ICTR-96-4-T, Trial Chamber, 2 September 1998.

88 William A. Schabas, 'The Crime of Genocide in the Jurisprudence of the International Criminal Tribunals for the Former Yugoslavia and Rwanda', in Horst Fischer, Claus Kress and Sascha Rolf Lüder (eds), *International and National Prosecution of Crimes Under International Law: Current Developments* (2001), 447, 450.

89 The Trial Chamber came to the following conclusion (para 516):
'[m]oreover, the Chamber considered whether the groups protected by the Genocide Convention, echoed in Article 2 of the [ICTR] Statute, should be limited to only the four groups expressly mentioned and whether they should not also include any group which is stable and permanent like the said four groups. In other words, the question that arises is whether it would be impossible to punish the physical destruction of a group as such under the Genocide Convention, if the said group, although stable and membership is by birth, does not meet the definition of any one of the four groups expressly protected by the Genocide Convention. In the opinion of the Chamber, it is particularly important to respect the intention of the drafters of the Genocide Convention, which according to the *travaux préparatoires*, was *patently to ensure the protection of any stable and permanent group*' (emphasis added).

90 In this regard, the Trial Chamber read the express terms of the definition of genocide beyond their ordinary meaning. Resort to the *travaux préparatoires* of a treaty for the purposes of

However, this 'expansive' view as to what constitutes a relevant 'group' for the purposes of the crime of genocide has not been followed in the subsequent jurisprudence of the *ad hoc* Tribunals. The specified list of four groups is regarded as exhaustive.[91]

In *Jelesić*, for example, Trial Chamber I of the ICTY confirmed that:[92]

> the preparatory work of the [1948 Genocide] Convention demonstrates that a wish was expressed to limit the field of application of the Convention to protecting "stable" groups *objectively defined* and to which individuals belong regardless of their own desires

Having said this, however, the Trial Chamber in *Jelesić* did recognize the difficulties associated with specifying precisely the meaning of the various groups that had been 'objectively defined' for the purposes of the crime. In the immediately following paragraph of its judgement, it noted that:[93]

> [a]lthough the objective determination of a religious group still remains possible, *to attempt to define a national, ethnical or racial group today using objective and scientifically irreproachable criteria would be a perilous exercise* whose result would not necessarily correspond to the perception of the persons concerned by such categorisation

As a consequence, the Chamber preferred to adopt a 'subjective criterion' – the perception of the person(s) who intends to target that group – to determine the existence of a national, ethnical or racial group for the purposes of the definition of the crime of genocide.

---

interpretation should only be undertaken to clarify 'ambiguous or obscure' terms, or those where the result would otherwise be 'manifestly absurd or unreasonable': Vienna Convention on the Law of Treaties, opened for signature 23 May 1969, 1155 UNTS 331; 8 ILM 679 (entered into force 27 January 1980), article 32 (see also chapter 2). Whilst the ICTR Statute is not a 'treaty' as such, and that therefore the VCLT rules for treaty interpretation do not *stricto sensu* apply to that instrument, the definition of the crime of genocide in the ICTR Statute is, as noted above, taken from a treaty – the Genocide Convention. There is no reason to read the (exhaustive) list of groups specified in the ICTR Statute in any other manner than would apply to interpreting them in that treaty.

91 See, for example, Judgement, *Prosecutor v Radislav Krstić*, Case No. IT-98-33-T, Trial Chamber, 2 August 2001, para 554, where the ICTY Trial Chamber stated that:
'the Genocide Convention does not protect all types of human groups. Its application is confined to national, ethnical, racial or religious groups'.

92 Judgement, *Prosecutor v Goran Jelesić*, Case No. IT-95-10-I, Trial Chamber, 14 December 1999, para 69 (emphasis added).

93 *Ibid*, para 70 (emphasis added).

While this subjective approach has not always been followed,[94] and has been criticised by some commentators,[95] the point to be taken for present purposes is that there may be circumstances where the definition of genocide might, in practice, be somewhat 'broader' than what might appear from a strict reading of the wording as set out in the relevant instruments. Indeed, in *Rutaganda*, Trial Chamber I of the ICTR (which comprised the same Judges as in *Akayesu*), when discussing the various groups specified in the definition, noted that '[e]ach of these concepts must be assessed in the light of a particular political, social and cultural context'.[96]

Nevertheless, irrespective of the expansive approach taken by the ICTR Trial Chamber I in *Akayesu* and *Rutaganda*, as well as the uncertainties highlighted by the ICTY in *Jelesić*, it appears highly unlikely that the intentional destruction of the environment during armed conflict would *per se* fall within the (current) definition of genocide. To assert otherwise would, in effect, impose a 'double intent' requirement to what is already a very complex crime. Not only would it be necessary to show the existence of the current *dolus specialis*, but an additional intent – to target the environment as a victim or use it as a weapon – would also be necessary to prove genocide for such acts. This would constitute a redefinition of the crime (at least in relation to circumstances involving the intentional destruction of the environment) and would mean that, for practical purposes, it would be virtually impossible to prove beyond a reasonable doubt the commission of what is already a very difficult crime to prove.

In any event, given the 'stigma' associated with this crime of crimes, it would be unlikely that circumstances pointing 'just' to the intentional targeting of the environment alone would be prosecuted as genocide. Consequently, even if acts of this type were such as to (possibly) constitute an act of genocide, it is probable that the Prosecutor of the ICC would take a very conservative approach. To date, prosecution in the international tribunals for the crime of genocide has generally been 'reserved' for 'high profile' acts that have resulted in, or have been intended to directly cause very considerable harm to a relevant group.

It is suggested, therefore, that it would, under current circumstances, be a 'bridge too far' to expect the Prosecutor to run the risk of being seen (rightly

94 See, for example, Judgement and Sentence, *Prosecutor v. Georges Anderson Nderubumwe Rutaganda*, Case No. ICTR-96-3, Trial Chamber I, 6 December 1999, where the Trial Chamber concluded (at para 57) that:
'a subjective definition alone is not enough to determine victim groups, as provided for in the Genocide Convention'.

95 See, for example, Schabas, *supra* note 88, 455–6.

96 Judgement and Sentence, *Prosecutor v. Georges Anderson Nderubumwe Rutaganda*, Case No. ICTR-96-3, Trial Chamber I, 6 December 1999, para 56.

or wrongly) to be 'stretching' the scope of genocide to incorporate actions that do not, on their face, appear to be a genocidal acts. Moreover, the Elements of Crimes appear to limit any scope for considering intentional environmental destruction as an act of genocide, by stipulating that, for each of the acts that might constitute an act of genocide under the Rome Statute, that act *inter alia* must have taken place:[97]

> in the context of a manifest pattern of similar conduct directed against that group or was conduct that could itself effect such destruction

This is not to say that the range of acts of genocide may not expand in the future – the decision in *Akayesu* that, as well as a crime against humanity, rape and sexual violence can constitute an act of genocide in particular circumstances,[98] demonstrates how even this crime is likely to be interpreted more expansively in the future in order to reflect contemporary values.[99] In this regard, Cherif Bassiouni has suggested that it is valid to consider the scope of the crime of genocide 'as susceptible of progressive interpretation in the light of new techniques that nefarious planners devise to achieve their evil goals'.[100] However, at least in the relatively early stages of the judicial activities of the ICC, there is a need for the Prosecutor, and the Court itself, not to be seen to be 'creating' crimes that are not clearly set out in the Rome Statute, as this flies in the face of the political concerns that lead to the compromised terms of the Rome Statute in 1998.

If, for example, an extension of the relevant groups within the definition of genocide was eventually to be accepted, it might be possible to apply it more easily to circumstances of 'cultural genocide' perpetrated through the destruction of the natural habitat or resources upon which indigenous or minority populations are dependent. However, this does not readily fall within the current framework of genocide under the Rome Statute and, as such, the prosecution of such acts therefore does not appear either legally sustainable or politically feasible at this stage.

---

97 See Elements of Crimes, articles 6(a)(4), 6(b)(4), 6(c)(5), 6(d)(5) and 6(e)(7).

98 Judgement, *Prosecutor v. Jean-Paul Akayesu*, Case No. ICTR-96-4-T, Trial Chamber, 2 September 1998, para 731. See also de Vito, Gill and Short, *supra* note 68.

99 In this regard, the *Akayesu* decision was described at the time as 'the most important decision rendered thus far in the history of women's jurisprudence': Kelly Askin, 'Women's Issues in International Criminal Law: Recent Developments and the Potential Contribution of the ICC', in Dinah Shelton (ed), *International Crimes, Peace, and Human Rights: The Role of the International Criminal Court* (2000), 47, 52. For a description of the circumstances leading to the inclusion of charges of rape and sexual violence against Jean-Paul Akayesu, see Cherie Booth, 'Prospects and issues for the International Criminal Court: lessons from Yugoslavia and Rwanda', in Philippe Sands (ed), *From Nuremberg to The Hague: The Future of International Criminal Justice* (2003), 157, 167–72.

100 Bassiouni, *supra* note 71, 10.

4.4.2. INTENTIONAL DESTRUCTION OF THE ENVIRONMENT AS A CRIME AGAINST HUMANITY?

The Rome Statute defines a crime against humanity as any of a number of specified acts 'committed as part of a widespread or systematic attack directed against any civilian population, with knowledge of the attack'.[101] These include:[102]

> [p]ersecution against any identifiable group or collectivity on political, racial, national, ethnic, cultural, religious, gender … or other grounds that are universally recognized as impermissible under international law, in connection with any act referred to in this paragraph or any crime within the jurisdiction of the Court

and:[103]

> [o]ther inhumane acts of a similar character intentionally causing great suffering, or serious injury to body or to mental or physical health

Although 'crimes against humanity' was not formally categorized as a separate crime until after the Second World War, the concept had by that time already been recognised. On 28 May 1915, the Governments of France, Great Britain and Russia made a declaration regarding the massacres of the Armenian population in Turkey, denouncing them as:[104]

> crimes against humanity and civilisation for which all the members of the Turkish government will be held responsible together with its agents implicated in the massacres

Crimes against humanity were recognized (and codified) in the Nuremberg Charter,[105] and considered in the Judgment of the Nuremberg Military Tribunal, as well as in Law No. 10 of the Control Council for Germany,[106] and the Tokyo Military Tribunal Charter.[107] Its inclusion within the jurisdiction of the Nuremberg

---

101 Rome Statute, chapeau article 7(1).

102 *Ibid*, article 7(1)(h).

103 *Ibid*, article 7(1)(k).

104 See Judgement, *Prosecutor v. Jean-Paul Akayesu*, Case No. ICTR-96-4-T, Trial Chamber, 2 September 1998, para 565 and the corresponding footnote.

105 See Nuremberg Charter, article 6(c).

106 See Law No. 10 of the Control Council for Germany, Punishment of Persons Guilty of War Crimes, Crimes Against Peace and Against Humanity (20 December 1945) 36 ILR 31, article II. This law was enacted to establish a legal basis in Germany for the trial of war criminals who were not prosecuted by the Nuremberg Military Tribunal: Jørgensen, *supra* note 57, 20.

107 See Charter of the International Military Tribunal for the trial of the major war criminals in the Far East (19 January 1946) TIAS 1589; 4 Bevans 20 (Tokyo Charter), article 5(c).

and Tokyo Military Tribunals was an important milestone in the formal recognition of crimes of international concern that violated fundamental human rights.[108]

Since the Nuremberg and Tokyo Military Tribunal trials, the concept of crimes against humanity has continued to undergo a gradual evolution, firstly in national cases such as *Eichmann*[109] and *Barbie*,[110] and subsequently as it has been defined in the Statutes of the *ad hoc* international tribunals[111] and, ultimately, in the Rome Statute itself. As noted, the definition of crimes against humanity in the Rome Statute is broader than any previous formulation in several important respects.

Despite this process of evolution and expansion, there is (still) no specific reference to the environment in the definition of crimes against humanity. It is true, however, that certain acts that constitute crimes against humanity might also relate to circumstances where the environment has been intentionally targeted, in particular those specified in articles 7(1)(h) and 7(1)(k) of the Rome Statute quoted above.

Article 7(1)(h) identifies persecution against 'any identifiable group' or 'collectively' on a very broad basis ('any other grounds') – the characterization of the targeted groups for this crime is therefore significantly wider than for the

108 Flavia Lattanzi, 'Crimes Against Humanity in the Jurisprudence of the International Criminal Tribunals for the Former Yugoslavia and Rwanda', in Horst Fischer, Claus Kress and Sascha Rolf Lüder (eds), *International and National Prosecution of Crimes Under International Law: Current Developments* (2001), 473, 475.

109 *Attorney-General of the Government of Israel v Eichmann* (1961) 36 ILR 5. Eichmann was prosecuted under Israeli law (1951 Nazi and Nazi Collaborators (Punishment) Law) for war crimes, crimes against the Jewish people (the definition of which was modeled on the definition of genocide in the 1948 Genocide Convention) and crimes against humanity. He was convicted by the District Court of Jerusalem and sentenced to death. His appeal to the Supreme Court of Israel was dismissed: *Eichmann v Attorney-General of the Government of Israel* (1962) 36 ILR 277.

110 In 1987, Klaus Barbie, who had been the head of the Gestapo in Lyon from November 1942 to August 1943 and was known as the 'Butcher of Lyon', was convicted by the Rhone *Cour d'assises* of 17 counts of crimes against humanity. His appeal was dismissed by the French Court of Cassation: *Féderation National des Déportées et Internés Résistants et Patriots and Others v. Barbie* 100 ILR 330.

111 Article 3 of the ICTR Statute defines crimes against humanity as any one of a number of enumerated acts '... when committed as part of a widespread or systematic attack against any civilian population on national, political, ethnic, racial or religious grounds'. Article 5 of the ICTY Statute defines crimes against humanity to include the same acts, but '... when committed in armed conflict, whether international or internal in character, and directed against any civilian population'. However, in practice, the ICTY has adopted the criteria mentioned in the chapeau of article 3 of the ICTR Statute: see, for example, Judgement, *Prosecutor v. Zoran Kupreskic, Mirjan Kupreskic, Vlatko Kupreskic, Drago Josipovic, Dragan Papic and Vladimir Santic*, Case No. IT-95-16-T, Trial Chamber, 14 January 2000, para 544; Judgement, *Prosecutor v. Tihomir Blaskic*, Case No. IT-95-14-A, Appeals Chamber, 29 July 2004, para 96–126. This practice has been criticised by some commentators: see, for example, Lattanzi, *supra* note 108, 478–82.

crime of genocide. 'Persecution' is defined for the purposes of article 7(1) of the Rome Statute as:[112]

> … the intentional and severe deprivation of fundamental rights contrary to international law by reason of the identity of the group or collectively

The deliberate destruction of a habitat[113] or of access to clean and safe water or food on a significant scale could, in certain circumstances, represent a breach of the fundamental human rights of the individuals within a targeted group, as perhaps might some other acts of environmental destruction. The right to 'adequate food' is, for example, expressly included in article 11 of the ICESCR,[114] one of the principal human rights instruments that makes up the so-called 'International Bill of Human Rights'.[115] The right to water is, in addition, guaranteed in articles 11 and 12 of the ICESCR.[116] The duty to provide water is also expressly specified in article 24 of the Convention on the Rights of the Child,[117] and article 14 of the Convention of All Forms of Discrimination against Women.[118]

In relation to the right to water, the United Nations Committee on Economic, Social and Cultural Rights has stated that:[119]

---

112 Rome Statute, article 7(2)(g).

113 As noted in chapter 1, *conservation* regimes such as those specified in the European Council Directive 92/43/EEC of 21 May 1992 on the conservation of natural habitats and of wild fauna and flora are not expressly designed, nor appropriate to deal with the intentional destruction of the environment during armed conflict.

114 International Covenant on Economic, Social and Cultural Rights, opened for signature 16 December 1966, 993 UNTS 3 (entered into force 3 January 1976) (ICESCR).

115 The other two instruments that, together with the ICESCR make up the 'International Bill of Human Rights', are the Universal Declaration of Human Rights, adopted by United Nations General Assembly Resolution 217A (III) (10 December 1948) UN Doc A/810 on the Universal Declaration of Human Rights, (UDHR), and the International Covenant on Civil and Political Rights, opened for signature 16 December 1966, 999 UNTS 171 (entered into force 23 March 1976) (ICCPR).

116 United Nations Committee on Economic, Social and Cultural Rights, 'Substantive Issues arising in the Implementation of the International Covenant on Economic, Social and Cultural Rights: General Comment No. 15 (2002) – The right to water (arts. 11 and 12 of the International Covenant on Economic, Social and Cultural Rights)' (November 2002) UN Doc E/C.12/2002/11, para 3. See also Horn and Freeland, *supra* note 31.

117 See Convention on the Rights of the Child, opened for signature 20 November 1989, 1577 UNTS 3 (entered into force 2 September 1990) (CROC), article 24(2)(c).

118 See Convention on the Elimination of All Forms of Discrimination against Women, opened for signature 1 March 1980, 19 ILM 33 (entered into force 3 September 1981) (CEDAW), article 14(h).

119 United Nations Committee on Economic, Social and Cultural Rights, 'Substantive Issues arising in the Implementation of the International Covenant on Economic, Social and Cultural Rights: General Comment No. 15 (2002) – The right to water (arts. 11 and 12 of the International Covenant on Economic, Social and Cultural Rights)' (November 2002) UN Doc E/C.12/2002/11, para 1 and 3.

> [w]ater is a limited natural resource and a public good fundamental for life and health. The human right to water is indispensable for leading a life in human dignity. It is a prerequisite for the realization of other human rights
> (...)
> [t]he right [to water] should also be seen in conjunction with the other rights enshrined in the International Bill of Human Rights, foremost amongst them the right to life and human dignity

Thus, in theory, the terms of article 7(1)(h) of the Rome Statute might extend to certain acts that intentionally target the environment during armed conflict. However, of course, it would also be necessary to prove the other elements of the crime of crimes against humanity – as set out in the chapeau of article 7 – for this provision to be applicable to such acts. This is discussed below; however, before doing so, one should note the 'catch all' provision, article 7(1)(k) of the Rome Statute, which is also referred to above. Once again, one could envisage the possibility of acts that constitute environmental crimes perhaps falling within the terms of this provision.

As noted, the chapeau of article 7 sets out a number of overarching elements of the crime of crimes against humanity, which must also be present in order to support a conviction. This includes a 'widespread or systematic attack directed against any civilian population'. There has been considerable jurisprudence in relation to the meaning of the concepts of 'widespread' and 'systematic' in this context. The Appeals Chamber of the ICTY has confirmed that:[120]

> 'widespread' refers to the large-scale nature of the attack and the number of targeted persons, while the phrase 'systematic' refers to the organized nature of the acts of violence and the improbability of their random occurrence

In similar vein, in 2009, a Pre-Trial Chamber of the ICC confirmed that:[121]

> [a]lthough the terms "widespread" and "systematic" are not specifically defined in the Statute, the Chamber has previously held that this language excludes random or isolated acts of violence, and that the term "widespread" refers to the large-scale nature of the attack, as well as to the number of victims, while the term "systematic" pertains to the organised nature of the acts of violence and to the improbability of their random occurrence

---

120 Judgement, *Prosecutor v. Tihomir Blaskic*, Case No. IT-95-14-A, Appeals Chamber, 29 July 2004, para 101, referring to Judgement, *Prosecutor v. Dragoljub Kunarac, Radomir Kovac and Zoran Vukovic*, Case No. IT-96-23 and IT-96-23/1-A, Appeals Chamber, 12 June 2002, para 94.

121 Decision on the Prosecution's Application for a Warrant of Arrest against Omar Hassan Ahmad Al Bashir, *The Prosecutor v. Omar Hassan Ahmad Al Bashir ('Omar Al Bashir')*, Case No. ICC-02/05–01/09, Pre-Trial Chamber I, 4 March 2009, para 81.

The two expressions are set out in the chapeau in the disjunctive ('or') form, and are therefore alternate requirements. It is not necessary to demonstrate that the relevant attack meets both criteria – one would be sufficient to support a conviction. Be that as it may, both expressions at least suggest on their face that a crime against humanity will generally only occur in the context of a multiplicity of actions, an observation that is reinforced by article 7(2)(a) of the Rome Statute, which specifies that an 'attack directed against any civilian population' means:[122]

> a course of conduct involving the *multiple commission of acts* referred to in [article 7(1) of the Rome Statute] against any civilian population, pursuant to or in furtherance of a State or organizational policy to commit such attack

However, it is still possible for a limited number of acts, or even a single act by an accused, to constitute a crime against humanity, assuming that all other elements of the crime are satisfied, if they are 'a part of [the] attack ... unless those acts may be said to be isolated or random'.[123] Nevertheless, the overall combined effect of these elements, coupled with the fact that there is a need to demonstrate the existence of a 'policy to commit such attack', would seem to make it very difficult to bring acts done with the intent to destroy the environment during armed conflict within the terms of crimes against humanity.

Further, and most significantly for the purposes of this discussion, any possible connection is made even more tenuous due to the requirement that the attack must be directed 'against any civilian population'. In essence, for a crime against humanity to be committed, it is necessary to demonstrate that the civilian population is the 'victim' or 'primary object' of such an attack.[124] Indeed, the 'status of the victim as a civilian' is one of the characterizing features of a crime against humanity.[125] In order to determine whether the attack has been directed

---

122 Rome Statute, article 7(2)(a) (emphasis added). It should be noted that, as has been confirmed in the jurisprudence of the *ad hoc* tribunals, the reference to a 'State ... policy' in this provision does not mean that the crime of crimes against humanity can be characterized as 'an act of criminal sovereignty' – the concept of individual criminal responsibility for such crimes still applies, and the *ad hoc* tribunals, as well as the ICC, only have jurisdiction over 'natural persons' (ICTY Statute, article 6; ICTR Statute, article 5; Rome Statute, article 25(1)): see also Judgement, *Prosecutor v. Tihomir Blaskic*, Case No. IT-95-14-T, Trial Chamber, 3 March 2000, para 205 and the corresponding footnotes.

123 Judgement, *Prosecutor v. Tihomir Blaskic*, Case No. IT-95-14-A, Appeals Chamber, 29 July 2004, para 101, referring to Judgement, *Prosecutor v. Dragoljub Kunarac, Radomir Kovac and Zoran Vukovic*, Case No. IT-96-23 and IT-96-23/1-A, Appeals Chamber, 12 June 2002, para 96.

124 Judgement, *Prosecutor v. Dragoljub Kunarac, Radomir Kovac and Zoran Vukovic*, Case No. IT-96-23 and IT-96-23/1-A, Appeals Chamber, 12 June 2002, para 91.

125 Judgement, *Prosecutor v. Tihomir Blaskic*, Case No. IT-95-14-A, Appeals Chamber, 29 July 2004, para 107.

against the civilian population in this way, the ICTY has stated that the following factors must be considered:[126]

> the means and method used in the course of the attack, the status of the victims, their number, the discriminatory nature of the attack, the nature of the crimes committed in its course, the resistance to the assailants at the time and the extent to which the attacking force may be said to have complied or attempted to comply with the precautionary requirements of the laws of war

Although the intentional destruction of the environment during armed conflict may often be intended as a method of harming specific populations, it is for this reason that the crime of crimes against humanity is not readily applicable in circumstances where it is the *environment* (as opposed to civilians) that is deliberately targeted as the victim or the primary object of such acts. Indeed, as outlined in chapter 1 and discussed further in the next chapter, although the concept of crimes against the environment is not to be regarded as an environmental protection measure *per se*, the principal focus of this book is to more effectively address actions that specifically *target the environment*. It is the intent to destroy the environment during warfare that is the distinguishing feature of the crime of crimes against the environment. It is this specific victim perspective that differentiates any notion of a crime *against* the environment from the other core international crimes in the Rome Statute and, as a consequence, such acts will generally not fall within crimes against humanity.

For all of these reasons, it is considered that crimes against humanity, as defined in the Rome Statute, would at most only have indirect relevance to such actions, although of course this has not been specifically considered by the Court. Consequently, the crime of crimes against humanity does not represent a satisfactory mechanism for the prosecution before the ICC of intentional destruction of the environment during armed conflict.

4.4.3. INTENTIONAL DESTRUCTION OF THE ENVIRONMENT AS A WAR CRIME?

4.4.3.1. Article 8(2)(b)(iv)

As noted, the 'natural environment' is expressly referred to in only one provision of the Rome Statute, this being a specific act within the definition of war

126 Judgement, *Prosecutor v. Dragoljub Kunarac, Radomir Kovac and Zoran Vukovic*, Case No. IT-96-23 and IT-96-23/1-A, Appeals Chamber, 12 June 2002, para 91: see also Judgement, *Prosecutor v. Tihomir Blaskic*, Case No. IT-95-14-A, Appeals Chamber, 29 July 2004, para 106.

crimes.[127] Article 8(2)(b)(iv) specifies that, 'in particular when committed as part of a plan or policy or as part of a large-scale commission of such crimes',[128] a war crime within the jurisdiction of the Court includes:[129]

> (b) ... serious violations of the laws and customs applicable in international armed conflict, within the established framework of international law, namely, any of the following acts:
> (...)
> (iv) Intentionally launching an attack in the knowledge that such attack will cause incidental loss of life or injury to civilians or damage to civilian objects or widespread, long-term and severe damage to the natural environment which would be clearly excessive in relation to the concrete and direct overall military advantage anticipated

As can be seen, the article specifies three 'variants of the offence'[130] – relating to i) incidental loss of life or injury to civilians; ii) damage to civilian objects; or iii) damage to the natural environment. Given the focus of this book, it is only the third variant that is the subject of the following discussion.

Before assessing the scope of this aspect of the provision, it is necessary to consider its drafting history. Although, as noted in chapter 1, the notion of the 'criminality' of States for significant environmental damage was, for a period, included in the ILC Draft Articles on State Responsibility,[131] and the concept of individual criminal responsibility for 'widespread, long-term and severe damage to the natural environment' was included in the Draft Code of Crimes against

---

127 Of course, other 'war crimes' defined in article 8 of the Rome Statute may also relate to conduct that might indirectly involve damage to the natural environment: Mark A. Drumbl, 'Waging War against the World: The Need to Move from War Crimes to Environmental Crimes', in Jay E. Austin and Carl E. Bruch (eds), *The Environmental Consequences of War: Legal, Economic, and Scientific Perspectives* (2000), 620, 622, footnote 8. These are discussed in 4.4.3.2.

128 Rome Statute, article 8(1).

129 *Ibid*, article 8(2)(b)(iv).

130 Gerhard Werle and Florian Jessberger, *Principles of International Criminal Law* (3rd ed, 2014), 493.

131 See Report of the International Law Commission to the General Assembly on its work of the thirty-second session, [1980] 2 *Yearbook of the International Law Commission* Part II, 32, UN Doc A/CN.4/SER.A/1980/Add.l (Part 2). As noted in chapter 1, draft article 19 of the 'Draft Articles on State Responsibility for International Wrongful Acts' had provided that an 'international crime' included:
'[a]n internationally wrongful act which results from the breach by a State of an international obligation so essential for the protection of fundamental interests of the international community that its breach is recognized as a crime by that community as a whole constitutes an international crime', and included (draft article 19(3)(d)) 'a serious breach of an international obligation of essential importance for the safeguarding and preservation of the human environment, such as those prohibiting massive pollution of the atmosphere or of the seas'.

the Peace and Security of Mankind,[132] such actions did not form part of the 1994 ILC Draft Statute.

Instead, it was suggested that such damage would usually fall within the scope of other crimes already included in the draft text, and that, if that were not the case in a specific situation, it might be that the act does not meet the 'threshold of gravity for an international crime'.[133]

The revised draft version of the Statute, which subsequently formed the basis for the final discussions at the 1998 Rome Conference, contained three additional options, other than the version that was ultimately adopted as article 8(2)(b)(iv), namely:[134]

(1) 'Intentionally launching an attack in the knowledge that such attack will cause incidental loss of life or injury to civilians or damage to civilian objects or widespread, long-term and severe damage to the natural environment, which is not justified by military necessity';
(2) 'Intentionally launching an attack in the knowledge that such attack will cause incidental loss of life or injury to civilians or damage to civilian objects or widespread, long-term and severe damage to the natural environment';
(3) 'No paragraph' – which would have meant that there would be no prohibition included in the Rome Statute in relation to widespread, long-term and severe damage to the natural environment.

In the end, a compromise was reached. As discussed in chapter 2, and as yet another illustration of the political concerns associated with criminalizing acts that damage the environment in the context of armed conflict, the final provision requires a balancing of the damage as against military advantage. In practical terms, this means that environmental issues are 'made secondary' to interests of military importance.[135] In so doing, the provision therefore does not focus on the issues that arise from the intentional targeting of the environment during the course of armed conflict. It fails to recognize the importance of the environment *as such*, instead falling back on the traditional and outdated approach that environmental harm is to be regarded as an unfortunate 'bi-product' of warfare, even in circumstances where the damage is deliberate and intentional.

---

132 [1991] 1 *Yearbook of the International Law Commission* 234, UN Doc A/CN.4/SER.A/1991, draft article 26.

133 Antonio Cassese, Albin Eser, Giorgio Gaja, Philip Kirsch, Alain Pellet and Bert Swart (eds), *The Rome Statute of the International Criminal Court: A Commentary* (2002), 522–3.

134 See Draft Statute for the International Criminal Court (14 April 1998) UN Doc A/CONF.183/2/Add.1, Part 2 'War Crimes', ss B(b), as quoted in Drumbl, *supra* note 127, 622–3.

135 Drumbl, *ibid*, 623.

Article 8(2)(b)(iv) of the Rome Statute has been discussed in general terms in earlier chapters, particularly in the context of the similar language used in article 55(1) of the 1977 Additional Protocol I,[136] and article I(1) of the ENMOD Convention.[137] However, in addition to the issues raised above, several points bear further elaboration here. First, it is apparent that article 8(2)(b)(iv) demands a very high threshold of injury to the environment before an act would fall within the scope of the crime. The use of the conjunctive ('and') between the words widespread, long-term and severe, rather than the disjunctive form, has, at least from an environmental integrity viewpoint, effectively meant a 'regression' from the standard that had been specified in the ENMOD Convention (where the disjunctive form is used).[138] Gerhard Werle and Florian Jessberger explain this variance as follows:[139]

> [s]ince environmental damage can be expected as a collateral consequence in any type of warfare, the requirements were raised in comparison to those in ENMOD.

Moreover, a comparison of article 8(2)(b)(iv) with article 55(1) of the 1977 Additional Protocol I also indicates how the degree of culpable action necessary to amount to a war crime under the Rome Statute appears to have been increased. Acts that would contravene article 55(1) would not necessarily constitute a war crime under article 8(2)(b)(iv) of the Rome Statute, since this latter provision includes the need for the damage to be 'clearly excessive'. The difficulties relating to the requirement of 'excessive' damage (let alone clearly excessive) have already been canvassed in chapter 3 in relation to the 2000 Committee Report examining NATO's actions during Operation Allied Force,[140] although it is generally agreed that, under customary international law, the parties to an armed conflict are prohibited from causing 'excessive damage to the environment during military operations'.[141]

In the case of article 8(2)(b)(iv), not only must the damage be 'excessive', but it must also be 'clearly' so. Irrespective of any uncertainties as to what 'excessive' means, the inclusion of the descriptor 'clearly' suggests an intention to set as

---

136 Protocol I Additional to the Geneva Conventions of August 12, 1949, and relating to the Protection of Victims of International Armed Conflicts, opened for signature 8 June 1977, 1125 UNTS 3, 16 ILM 1391 (entered into force 7 December 1978) (1977 Additional Protocol I).

137 Convention on the Prohibition of Military or Any Other Hostile Use of Environmental Modification Techniques, opened for signature 10 December 1976, 1108 UNTS 151; 16 ILM 88 (entered into force 5 October 1978) (ENMOD Convention).

138 Drumbl, *supra* note 127, 624.

139 Werle and Jessberger, *supra* note 130, 493.

140 Final Report to the Prosecutor of the International Criminal Tribunal for the former Yugoslavia by the Committee Established to Review the NATO Bombing Campaign Against the Federal Republic of Yugoslavia, 13 June 2000, 39 ILM 1257 (Committee Report).

141 See Werle and Jessberger, *supra* note 130, 492 and the references in the corresponding footnote. See also chapter 3.

the requisite threshold an even higher level of damage, since there must be a difference between damage that is *excessive* and damage that is *clearly excessive*, although how this is to be determined is unclear.

There is also no guidance provided as to the meaning of 'widespread', 'long-term' or 'severe' as they appear in article 8(2)(b)(iv). Reference was made in chapter 2 to the interpretative guidance provided in relation to their meaning in the ENMOD Convention. As noted, these three terms are defined, for the purposes of the ENMOD Convention, as follows:[142]

(a) 'widespread': encompassing an area on the scale of several hundred square kilometres;
(b) 'long-lasting': lasting for a period of months, or approximately a season;
(c) 'severe': involving serious or significant disruption or harm to human life, natural and economic resources or other assets.

The breadth of these interpretations, particularly as *all* of the variables must be satisfied ('and'), already sets the bar at a very high point in terms of what level of environmental damage will be necessary before article 8(2)(b)(iv) might apply.[143] Yet, in relation to the use of these words in article 55(1) of the 1977 Additional Protocol I, various commentaries have provided even more restrictive interpretations. In a 1993 report to the United Nations General Assembly, the Secretary-General stated that:[144]

> [t]here are substantial grounds, including from the *travaux préparatoires* of [the 1977 Additional] Protocol I, for interpreting 'long-term' to refer to *decades rather than months*. On the other hand, it is not easy to know in advance exactly what the scope and duration of some environmentally damaging acts will be.

Naturally, it will be important to develop an understanding as to the scope of these words as they are used in any specific provision of the Rome Statute. The interpretations offered for the purposes of ENMOD 'should not be transferred mechanically' to article 8(2)(b)(iv).[145] Moreover, the words will not necessarily have the same meaning in different articles of the same instrument – obviously,

142 1976 CCD Understanding Relating to Article I of ENMOD, 31 United Nations General Assembly Official Records Supp. No. 27 (A/31/27), Annex I.

143 As noted previously, William Fenrick has suggested, for example, that the threshold would probably not be reached even by 'the sort of damage caused by heavy shelling during World War I battles on the Western Front': William J. Fenrick, 'Article 8: War Crimes – article 8(2)(b)(iv)', in Otto Triffterer (ed), *Commentary on the Rome Statute of the International Criminal Court: Observers' Notes, Article by Article* (1999), 197, 197.

144 United Nations Secretary-General, 'Report of the Secretary-General on the Protection of the Environment in Times of Armed Conflict' (29 July 1993) UN Doc A/48/269, para 34 (emphasis added).

145 Werle and Jessberger, *supra* note 130, 493.

the context in which a particular word appears is relevant to its precise meaning. For example, as will be detailed in the following chapter, the meaning of the word 'widespread' as it appears in the *chapeau* of article 7(1) (crimes against humanity) will not be the same as its meaning in article 8(2)(b(iv). Without a specific definition of such words, or any agreements or discussions that may provide such guidance, it would ultimately fall to the Court itself to determine their precise meaning, thus leaving the issue unclear at least until that point.

In the absence of an express definition (or clarification) within the Rome Statute, for the ICC Judges to determine a different (lower) interpretation of these words in the context of environmental damage than the thresholds in the ENMOD Convention and the 1977 Additional Protocol I, it would be necessary for them to base their conclusions, at least partially, on environmental concerns. This might possibly eventuate, but such an outcome would require both the involvement of environmental groups in the proceedings, although it is not entirely clear how this would work in practice,[146] and a 'brave' bench of Judges, given the reality that environmental damage invariably occurs in armed conflict, and the fact that there has traditionally been political reluctance to extend the express sanctions against environmental damage beyond what is specifically provided for in the Rome Statute.

Until such time, there will be a great deal of uncertainty as to precisely what level of environmental damage would be required for the provision to be applicable, or even relevant. As a consequence, it is suggested that the Prosecutor of the Court would be hesitant to initiate a prosecution alleging a violation of this war crime, at least in the absence of further clarity on these points, for fear of falling short of demonstrating the requisite degree of environmental harm.

In addition, as noted, article 8(2)(b)(iv) requires an assessment of the proportionality of the environmental damage seen in the context of the military contingencies surrounding such actions. The requirement that the anticipated military advantage must be taken into account when looking at the damage to the environment – which was not the case with respect to either of articles 35(3) or 55(1) of the 1977 Additional Protocol I – adds to the uncertainty, subjectivity and difficulty of applying the provision. Even if the requisite threshold of 'widespread, long-term and severe' damage is found to have resulted from a particular act – and of course this would assume that the scope of this threshold is itself clear in practice – a war crime still would not have been committed if this falls within what was acceptable in the light of the anticipated military advantage.

146 One possibility might be for the Judges to require that specific information relating to the relevant environmental concerns be presented, pursuant to article 64(6)(d) of the Rome Statute, which empowers a Trial Chamber to '[o]rder the production of evidence in addition to that already collected prior to the trial or presented during the trial by the parties'.

The Elements of Crimes, when referring to article 8(2)(b)(iv), emphasize that the military considerations that are to be weighed in determining the proportionality of the act are to be determined on a subjective basis. It provides that:[147]

> [t]he expression 'concrete and direct overall military advantage' refers to a military advantage that is foreseeable by the perpetrator at the relevant time. Such advantage may or may not be temporally or geographically related to the object of the attack ... It reflects the proportionality requirement inherent in determining the legality of any military activity undertaken in the context of an armed conflict.

The military 'value' of any particular act(s) is to be determined on the basis of 'the broader purpose' of the particular operation.[148] As has also been noted in previous chapters, a determination of whether the relevant act is acceptable will be dependent upon the knowledge of the alleged perpetrator, based on his/her ('foreseeable') perceptions at the time. The Elements of Crimes confirm this as follows:[149]

> ... this knowledge element requires that the perpetrator make the *value judgement* as described therein. An evaluation of that value judgement must be based on the *requisite information available to the perpetrator at the time.*

It is highly likely that the terms of article 8(2)(b)(iv), when read together with the guidance provided by the Elements of Crimes, would 'excuse' many (all?) decisions made by military commanders to intentionally target the environment.[150] Taking these considerations into account, the terms of article 8(2)(b)(iv) are therefore, in the words of Phoebe Okowa, 'heavily tilted in favour of military advantage and against environmental protection'.[151]

In addition, article 30 of the Rome Statute applies as the *mens rea* standard for article 8(2)(b)(iv).[152] The dual requirement of both intention and, in particular, knowledge further restricts any possible practical application of article 8(2)(b)(iv). For example, the Committee Report concluded that:[153]

---

147 Elements of Crimes, article 8(2)(b)(iv), footnote 36.

148 Knut Dörmann, *Elements of War Crimes under the Rome Statute of the International Criminal Court: Sources and Commentary* (2004), 173.

149 Elements of Crimes, article 8(2)(b)(iv), footnote 37 (emphasis added).

150 Knut Dörmann, 'War Crimes in the Elements of Crimes', in Horst Fischer, Claus Kress and Sascha Rolf Lüder (eds), *International and National Prosecution of Crimes Under International Law: Current Developments* (2001), 95, 127.

151 Phoebe Okowa, 'Environmental Justice in Situations of Armed Conflict', in Jonas Ebbesson and Phoebe Okawa (eds), *Environmental Law and Justice in Context* (2009), 231, 248.

152 Werle and Jessberger, *supra* note 130, 494.

153 Committee Report, para 23 (emphasis added).

> the requisite *mens rea* [under article 8(2)(b)(iv)] on the part of a commander would be *actual or constructive* knowledge as to the grave environmental effects of a military attack; a standard which would be *difficult to establish for the purposes of prosecution* and which may provide an insufficient basis to prosecute military commanders inflicting environmental harm in the (mistaken) belief that such conduct was warranted by military necessity

It must also be noted that article 8(2)(b)(iv) is only applicable 'in international armed conflict, within the established framework of international law'. This express limitation by definition further restricts its effectiveness in dealing with the environmental aspects of armed conflict. Earlier drafts of the instrument had included wording similar to article 8(2)(b)(iv) within articles 8(2)(c), and 8(2)(e), which both deal with war crimes committed in an armed conflict 'not of an international character'; however, this was omitted during the negotiation process and was not included in the final version of the Rome Statute.[154]

As mentioned below, a number of other provisions in the Rome Statute that originally only applied to international armed conflicts have more recently been extended to the war crimes provisions applying to non-international armed conflicts (see 4.4.3.2). However the scope of article 8(2)(b)(iv) was never part of that debate. In any event, the definition of a non-international armed conflict in the Rome Statute is itself quite limited.[155]

This is a particularly notable omission given that, as pointed out in earlier chapters, an increasing number of armed conflicts in the world are of a non-international nature and, in any event, there is no overwhelming logic preventing the applicability of such a provision in relation to internal conflicts. The environmental damage that occurred during the Rwandan and Yugoslav conflicts (the latter being regarded as having the characteristics of both an international and a non-international armed conflict),[156] indicate the extent of environmental destruction that can take place during the course of a 'civil war'. This would be the case, for example, in circumstances where an insurgency group uses a tropical forest as its 'base' and the Government forces, believing that such acts are 'legitimate theaters of operations', deliberately destroy the forest and/or poison water and river systems as part of its attempt to defeat the insurgents.[157]

---

154 Drumbl, *supra* note 127, 631.

155 See Rome Statute, article 8(2)(f). See also Decision on the Prosecution's Application for a Warrant of Arrest against Omar Hassan Ahmad Al Bashir, *The Prosecutor v. Omar Hassan Ahmad Al Bashir ('Omar Al Bashir')*, Case No. ICC-02/05-01/09, Pre-Trial Chamber I, 4 March 2009, para 59.

156 See, for example, Decision on Defence Motion for Interlocutory Appeal on Jurisdiction, *Prosecutor v Duško Tadić*, Case No. IT-94-1, Appeals Chamber, 2 October 1995, para 73.

157 Drumbl, *supra* note 127, 631.

There is also another issue with respect to article 8(2)(b)(iv), albeit in practical terms, relatively minor. Under the obliquely sanitized heading 'Transitional Provision,' article 124 of the Rome Statute enables States to 'opt out' of the war crimes provisions for seven years.[158] The inclusion of article 124 was regarded as yet another necessary compromise to allow for the acceptance of the Rome Statute, given the political sensitivities associated with any restriction on the acts of a State's official armed forces (war crimes are typically – though not exclusively – committed by military personnel). France and Colombia indicated early on that they would consider invoking this provision,[159] and subsequently proceeded to do so.[160] Unless and until this provision is amended,[161] there is nothing – apart from (the admittedly not insubstantial) political pressure – to stop a new State Party to the Rome Statute from making such a declaration in relation to war crimes, which would, of course, include article 8(2)(b)(iv). However, as noted, this is not likely to be a major concern.

In sum, therefore, it seems that there is a real risk that, in reality, resort will not be made to the environmental damage variant in article 8(2)(b)(iv). This provision is but one of a multitude of different war crimes set out in the Rome Statute, and the requirements necessary for it to be applied appear virtually impossible to satisfy in practical terms. Not only are there very significant legal hurdles to overcome in this regard, but this is also a reflection of the resistance towards the issue at the political level, given the general reluctance of States to limit the actions of their own military personnel.

It is true in one respect that express reference to environmental damage as a war crime is to be regarded as a step forward in the development of international criminal law, particularly since there is now a mechanism of international justice – the ICC – through which such acts can be prosecuted. At least the issue

---

158 Article 124 of the Rome Statute provides that:
'... a State, on becoming a party to this Statute, may declare that, for a period of seven years after the entry into force of this Statute for the State concerned, it does not accept the jurisdiction of the Court with respect to the category of crimes referred to in article 8 when a crime is alleged to have been committed by its nationals or on its territory.'

159 Michael Matheson, *Council Unbound: The Growth of UN Decision Making on Conflict and Postconflict Issues after the Cold War* (2006), 375.

160 In 2008, France withdrew its declaration. On several occasions Colombian authorities had publicly stated their intention to withdraw Colombia's declaration: Coalition for the International Criminal Court, 'Delivering on the Promise of a Fair, Effective and Independent Court – Review Conference of the Rome Statute: Article 124', October 2010 <www.iccnow.org/?mod=article124> at 24 January 2015. Although, in the end, no such withdrawal was initiated by Colombia, the effects of the declaration with respect to that country expired on 1 November 2009.

161 Article 124 was considered at the Review Conference of the Rome Statute (convened under article 123(1) of the Rome Statute) held in Kampala from 31 May to 11 June 2010; however it was decided at that time not to delete the provision, but rather to review it again five years later.

has been raised and discussed, and there is an acceptance, in relation to the jurisdiction of the ICC, that completely unfettered environmental destruction is no longer accepted – refer, for example, to the fact that the third possible 'no paragraph' option quoted above in the draft version of the Statute was ultimately rejected at the 1998 Rome Conference.

However, as indicated by the discussion above, and in the absence of a specific environmental crime, the issue of intentional environmental destruction during armed conflict still appears at best to be a narrowly defined 'add-on',[162] with very considerable – perhaps even insurmountable – legal hurdles that, for all practical purposes, will serve to curtail any effective prosecution. As Mark Drumbl has stated, the decision to criminalize widespread, long-term and severe damage to the natural environment as a war crime, as is specified in article 8(2)(b)(iv) is 'cause for limited celebration, considerable disappointment, and some concern'.[163]

Indeed, the high damage thresholds and very limited circumstances as to when the provision may be relevant, if anything, serve to reinforce traditional perceptions that environmental concerns remain very much a minor or subsidiary issue when planning and implementing a military action.

4.4.3.2. Other War Crimes within Article 8 of the Rome Statute

For the sake of completeness, it is appropriate to consider whether any other acts that might constitute a war crime within the terms of the Rome Statute may be helpful in addressing the issue of intentional destruction of the environment, even though, unlike article 8(2)(b)(iv), they make no specific reference to such damage at all. In doing so, it must be noted that the chapeau of each potentially relevant paragraph of article 8 suggests that the enumerated war crimes within that paragraph are to be regarded as exhaustive, since it has generally been accepted that the use of the words 'namely, any of the following acts'[164] implies exclusivity.[165] As a consequence, it does not appear possible to extend the scope of article 8 of the Rome Statute beyond the specific war crimes that have been expressly stipulated, notwithstanding the fact that article 21(1)(b) of the Rome Statute provides that the applicable law of the Court shall include:[166]

---

162 Drumbl, *supra* note 127, 632.

163 *Ibid*, 630.

164 See Rome Statute, chapeau of articles 8(2)(a), 8(2)(b), 8(2)(c) and 8(2)(e) respectively.

165 Drumbl, *supra* note 127, 633. For a contrary view, see Carl E. Bruch, 'All's Not Fair in (Civil) War: Criminal Liability for Environmental Damage in Internal Armed Conflict' (2001) 25 *Vermont Law Review* 695, 719.

166 Rome Statute, article 21(1)(b). For an apparently contrary view, however, at least in relation to the applicability of what are referred to as 'parent norms', see Werle, *supra* note 75, 107.

> [i]n the second place, where appropriate, applicable treaties and the principles and rules of international law, including the established principles of the international law of armed conflict

With this caveat in mind, one can list those war crimes provisions that criminalize particular acts, the implementation of which could have destructive effects on the environment. Within the 'grave breaches' provisions, articles 8(2)(a)(iii) and 8(2)(a)(iv) of the Rome Statute may be relevant in this regard.

In addition, within the context of an international armed conflict, articles 8(2)(b)(v), 8(2)(b)(xiii), 8(2)(b)(xvi), 8(2)(b)(xvii) and 8(2)(b)(xviii) also appear to apply to acts that may impact negatively on the environment. The latter two of these provisions relate to the use of weapons that might already be prohibited by some of the international agreements discussed in chapter 2.[167] With respect to the prosecution of environmentally destructive acts within the context of a non-international armed conflict, the only possible provisions of relevance are articles 8(2)(e)(v), 8(2)(e)(xii), and the more recently included articles 8(2)(e)(xiii) and 8(2)(e)(xiv).

However, each/all of these provisions, to the extent they may be relevant, do not, either individually or collectively address adequately all of the fundamental issues associated with the intentional destruction of the environment during warfare.[168] The structure of article 8 of the Rome Statute thus creates somewhat of a dilemma; in the event that the Prosecutor wishes to proceed against an accused in relation to such damage, he/she would in all likelihood look at the first instance to article 8(2)(b)(iv) as the 'standard' for such a war crime. However, as discussed above, there are very significant difficulties associated with proving that the requisite threshold requirements of that provision could in reality ever be met.

---

167 See, for example, Geneva Protocol for the Prohibition of the Use in War of Asphyxiating, Poisonous and Other Gases, and of Bacteriological Methods of Warfare, opened for signature 17 June 1925, 26 Martens (3rd) 643 (entered into force 8 February 1928); Convention on the Prohibition of the Development, Production and Stockpiling of Bacteriological (Biological) and Toxic Weapons and on their Destruction, opened for signature 10 April 1972, 1015 UNTS 163 (entered into force 26 March 1975); Convention on Prohibitions or Restrictions on the Use of Certain Conventional Weapons Which May be Deemed to be Excessively Injurious or to Have Indiscriminate Effects, opened for signature 10 April 1981, 1342 UNTS 137 (entered into force 2 December 1983); Convention on the Prohibition of the Development, Production, Stockpiling and Use of Chemical Weapons and on Their Destruction, opened for signature 13 January 1993, 32 ILM 800 (entered into force 29 April 1997).

168 For a discussion of the limited scope to which these provisions might possibly be applied in relation to intentional environmental destruction during armed conflict, see Aaron Ezekiel, 'The Application of International Criminal Law to Resource Exploitation: Ituri, Democratic Republic of the Congo' (2007) 47 *Natural Resources Journal* 225, 237–9.

On the other hand, a 'fall back' position of resorting to other war crimes, assuming that they may be applicable, and that all of the relevant elements of those crimes have been met, would in most circumstances not cover all of the elements of such damage and, in any event, would not address the specific 'intent' aspect of a crime against the environment – the deliberate targeting of the environment as a victim, and/or its use as a weapon, during armed conflict.

4.5. CONCLUSION

One of the principal goals behind the establishment of the ICC has been the deterrence and punishment of the most serious international crimes, which also 'threaten the peace, security and well-being of the world'.[169] Given the rapid development of technology and the sometimes bewildering shift in geopolitical alliances, such threats are themselves 'fluid' in nature and will almost certainly vary (expand) in the future.

The intentional destruction of the environment during the course of armed conflict now falls plainly within this description, particularly given the catastrophic environmental damage that may result. However, the limitations of, and uncertainties with the definitions of the existing crimes within the jurisdiction of the Court significantly restrict any attempt to utilize them in such a way so as to comprehensively apply to such acts.

It might be suggested by some that the inclusion of article 8(2)(b)(iv) of the Rome Statute, which makes express reference to the natural environment, may be sufficient in this regard. Indeed, the fact that there currently exists a specific war crime dealing with the issue might reinforce to most military personnel and others engaged in armed conflict that they cannot act with complete disregard of the environmental impact of their actions. In this regard, it could even be argued that, on a cursory reading, this provision provides some protection against such acts, and that therefore the 'need' to impose an enforcement mechanism has somehow been satisfied.

Yet, this has been shown to be both simplistic and inaccurate. The intentional destruction of the environment during armed conflict represents behaviour that is no longer acceptable, as it represents a blatant disregard for the environment and for the (potential) consequences of such acts, including on human populations but also more broadly. In certain circumstances, such acts can therefore also 'shock the conscience of humanity' and give rise to outrage and calls for appropriate criminal accountability. Article 8(2)(b)(iv) is insufficient

[169] Rome Statute, preamble para 3.

and inadequate to address the problem, due to the uncertainties of the provision and the inordinately high threshold level of damage that it requires, and even then only after taking account of the military contingencies. For the reasons outlined above, the elements of the crime appear to be almost unattainable as far as proof of violation is concerned.

Moreover, as discussed in this chapter, none of the core international crimes within the jurisdiction of the ICC are in terms that would adequately regulate such acts. In relation to these crimes, it will no doubt be important, particularly in the relatively early days of the Court's judicial activities, that the Court and the Prosecutor of the ICC avoid claims (whether or not they are justified) that they are perhaps 'overreaching' the boundaries of their respective powers, particularly given the highly political nature of the Court's mandate. It would almost certainly be unrealistic to expect that the Court would seek to exercise a role unless and until any act can quite readily be 'pigeon-holed' into the crimes within the Court's jurisdiction.

Yet, it is argued that acts done with the intent to cause significant environmental destruction during armed conflict should be prosecuted at the international level in particular circumstances, and certainly in broader terms than appear possible within the existing structure of the Rome Statute. This flows from two important considerations – first, the need to properly formalize the criminalization of such acts through a mechanism (a clearly defined crime) that appropriately addresses their heinous nature; and secondly, that this mechanism should be included within the jurisdiction of the ICC, given the functions mandated to that institution by the broader international community, and the level of seriousness of the actions it addresses.

As regards the latter point, it is appropriate that these prosecutions should be undertaken by an international mechanism of criminal justice that has been established according to diplomatic, legal and political processes. The need to ensure the integrity of such prosecutions means that they must be carried out by a body that has been created with the general acceptance of States. As the first and only *permanent* international criminal court, the ICC represents the appropriate judicial 'forum' through which to prosecute such acts.

Hence, we are now at a crossroads – there is an imperative to address the intentional targeting of the environment during armed conflict, and an appropriate judicial institution exists for that to be done. Yet, article 8(b)(2)(iv), and indeed the Rome Statute as a whole, is simply not performing the role that it should with respect to such acts and, frankly, is incapable of doing so as presently constituted. The current legal framework fails to provide sufficient protection against such acts and thus fails humanity on this issue.

The following chapter therefore elaborates on the elements that are required for more effective regulation of the intentional destruction of the environment during armed conflict. It offers a legally consistent, practical and politically acceptable codification of the form of international criminal accountability that should apply and which remedies the *lacunae* within existing international (criminal, as well as humanitarian and environmental) law that have been highlighted in this book.

CHAPTER FIVE

INCORPORATING CRIMES AGAINST THE ENVIRONMENT INTO THE ROME STATUTE

5.1. THE NEED FOR A *SUI GENERIS* CRIME OF CRIMES AGAINST THE ENVIRONMENT

5.1.1. REVISITING THE IMPERATIVE TO MORE EFFECTIVELY ADDRESS THE INTENTIONAL DESTRUCTION OF THE ENVIRONMENT DURING WARFARE UNDER INTERNATIONAL CRIMINAL LAW

In 1997, the International Court of Justice, when dealing with a dispute that involved issues relating to the environment and the concept of sustainable development, made the following observation:[1]

> Throughout the ages, mankind has, for economic and other reasons, constantly interfered with nature. In the past, this was often done without consideration of the effects upon the environment. *Owing to new scientific insights and to a growing awareness of the risks for mankind – for present and future generations – of pursuit of such interventions at an unconsidered and unabated pace, new norms and standards have been developed, set forth in a great number of instruments during the last two decades.* Such new norms have to be taken into consideration, and such new standards given proper weight, not only when States contemplate new activities but also when continuing with activities begun in the past.

In other words, in the Court's view, when it comes to environmental concerns, it has been necessary from time to time to recognize 'new norms and standards' to react to, and counter, the threats posed by the increasingly destructive effects of various activities undertaken by humans. There is an 'expanding need' for the development of appropriate universal norms by the international

---

1 *Gabcíkovo-Nagymaros Project (Hungary v. Slovakia)* (Judgment) [1997] ICJ Rep 7, para 140 (emphasis added).

community,[2] in order to address issues of global concern, including the state of the environment. As a consequence, acts that may represent significant threats to the environment must be effectively regulated through appropriate legal mechanisms.

Not only is this important given that our values and perspectives with respect to the environment have altered and been reshaped over time, but also because of our increasing 'ability' to cause (and to understand the effects of) catastrophic and potentially permanent environmental damage.

In the case of environmental treaty norms, their development and codification has been a relatively recent phenomenon. As noted in chapter 2, it was only in 1972 that the first significant statement of fundamental international principles relating to the protection of the environment was elaborated in a multilateral instrument.[3] The development of environmental norms is as much an ongoing process today as it has been over the previous four decades.[4]

Yet, as discussed in earlier chapters, the development of Multilateral Environmental Agreements (MEAs), and more particularly the *jus in bello* treaty and customary principles, has not paid significant attention to the impact upon the environment of acts undertaken during warfare. While the *humanitarian* consequences of armed conflict have been increasingly addressed in the relevant *jus in bello* instruments, the international community has been far more resistant to the notion of effectively regulating the *environmental* consequences,[5] with the result that the interaction between the conduct of armed conflict and the natural environment is largely under-explored.

The same could be said about the relationship between international criminal law, which increasingly appears to be the global community's chosen mode

---

2 Jonathan I. Charney, 'Universal International Law' (1993) 87 *American Journal of International Law* 529, 529.

3 See Declaration of the United Nations Conference on the Human Environment (16 June 1972) UN Doc A/CONF.48/14/Rev.1 (1972 Stockholm Declaration).

4 For example, the International Environmental Agreements (IEA) Database Project maintained by the University of Oregon has calculated that, in the period from January 2000 to January 2015, there were 321 multilateral IEAs and modifications concluded (comprising 97 New Agreements, 176 Amending Agreements and 48 Protocols), as well as 100 bilateral agreements and modifications (91 New Agreements, 3 Amending Agreements and 6 Protocols): University of Oregon, 'International Environmental Agreements (IEA) Database Project' <http://iea.uoregon.edu/page.php?query=summary&type=MEA> at 25 January 2015.

5 Mark A. Drumbl, 'Waging War against the World: The Need to Move from War Crimes to Environmental Crimes', in Jay E. Austin and Carl E. Bruch (eds), *The Environmental Consequences of War: Legal, Economic, and Scientific Perspectives* (2000), 620, 620.

of enforcing some of its most cherished norms,[6] and concerns for the environment during armed conflict. Addressing atrocities perpetrated against humans in the conduct of warfare has totally overshadowed any environmental concerns. In some senses, this primary anthropocentric focus is perfectly understandable as international criminal law has been increasingly utilized as a legal tool for the regulation of conduct during warfare. However, this has also meant that, thus far, relatively little effective focus has been directed towards the real threats to the environment that arise in the context of armed conflict (and beyond).

To the extent that these environmental concerns *have* been dealt with within either the *jus in bello* or the principles of international criminal law – either directly or (more typically) indirectly – other factors have still tended to dominate the discussions, in particular those relating to military necessity and military advantage, and/or the traditional reluctance of States to 'unduly' restrict the conduct of their forces on the battlefield by referring to the environmental consequences that may result therefrom. Such institutionalized reluctance stems from the fact that warfare almost invariably results in environmental damage, thus generating a concern among States that *any* action taken during armed conflict may potentially attract some form of sanction, unless certain 'legal safeguards' – typically in the form of minimum thresholds of culpability – are put into place.

Thus, any possible regulation of armed conflict that might otherwise have been driven primarily by environmental concerns has historically been perceived as unnecessarily hindering the conduct of military operations, with the consequence that those thresholds necessary for the application of any criminal accountability that have been codified are deliberately set at extremely high levels and, consequently, are of little practical or 'deterrence' value.

As a result, the relevant *jus in bello* principles, and the more recently developed standards codified within international criminal law instruments, do not adequately address the intentional targeting of the environment during warfare, as has been demonstrated in earlier chapters, and have probably never done so. This reality stands in stark contrast to the growing movement in other areas towards the universal recognition of environmental rights[7] and, more fundamentally, the overarching importance of the environment.

---

6 Frédéric Mégret, 'The Problem of an International Criminal Law of the Environment' (2011) 36:2 *Columbia Journal of Environmental Law* 195, 201.

7 Peggy Rodgers Kalas, 'International Environmental Dispute Resolution and the Need for Access by Non-State Entities' (2001) 12 *Colorado Journal of International Environmental Law and Policy* 191, 243.

With the (possible) exception of article 8(2)(b)(iv) of the Rome Statute (but see further below), the international crimes that have been included within the jurisdictional mandate of the international criminal tribunals have therefore not focused in any specific way on addressing the intentional destruction of the environment during armed conflict. The threats posed to the natural environment have largely not had a role to play in the codification of unacceptable behaviour warranting individual criminal responsibility under international law.

In this context, it is no surprise that the concept of an international environmental 'crime' has not been seriously considered by States within an international criminal law context. Rather, to the degree that environmental issues have arisen, they have, as noted earlier, typically only been treated as 'at most … an add-on in narrowly circumscribed areas'.[8]

This notion that environmental issues are only of supplementary concern has restricted the application of principles relating to the concern for the environment in international criminal law and, as was discussed with regard to article 8(2)(b)(iv) of the Rome Statute, limits the effectiveness of regulating such damage within the context of armed conflict. In other words, the previous analysis suggests that, if anything, the possibility of criminal accountability for intentional destruction of the environment has *decreased* due to the terms of article 8(2)(b)(iv), given the near impossibility of ever proving that the requisite threshold of damage required under that provision, to be balanced against various military considerations, has been perpetrated.

In chapter 1, a proposition was put forward for assessment; that there is an imperative to effectively prosecute the intentional destruction of the environment during warfare under international criminal law. It is submitted that this case has been demonstrated in this book. The analysis of the *jus in bello* treaty instruments and customary principles, the international environmental law treaty instruments, and the jurisdictional mandates of the international criminal justice mechanisms clearly point to the fact that the current international regulatory framework is not designed to meet this imperative. The *jus in bello* in particular and, to a far lesser extent, the international environmental law principles, do touch on certain aspects of this issue, but in neither a comprehensive, consistent or focused manner. In terms of international criminal law, the Rome Statute also does refer to environmental damage, but, as discussed, in terms that are, for practical purposes, largely ineffective.

8 Drumbl, *supra* note 5, 632. This is also illustrated by the fact that damage to the environment is currently addressed in the Rome Statute within the provision (article 8(2)(b)(iv)) that also relates to *incidental* loss or damage to civilians or civilian objects.

Simply put, to proscribe certain behaviour (the intentional destruction of the environment during armed conflict) in terms, and with conditions, that imply that any relevant sanctions will not be enforced because the crime can in practical terms never be proven, may at a minimum render any possible criminal accountability as an irrelevant consideration. Ironically, therefore, just as the need for effective regulation continues to increase, the institutional response has had a regressive effect.

Thus, if one of the important aims of criminalizing certain behaviour within the Rome Statute is to deter such actions from being perpetrated – so as 'to contribute to the prevention of such crimes'[9] – then article 8(2)(b)(iv) does not function as it should. In crude terms, it simply does not 'work' to any effective degree and fails to encourage those involved in armed conflict to consider the environmental consequences of their actions in a significant way. This is a situation that must be redressed. As discussed below, it is (and has been for some time) generally agreed that the intentional destruction of the environment during armed conflict on a significant scale exhibits 'elements' that characterize it as an international crime, and thus represents an appropriate subject matter for international criminal law.

This must also, as noted, be seen in the light of the continued development of weapons technology, with its ever-increasing destructive capabilities. A number of studies indicate that, as this technology continues to develop, the amount of (and potential for) environmental damage caused during armed conflict has also escalated.[10] As noted in chapter 1, already in 1976, the International Law Commission (ILC), in its commentary on the then draft article 19 of the Articles on State Responsibility for Internationally Wrongful Acts, had observed that:[11]

> [t]he astounding progress of modern science, although it has produced and continues to produce marvellous achievements of great benefit to mankind, nevertheless imparts a capacity to inflict kinds of damage which would be fearfully destructive not only of man's potential for economic and social development but also of his health and of the very possibility of survival for the present and future generations.

---

9 Rome Statute of the International Criminal Court, opened for signature 17 July 1998, 2187 UNTS 3; 37 ILM 999 (entered into force 1 July 2002) (Rome Statute), preamble para 5.

10 See, for example, Eric Talbot Jensen, 'The International Law of Environmental Warfare: Active and Passive Damage During Armed Conflict' (2005) 38 *Vanderbilt Journal of Transnational Law* 145, 184–5 and the corresponding footnote.

11 Commentary by the International Law Commission on Article 19 'Articles on State Responsibility for International Wrongfully Acts' [1976] 2 *Yearbook of the International Law Commission* Part II, 96, UN Doc A/CN.4/SER.A/1976/Add.l (Part 2), 108.

In this regard, the International Committee of the Red Cross has concluded that:[12]

> [i]n the study, development, acquisition or adoption of a new weapon, means or method of warfare, States are under an obligation to determine whether its employment would, in some or all circumstances, be prohibited by applicable rules of international law, *including those providing protection to the environment in times of armed conflict*

Moreover, the fact that environmental destruction is an inevitable occurrence in virtually all armed conflict cannot (or at least can no longer) justify its exclusion from otherwise attracting international criminal responsibility in appropriate circumstances. In this regard, parallels can be drawn from other developments in the evolution of international criminal law, a notable example being the fact that an act of rape (or other acts of sexual violence) can, as noted, now constitute a war crime, a crime against humanity and even an act of genocide.[13] In 1998, the then judge of the International Criminal Tribunal for Rwanda (ICTR),[14] Navi Pillay (until recently, the United Nations High Commissioner for Human Rights),[15] when interviewed about the landmark *Akayesu* decision,[16] was quoted as saying:[17]

> [f]rom time immemorial rape has been regarded as spoils of war ... now it will be considered a war crime. We want to send out a strong signal that rape is no longer a trophy of war.

12 International Committee of the Red Cross, 'Guidelines for Military Manuals and Instructions on the Protection of the Environment in Times of Armed Conflict' (1995), article IV(18) (emphasis added) <www.icrc.org/Web/Eng/siteeng0.nsf/html/57JN38#a2> at 25 January 2015. In support of this statement, the Guidelines refer to article 36 of Protocol I Additional to the Geneva Conventions of August 12, 1949, and relating to the Protection of Victims of International Armed Conflicts, opened for signature 8 June 1977, 1125 UNTS 3, 16 ILM 1391 (entered into force 7 December 1978).

13 See, for example, Rome Statute, articles 7(1)(g), 8(2)(b)(xxii) and 8(2)(e)(vi).

14 International Criminal Tribunal for the Prosecution of Persons Responsible for Genocide and Other Serious Violations of International Humanitarian Law Committed in the Territory of Rwanda and Rwandan Citizens Responsible for Genocide and Other Such Violations Committed in the Territory of Neighbouring States between 1 January 1994 and 31 December 1994 (ICTR).

15 Navanethem 'Navi' Pillay was a Judge of the ICTR from 1995–2003, a Judge of the International Criminal Court from 1 March 2003–31 August 2008, and United Nations High Commissioner for Human Rights from 1 September 2008–31 August 2014.

16 Judgement, *Prosecutor v. Jean-Paul Akayesu*, Case No. ICTR-96-4-T, Trial Chamber, 2 September 1998.

17 Bill Berkeley, 'Judgment Day' *The Washington Post*, 11 October 1998, at W10, quoted in Mark Ellis, 'Breaking the Silence: Rape as an International Crime' (2006–2007) 38 *Case Western Reserve Journal of International Law* 225, 236.

In other words, even though rape had always been associated with warfare, it was considered necessary and appropriate at some point in time (albeit much later than was justified) to deem the perpetration of such acts as sufficiently serious so as to constitute an international crime. Similarly, the time has now come to codify in appropriate terms that the intentional destruction of the environment during armed conflict constitutes such a crime (subject to minimum damage thresholds); and not only so, but that the elaboration of that crime, and the scope for criminal accountability for such acts, must bear due correlation to the prevailing standards of what constitutes (un)acceptable behaviour in the eyes of the international community.

In its oral submissions before the International Court of Justice during hearings related to that Court's Advisory Opinion regarding the legality of nuclear weapons, when referring to the 'Martens clause',[18] Australia made the following point:[19]

> neither the concept of "humanity", nor the "dictates of public conscience" are static. Conduct which might have been considered acceptable by the international community earlier ... might be condemned as inhumane by the international community today

In similar vein, in his Dissenting Opinion in the same case, Judge Shahabuddeen noted that the principles of humanity and the dictates of public conscience – both of which are specified in the Martens clause – should be ascertained:[20]

> in the light of changing conditions, inclusive of changes in the means and methods of warfare and the outlook and tolerance levels of the international community. The principles would remain constant, but their practical effect would vary from time to time: *they could justify a method of warfare in one age and prohibit it in another*

These sentiments – which were expressed by the Judge when considering the (potential) use of massively destructive weapons during warfare – apply equally to the context of acts targeting the environment on a significant scale during armed conflict. The 'tolerance threshold' of the international community

---

18 See, for example, the preamble of the Hague Convention IV Respecting the Laws and Customs of War, opened for signature 18 October 1907, 3 Martens (3rd) 461 (entered into force 26 January 1910) (Hague Convention IV). For a discussion of the Martens clause, see chapter 2.

19 Verbatim Record, Public Sitting held on Monday 30 October 1995 in the case in *Legality of the Use by a State of Nuclear Weapons in Armed Conflict* (Request for Advisory Opinion Submitted by the World Health Organization) and in *Legality of the Threat or Use of Nuclear Weapons* (Request for Advisory Opinion Submitted by the General Assembly of the United Nations), Statement of Australia, Foreign Minister Gareth Evans, para 10.

20 Dissenting Opinion of Judge Shahabuddeen, *Legality of the Threat or Use of Nuclear Weapons* (Advisory Opinion) [1996] ICJ Rep 226, 375, at 406 (emphasis added).

towards environmental damage suffered in warfare is now lower than what might previously have been the case. Acts committed during armed conflict that exceed acceptable levels in this regard should now render those responsible as criminally accountable, and enforcement mechanisms must be structured so as to make this possible in a practical sense. As Bassiouni has opined:[21]

> The object of the normative proscription of international criminal law is to specify conduct identified as harmful to a given world social interest whose protection is deemed to require the imposition of criminal sanctions on violators and which sanctions are enforced by the member states of the world community through international collective, cooperative or national action.

States did not for a long time establish a specific tool to allow for international criminal responsibility for intentional environmental destruction during armed conflict. However, given the increasing appreciation of the importance of the natural environment and the consequences of environmental destruction, past inaction is not a valid reason to prevent the elaboration and development of the rules of international criminal law in more recent times,[22] particularly where those rules are warranted. Indeed, this has been expressly recognized by the inclusion of article 8(2)(b)(iv) of the Rome Statute – however, for the reasons already discussed, this provision does not codify the criminalization of such acts in a way that properly addresses the realities of environmental destruction; nor does it conform with the increasing need for appropriate standards and enforcement mechanisms to project the significance of environmental concerns during the conduct of armed conflict.

5.1.2. A *SUI GENERIS* CRIME – CRIMES AGAINST THE ENVIRONMENT

5.1.2.1. Why a Stand-Alone Crime?

In order to achieve these goals and following on from the analysis in previous chapters, a more rigorous criminalization of the intentional destruction of the environment during armed conflict is required. The Rome Statute is the most appropriate instrument in which this should be included and through which it should be enforced. This is both due to the fact that the ICC is the best equipped of any of the existing mechanisms of international criminal justice to deal with

---

21 M. Cherif Bassiouni, as quoted in Mark Allan Gray, 'The International Crime of Ecocide' (1996) 26 *California Western International Law Journal* 215, 270.

22 Tullio Scovazzi, 'State Responsibility for Environmental Harm' (2001) 12 *Yearbook of International Environmental Law* 43, 67.

this crime, and also, as noted in chapter 1, that the Rome Statute itself plays 'a powerful trendsetting role'[23] in articulating and shaping the actions of States.

As demonstrated in chapter 4, this cannot, however, be achieved simply by trying to 'pigeon-hole' such environmental concerns into, or add them onto, the already existing core international crimes in the Rome Statute, not the least because those crimes largely do not have a focus directed towards the environment. To attempt to do so might, in certain circumstances, also unduly compromise the *nullum crimen sine lege* principle in relation to those crimes.

In addition, at a more general level, it would not be appropriate to manipulate or dilute the focus of the core international crimes away from the egregious violations of human rights with which they already deal. Such an approach may, in fact, compromise the terms of those crimes, and even then may (still) fail to more effectively address the issue of intentional environmental destruction during armed conflict.

One solution that was initially considered by this author was to (simply) amend the existing war crime provision, or perhaps to include a new war crime. After careful thought, however, it was concluded that such an approach would not be appropriate. First, amending article 8(2)(b)(iv) itself to alter the prohibitions relating to environmental damage would be complicated, particularly because the provision also deals with the other 'variants' of the crime – incidental loss of life or injury to civilians and damage to civilian objects. Those aspects of the provision are not the subject of this discussion and, therefore, the only amendment to article 8(2)(b)(iv) that is proposed is the deletion of the third variant dealing with damage to the natural environment (see Appendix I 5.3.2.2.2).

Moreover, adding a 'new' war crime within article 8 of the Rome Statute to address the intentional destruction of the environment does not adequately meet the imperatives in terms of environmental concerns during hostilities. Whilst, on the one hand, that approach might be more amenable to States than the approach suggested in this chapter, it would not sufficiently demonstrate the importance of instituting international justice mechanisms that address acts that target the environment *per se*. Apart from issues relating to the type

23 Mark A. Drumbl, *Reimagining Child Soldiers in International Law and Policy* (2012), 121. See also African Union, 'Draft Protocol on Amendments to the Protocol on the Statute of the African Court of Justice and Human Rights' AU Doc Exp/Min/IV/Rev.7, which was concluded at a meeting of Government Experts and Ministers of Justice/Attorneys General of the African Union States on 15 May 2012, draft article 28D(b)(iv) of which is identical in its terms to article 8(2)(b)(iv) of the Rome Statute: see <http://africlaw.files.wordpress.com/2012/05/au-final-court-protocol-as-adopted-by-the-ministers-17may. pdf> at 18 February 2014.

of armed conflict to which it would apply, and the sensitivity of States in respect of war crimes, categorizing such acts simply as a war crime and within the other restrictions that apply to war crimes, as well as the underlying military considerations, would not adequately stigmatize such acts – in simple terms, apart from the legal hurdles, it would send 'the wrong message' regarding the importance of the environment, even during hostilities.

What is required is an appropriate 'tool' that regulates such acts reaching a (specified) level of gravity by way of its codification as a crime *in its own right*. This therefore calls for an alternate approach; an amendment to the Rome Statute that would see the inclusion of a separate *sui generis* crime within the Rome Statute – 'crimes against the environment' – with a principal and direct focus on acts done with intent to significantly destroy the environment during armed conflict. Having said this, it is important to bear in mind that the Rome Statute is not an 'environmental' treaty, but rather an elaboration of fundamental aspects of international criminal law. This does not, however, rule out the inclusion of such a crime; rather what it emphasizes is that such acts have now reached a level of gravity and seriousness such that they will, in certain circumstances, constitute an international crime.

The inclusion, within the jurisdiction of a mechanism of international justice created by the international community, of a specific crime directed towards this type of act therefore represents an appropriate elaboration of law that, as Kittichaisaree notes, should be undertaken 'in light of the latest developments in law, morality, and the sense of criminal justice at the [relevant] time.'[24]

5.1.2.2. Why 'During Armed Conflict'?

In this context, this chapter seeks to set out a relatively modest proposal to more adequately address such acts – modest in the sense that it focusses solely on acts done with intent to cause significant environmental destruction *during armed conflict*. History has demonstrated that such attacks on the environment have often occurred in armed conflict. It is, however, recognized that this is not the only context in which they take place, since they might also be perpetrated in circumstances where there is not necessarily a clear nexus to an armed conflict, or even perhaps in the course of 'peaceful activities'.[25]

---

24 Kriangsak Kittichaisaree, *International Criminal Law* (2001), 3.

25 See, for example, Ludwik A. Teclaff, 'Beyond Restoration – the Case of Ecocide' (1994) 34 *Natural Resources Journal* 933, who (at 934) contended that the term 'ecocide' could be applied to *peaceful* activities that destroy the environment on a massive scale. See also Mansour Jabbari-Gharabagh, 'Type of State Responsibility for Environmental Matters in International Law' (1999) 33 *Revue Juridique Themis* 59, 86. For a contrary view, see Malcolm

In a sense, the potential range of circumstances in which a deliberate targeting of the environment might take place presents a fundamental choice to those seeking to 'define' such a crime. As someone with sympathies for the need to prevent such attacks in *all* contexts, this author could have chosen to put forward a broad, all-embracing proposal for a crime – covering actions that are perpetrated either in the context of armed conflict *or* occasioned during times of peace. In the end, however, the author has chosen not to take this path – despite its attractions – and has limited it to armed conflicts, for the reasons set out below.

First, as discussed in the following sections, the intentional targeting of the environment is addressed in this book not from an environmental protection perspective, but rather from a (international) criminal law viewpoint, giving rise to the need to elaborate appropriate and effective tools of criminal accountability for acts that represent unacceptable (criminal) behaviour, as opposed to environmental law sanctions. In other words, the proposed amendment to the Rome Statute set out in this chapter addresses acts that constitute an *international crime*, for which regulation is necessary and appropriate under international criminal law. In essence, it is the *intention* to target the environment that indicates the gravity of such acts, and this is to be seen within the context of international criminal law.

Whilst the inclusion of this crime within the mandate of the ICC will have a much greater deterrence value than the current *status quo*, and thus serve to (potentially) reduce the risk of significant environmental damage, it is therefore first and foremost to be considered under the ambit of international criminal law. Measures designed more generally to specifically protect the environment might take other forms, since not every act that has adverse consequences for the environment should be considered as falling within the scope of an international crime.

Secondly, it is important to recognize the boundaries under which the ICC operates. In general, the underlying contexts of the crimes that fall within the mandate of the Court revolve around armed conflicts. Although it may be possible – technically – for some of the crimes to be perpetrated in 'peace-time', since there is no (or no longer in the case of the latter) direct linkage or nexus required to an armed conflict in the case of the crimes of genocide and crimes against humanity, the reality is that even these crimes will typically take place during some form of armed conflict, even if this is not a definitional element of the crime.

To introduce into the Rome Statute a crime that would cover acts that clearly take place outside the scope of any conflict – for example, the deliberate dumping of nuclear waste into the ocean by a merchant vessel – would, it is submitted, 'stretch' the reach of the Court to situations beyond which it was principally

N. Shaw, 'Genocide and International Law', in Yoram Dinstein (ed), *International Law at a Time of Perplexity* (1989), 797, 810.

designed to address, notwithstanding the seriousness of such an action. This would potentially dilute the important role of the Court in dealing with crimes that take place (generally) within the context of conflict, and may also involve it in situations that are outside of its principal focus. Consequently, this would also give rise to political opposition, which might seriously hamper any acceptance of the need to codify this crime.

Thirdly, to the extent that the international community *has* thus far chosen to specifically address the deliberate destruction of the environment under international criminal law, it has done so as a *war crime* (article 8(2)(b)(iv) of the Rome Statute). Notwithstanding that this provision is flawed and ineffective, the fact remains that the concept of addressing environmental destruction has (only) been accepted in circumstances that have a clear nexus to an (international) armed conflict.[26] In this sense, it is already the case that environmental issues are relevant in the context of armed conflict for the purposes of the ICC, with the only issue (albeit a significant one, and a principal focus of this book) being *how* to effectively proscribe such acts. As noted, however, limiting any prohibition of intentional environmental destruction during warfare *only* to the classification of war crimes does not achieve all that is required. On the other hand, for the reasons outlined, a link between the crime and armed conflict is still considered to be necessary and appropriate, if the Rome Statute is to be the conduit by which such acts are to be prohibited.

Thus, criminalizing such acts as are perpetrated in an armed conflict context is an accepted approach under international criminal law, and the ICC is already regarded as an appropriate mechanism of international criminal justice to deal with them. In essence, to comply with already accepted enforceability mechanisms in respect of such acts is consistent with existing practice whilst not imposing systematic boundaries limiting the scope of any definition of crimes against the environment. It is acknowledged, however, that to require this 'relationship' with an armed conflict will necessarily exclude other examples of deliberate environmental destruction that do not indicate such a link.

Finally, from a political realist viewpoint, a broad all-embracing proposal would, it is submitted, have no chance of ever seeing the light of day, not only because its scope would be too wide-reaching to be politically acceptable, but also because it would most likely require the establishment of a new, specific judicial infrastructure – an international environmental (criminal) court[27] – to measure

---

26 See the opening words of article 8(2)(b) of the Rome Statute, which make direct reference to an (international) 'armed conflict'. This was raised in chapter 4 and will be discussed further in this chapter.

27 There have, over the past two decades, been a number of calls for the establishment of an international environmental court: see, for example, details of the 1992 draft convention for the Establishment of an International Court for the Environment proposed under the

compliance with the appropriate standards and also to regulate enforcement. Such a development does not appear likely, particularly in the light of the sensitivity of the international community towards increasing the number of international (criminal) tribunals,[28] given their very significant operational costs.[29]

For all of these reasons, this author has chosen to define crimes against the environment in terms that emphasize the need to better protect the environment *per se* and, by codifying it as a stand-alone crime, elevate its importance quite unambiguously in the context of armed conflict, so that it is no longer to be treated as a minor and incidental matter, clouded by overarching military considerations that are traditionally applied in the sphere of war crimes. Instead, it sends a strong and clear message that the destruction of the environment can no longer be treated simply as a 'spoil of war'.

At the same time, the terms of the crime as proposed below are not only workable in a functioning criminal law sense, but are consistent with current practice, far more effective than the existing legal regime in terms of addressing the fundamental environmental concerns, and politically more 'achievable' in the short-medium term. This is both because its scope is in line with the increasing acceptance by the international community of the principles of international criminal law as a tool for addressing the most egregious of acts, particularly (although, as noted above, not exclusively),[30] during armed conflict,[31] and also

auspices of the National Academy of Lincei in Rome, discussed in Kalas, *supra* note 7, 232–40. For another example of a proposed treaty for such a court, see Kenneth F. McCallion, 'International Environmental Justice: Rights and Remedies' (2003) 26 *Hastings International and Comparative Law Review* 427, 438–443. More recently, see Alliance des Démocrates et des Libéraux pour l'Europe, 'Charter of Brussels for the creation of European and International Criminal Court of the Environment and Health', 30 January 2014, article 3.

28 See, for example, the Completion Strategy for the International Tribunal for the Prosecution of Persons Responsible for Serious Violations of International Humanitarian Law Committed in the Territory of the Former Yugoslavia since 1991 (ICTY) and of the ICTR, discussed in chapter 4.

29 See, for example, Steven D. Roper and Lilian A. Barria, *Designing Criminal Tribunals – Sovereignty and International Concerns in the Protection of Human Rights* (2006), who calculated (at page 61) that, already by the end of 2005, the ICTY and ICTR had 'received over $4.5 billion through Chapter VII financing'.

30 In this regard, the crime of crimes against humanity is no longer formally linked to the existence of an armed conflict: compare, for example, the *chapeau* of the definition of crimes against humanity in article 5 of the Statute of the International Tribunal for the Prosecution of Persons Responsible for Serious Violations of International Humanitarian Law Committed in the Territory of the Former Yugoslavia since 1991, 32 ILM 1159 (entered into force 25 May 1993) (ICTY Statute) ('... when committed in armed conflict, whether international or internal in character') with article 7(1) of the Rome Statute, where no such nexus is specified. See chapter 4 for more details.

31 As mentioned in chapter 1, it is also noteworthy to observe that, at its 65th session in 2013, the International Law Commission decided to include the topic of 'Protection of the Environment in relation to Armed Conflicts' in its programme of work, on the basis of the

because an existing and operative enforcement mechanism – the ICC – has already been accepted as an appropriate institution to deal with such acts, and has (or is developing) much of the necessary expertise.

Moreover, designating an existing *permanent* (as opposed to *ad hoc*) mechanism of international criminal justice with this jurisdiction would allow for future accountability for acts perpetrated in as yet un-conducted armed conflicts, thus adding considerably to the deterrence effect of the provisions, assuming of course, that the terms of the crime are practical and workable.

All of these considerations address many issues that may otherwise have given rise to a general reluctance on the part of States to codify an additional international crime on the basis of the lack of existing enforcement machinery.[32] They also counter arguments as to any 'novelty' of the legal approach taken in this book to address the intentional destruction of the environment during armed conflict.

Although some may advocate that a far broader crime is required, it is submitted that the most appropriate approach to take is to codify the crime on the basis that it might realistically be accepted and acceptable within a relatively short time period. As Frédéric Mégret has noted:[33]

> [i]nternational criminalization ought not to attempt too much at the risk of failing to accomplish what can realistically be achieved.

This does not represent a lack of ambition; nor, it is submitted, is it an exercise in undue cynicism. Rather, it is an acceptance of the *realpolitik*, as well as recognition of the advantages of developing the regime of institutionalized international criminal law along the lines of existing accepted practice, thus reflecting a more seamless evolution of legal principles. There are no doubt many worthy and 'justifiable' calls for change and the extension of accountability mechanisms in a wide range of fields, including with respect to the Rome Statute itself.[34] However, some of these are unlikely to have any practical relevance for, or impact on, the ongoing development of international law, due to the inherent

---

recommendations contained in a working group report: see International Law Commission, 'Protection of the environment in relation to armed conflicts' (5 June 2013) <www.un.org/law/ilc/index.html> at 21 January 2015.

32 Gray, *supra* note 21, 271.

33 Mégret, *supra* note 6, 211.

34 See, for example, Kamari Maxine Clarke, 'Treat Greed in Africa as a War Crime', *The New York Times* (New York), 29 January 2013 <www.nytimes.com/2013/01/30/opinion/treat-greed-in-africa-as-a-war-crime.html?emc=eta1> at 25 January 2015.

conservatism of States when it comes to submitting themselves to international legal regulation and enforcement.

A tangible consequence of this conservatism, at least since the 1990s, has been the creation of international criminal tribunals with mandates that, to a large degree, were intended to prosecute only those crimes that were 'doubtless part of customary international law.'[35] However, this limitation only to customary international law, as discussed in the previous chapter, may not always be the case.[36] Seen in this context, it is argued that the inclusion of the *sui generis* international crime of crimes against the environment is a necessary and appropriate step at this point, even though existing customary law may not (yet) support every aspect of the definition set out in 5.3. As discussed in 5.2, a requirement that every one of its elements represents customary international law does not appear to be a necessary criterion for the legal characterization of an act as an international crime.

In any event, the suggestion that the targeting of the environment during armed conflict may attract international criminal prosecution has, as noted, already been accepted.[37] International criminal law already deals with acts directed against victims and addresses prohibitions on certain weapons. Thus, the inclusion of a crime that encompasses situations where the environment is targeted as a victim or utilized as a weapon is not at odds with existing accepted practice with respect to codifying acts that give rise to international criminal responsibility.

As mentioned, given the status of the ICC as a permanent international criminal court, that institution is the appropriate judicial body to have jurisdiction over this crime. This is so even though the Court undoubtedly faces some

---

35 United Nations Security Council, 'Report of the Secretary-General Pursuant to Paragraph 2 of Security Council Resolution 808' (3 May 1993) UN Doc S/25704, para 33–5. As noted in chapter 1, this can be contrasted to arguments raised before the Nuremberg Military Tribunal to the effect that the crimes prosecuted in that tribunal offended the *nullum crimen sine lege* principle. In respect of the ICTY, one commentator suggests that it also adopted a 'relaxed attitude' to the *nullum crimen* principle quite early on in its judicial operation, an approach that he considered was 'certainly consistent' with that also taken by the European Court of Human Rights: see, William Schabas, 'Commentary on Judgement, *Prosecutor v Furundžija*', in André Klip and Göran Sluiter (eds), *Annotated Leading Cases of International Criminal Tribunals, Volume III, The International Criminal Tribunal for the Former Yugoslavia, 1997–1999* (2001), 753, 757. See also Steven Freeland, 'Judicial Decision-Making in International Criminal Courts: 'Effective' Justice?' (2014) 3(1) *Griffith Journal of Law and Human Dignity* (forthcoming).

36 See, for example, Kenneth S. Gallant, 'Jurisdiction to Adjudicate and Jurisdiction to Prescribe in International Criminal Courts' (2003) 48 *Villanova Law Review* 764, 783.

37 See also Byung-Sun Cho, 'Emergence of an International Environmental Criminal Law?' (2001) 19 *UCLA Journal of Environmental Law and Policy* 11, 26.

considerable challenges, and there remain imperfections in this still-evolving system of international justice.[38] Yet, it has continued to develop expertise in the prosecution of serious international crimes, notwithstanding difficulties along the way.[39] Incorporating a crime of crimes against the environment into mandate of the ICC will, it is submitted, further emphasize the criminal aspects associated with such acts, and also allow for a relatively seamless adoption of the crime into an existing and functioning mechanism of international criminal justice.

For all of these reasons, therefore, this book argues for the introduction of a separate and independent international crime of crimes against the environment, as defined in section 5.3, to be included in the Rome Statute, and thus within the jurisdiction of the ICC.

5.2. CRIMES AGAINST THE ENVIRONMENT – AN INTERNATIONAL CRIME?

As noted, the Rome Statute is intended to address 'the most *serious* crimes of concern to the international community as a whole.'[40] The core crimes within the Rome Statute are regarded as international crimes, given their degree of 'seriousness', the level of which is reflected in the nature of the act(s) in question, the extent of its (their) effects, the motives of the perpetrator, or a combination of these factors.[41] In addition, the ILC has suggested that '[t]he more important the subject-matter [of an offence], the more serious the transgression'.[42] At issue here is whether acts that would constitute a crime against the environment meet the criteria as to constitute an international crime.

---

38 For a discussion of some of the challenges facing the ICC, see Steven Freeland, 'International Criminal Governance: Will the International Criminal Court be an 'Effective' Mechanism for Justice?', in Michael Head, Scott Mann and Simon Kozlina (eds), *Transnational Governance: Emerging Models of Global Legal Regulation* (2012), 213.

39 See, for example, the very pointed criticism of the Prosecution by the Trial Chamber of the ICC in Decision on Sentence pursuant to Article 76 of the Statute, *Prosecutor v. Thomas Lubanga Dyilo*, Case No. ICC-01/04-01/06-2901, Trial Chamber I, 10 July 2012, at *inter alia* para 60, 75, 89, 91 and 97.

40 Rome Statute, preamble para 4 (emphasis added).

41 See International Law Commission, 'Draft Code of Offences Against the Peace and Security of Mankind' [1987] 2:2 *Yearbook of the International Law Commission* 7, UN Doc A/CN.4/SER.A/1987/Add.1 (Part 2), 13, para 66, where the ILC concluded that 'it is this seriousness which constitutes the essential element of a crime against the peace and security of mankind'. Mohammed Saif-Alden Wattad goes even further, believing that the notion of international crimes 'is not necessarily limited to the most serious harms, but rather the most outrageous actions': Mohammed Saif-Alden Wattad, 'The Rome Statute & Captain Planet: What lies between 'Crimes Against Humanity' and the 'Natural Environment'' (2009) XIX *Fordham Environmental Law Review* 265, 268.

42 Nina H.B. Jørgensen, *The Responsibility of States for International Crimes* (2000), 107.

At the very least, the intentional targeting of the environment during armed conflict may, in certain circumstances, meet (and exceed) the relevant threshold level of seriousness, principally in terms of its potential effects; and particularly because the intent of the perpetrator is to *cause* such damage. In this respect, it should also be recalled that a Special Rapporteur of the ILC dealing with the Draft Code of Offences against the Peace and Security of Mankind had stated already in 1985 that:[43]

> ... all international crimes are characterized by the breach of an international obligation that is essential for safeguarding the fundamental interests of mankind. But some interests should be placed at the top of the hierarchical list. These are international peace and security, the right of self-determination of peoples, the safeguarding of the human being and the *preservation of the human environment.* Those are the four cardinal points round which the most essential concerns revolve, and these concerns constitute the summit of the pyramid on account of their primordial importance.

In general terms, it follows from this that such crimes attain the 'status' of an international crime because they are regarded as an affront to us all. In this sense, international criminal law since the 1990s has generally taken an anthropocentric approach to accountability. It is the perspective of the international community as to the abhorrent or heinous nature of the relevant act(s) directed towards humans that determines whether an international criminal response is warranted.[44]

Whilst this will remain the predominant motivation for the codification of crimes in the constituent documents of the international criminal tribunals, there are signs that there are also other factors at play – as indicated, for example, by the incorporation into the Rome Statute in 2010 of the definition of the crime of 'aggression', which is a more State-centered crime, and thus not so much a response to human rights concerns but rather to a breakdown in inter-State relationships.

As a consequence, there is no inherent impediment, based on any assertion that not all '[a]ttacks on the environment ... categorically "shock the conscience of mankind"', to prevent the conclusion that crimes against the environment constitutes an international crime for which specific provision should be made within the Rome Statute in a manner that gives rise to effective enforcement.[45]

---

43 Third Report on the Draft Code of Offences against the Peace and Security of Mankind, by Mr. Doudou Thiam, Special Rapporteur, International Law Commission, 8 April 1985, UN Doc. A/CN.4/387 and Corr.1 and Corr.2, para 61 (emphasis added).

44 Mégret, *supra* note 6, 208; Wattad, *supra* note 41, 268.

45 *Id* Mégret. See also M. Cherif Bassiouni (ed), *A Draft International Criminal Code and Draft Statute for an International Criminal Tribunal* (1987), who (at page 30) rates environmental crimes as 5 out of 10 – with 10 being the 'highest degree of prevalence' and 0 meaning 'non-existence' – under the element headed '"Shocking" to the Conscience of Humanity' (discussed below).

What is important in this respect is to provide a mechanism to determine the appropriate level of gravity, reflected principally in the intent to cause a certain threshold of damage, to constitute such a crime. In any event, it is clear that the perception of the international community as to what constitutes a serious crime of international concern may change over time and, as noted, this also applies to the intentional destruction of the natural environment, given our changing attitudes towards the state of the environment and acknowledgement of the need to better protect it during armed conflict.

Other criteria have also been suggested as indicative characteristics of what might constitute an international crime. Particular acts are, for example, categorized as international crimes because they are regarded as of such importance to the international community as to warrant accountability in the form of individual criminal responsibility.[46] As noted in chapter 1, in the context of the general international law principles of State responsibility, the ILC had earlier suggested that:[47]

> [a]n internationally wrongful act which results from the breach by a State of an international obligation *so essential for the protection of fundamental interests of the international community* that its breach is *recognized* as a crime by the community as a whole constitutes an international crime.

This definition incorporated both an objective (the type of obligation specified) and a subjective factor (the recognition by the international community). To provide further clarity as to what might constitute such a crime, the draft article then went on to indicate that one example might be:[48]

> a serious breach of an international obligation of essential importance for the safeguarding and preservation of the human environment ...

The draft article thus indicated that damage to the human environment on a significant scale was sufficiently serious as to be recognized as an international crime. It is true that draft article 19 was not directed towards the 'criminalization' of such acts, but was rather focused on the responsibility of States. However, applying the nomenclature of an 'international crime', even (or perhaps more so) within the concept of State responsibility under the general principles of public

46 Dinah Shelton, 'Centennial Essay: In Honor of the 100th Anniversary of the AIJL and the ASIL: Normative Hierarchy in International Law' (2006) 100 *American Journal of International Law* 291, 318.

47 See Report of the International Law Commission to the General Assembly on its work of the thirty-second session, [1980] 2 *Yearbook of the International Law Commission* Part II, 32, UN Doc A/CN.4/SER.A/1980/Add.l (Part 2), draft article 19(2) (emphasis added).

48 *Ibid*, draft article 19(3)(d).

international law, was indicative of what has become a trend towards regarding such acts from a criminal law perspective.

In 1987, Cherif Bassiouni suggested that an international crime exhibits one or more of the following characteristics:[49]

(a) An 'International Element', in that it:
   (i) threatens the peace and security of mankind, either directly or indirectly;
   (ii) is conduct that is 'shocking to the conscience of the world community' and is thus contrary to its 'shared values';
(b) A 'Transnational Element', in that it:
   (iii) is 'transnational' in nature, affecting public safety and economic interests in more than one State, involves citizens of more than one State, and transcends national borders; or
(c) A 'Necessity of International Cooperation Element', in that it:
   (iv) requires international cooperation for its prevention, suppression and control.

Bassiouni then goes on to assess crimes involving damage to the environment against these elements, and concludes that they display characteristics of all of them apart from factor (a)(i) above.[50]

In this author's view, the intentional destruction of the environment during warfare does, in fact, meet *all* of the elements outlined by Bassiouni, *including* the link to a threat to international peace and security, which is now regarded as a 'firmly established element of a theory of international criminal law'.[51] As discussed in chapter 1, there is a strong body of thought that also subscribes to this view. It is noteworthy that the ILC, in providing examples of what might constitute an international crime in its then draft article 19, 'equated'[52] a breach threatening the human environment (as quoted above) with:[53]

> (a) a serious breach of an international obligation of essential importance for the maintenance of international peace and security…
> (b) (…)

49 Bassiouni, *supra* note 45, 36–40.
50 *Ibid*, 39.
51 Florian Jessberger, Report to Chairman of Degree Committee regarding dissertation submitted by Steven Robert Freeland to the University of Maastricht, 23 July 2014, 7 (copy with author).
52 Jabbari-Gharabagh, *supra* note 25, 88.
53 See Report of the International Law Commission to the General Assembly on its work of the thirty-second session, [1980] 2 *Yearbook of the International Law Commission* Part II, 32, UN Doc A/CN.4/SER.A/1980/Add.l (Part 2), draft articles 19(3)(a), 19(3)(c) (emphasis in original).

(c) a serious breach on a widespread scale of an international obligation of essential importance for safeguarding the human being, such as those prohibiting slavery, genocide and *apartheid*

If anything, this further reinforces the view that crimes against the environment display the characteristics that are normally associated with the select group of international crimes, and thus is suitable for inclusion as a *sui generis* crime within the terms of the Rome Statute.

The importance of the natural environment – and thus the 'seriousness' of any significant damage caused to it – also stems from the fact that it covers a wide range of concepts, each of which is vital for the existence of not only humankind, but also the other co-existing and interdependent elements of the environment. Within the context of determining the degree to which damage to the environment should be criminalized, the ILC regarded it as encompassing:[54]

> the environment of the human race and where the human race develops, as well as areas the preservation of which is of fundamental importance in protecting the environment. These words therefore cover the seas, the atmosphere, climate, forests and other plant cover, fauna, flora and other biological elements.

In 1994, at its 15th International Congress on Penal Law, the International Association of Penal Law adopted a resolution relating to the concept of crimes against the environment. The discussions did not differentiate between such crimes occurring during armed conflict and those committed in times of peace – and thus were *broader* than the codified crime proposed in this book, which is just limited to acts during armed conflict. Even so, the resolution, which largely dealt with prosecution under national law, contained the following paragraph:[55]

> Core crimes against the environment affecting more than one jurisdiction or affecting the global commons outside any national jurisdiction *should be recognized as international crimes* under multilateral conventions

In sum, therefore, it is generally agreed that, in appropriate circumstances, acts done with intent to destroy the environment during armed conflict 'qualify' as an international crime. This appears to have been widely accepted by academic scholars already in the 1990s (most of the examples referred to above were

---

54 International Law Commission, 'Draft Code of Crimes Against the Peace and Security of Mankind' [1991] 2:2 *Yearbook of the International Law Commission* 107, UN Doc A/CN.4/SER.A/1991/Add.1 (Part 2).

55 See International Association of Penal Law, Resolutions of the XVth International Congress on Penal Law (1995) 66 1:2 *International Review of Penal Law* 52, 52–3 (emphasis added).

from that period). It is submitted that, if anything, there is an even stronger case for asserting this now, given the increased significance attributed by the international community to issues associated with the environment.[56] Of course, this will depend on a range of factors and will therefore not apply to all such damage – the extent of (potential) damage will be one, among several, relevant circumstances to take into account, along with the intent of the alleged perpetrator.

The following section therefore sets out a proposed definition of the crime of crimes against the environment, drafted on the basis as discussed in 5.1 (that is, limited to armed conflicts) and as a *sui generis* crime within the Rome Statute.

5.3. CRIMES AGAINST THE ENVIRONMENT – A DEFINITION FOR THE ROME STATUTE

As noted, this section sets out a definition of the crime in a form that would be consistent with its inclusion as a separate crime within the Rome Statute. It also discusses the most important aspects of the definition. In addition to the terms of the crime, the applicable updated Elements of Crime are also included. Other necessary associated changes in the Rome Statute as a consequence of the inclusion of this new provision are set out in Appendix I.

Section 5.4 will then briefly outline the process for amending the Rome Statute to include these provisions, although this is not particularly controversial and will only be dealt with in relatively brief terms.

5.3.1. THE APPLICABLE APPROACH TO DEFINING CRIMES AGAINST THE ENVIRONMENT

An obvious starting point for this exercise is to establish what the proposed new provision would seek to 'accomplish'[57] and, equally important, what it would not. Moreover, it is important to bear in mind that, as noted, the Rome Statute is an instrument directed towards the prevention of international crimes and enforcement by way of international criminal responsibility. It is not an environmental protection treaty, as demonstrated by the fact that the scope of

---

56 See, for example, 'End Ecocide', (2013), which claims that '[o]ver the course of 2012 making Ecocide a crime has developed into a global movement' <http://eradicatingecocide.cmail5.com/t/ViewEmail/r/9F38DE90606F959E/9E223A7C45EC01860CC2E775D3CF5869> at 25 January 2015.

57 Peter J. Richards and Michael N. Schmitt, 'Mars Meets Mother Nature: Protecting the Environment During Armed Conflict' (1999) 28 *Stetson Law Review* 1047, 1089.

the penalties that the Court may impose are i) relatively limited;[58] ii) do not easily lend themselves to an order for restitution, remediation or the 'clean up' of any environmental harm;[59] and iii) are clearly not specifically directed towards the 'curing' of environmental damage.[60]

Given this understanding of what the Rome Statute is and is not, debates about what constitutes the most appropriate requirements for 'crimes' under environmental treaties (and regional and national law)[61] might not always be directly relevant. Rather, the approach taken in this book, as emphasized previously, is from an international criminal law perspective *vis-à-vis* those who intend to damage the environment during armed conflict.[62] That said, as noted, the inclusion of the crime of crimes against the environment into the Rome Statute will have the effect of better protecting the interests of the environment during armed conflict.

As will be seen, for the crime of crimes against the environment, it is not necessary that the *actual* harm, in the sense of what is *intended* by the perpetrator, is 'achieved'. This is also the case with respect to other crimes under the Rome Statute. For example, for a crime of genocide to have been perpetrated, it is (only) necessary that one of the listed acts be committed, that it is directed towards a specific group, and that it is committed with the requisite intention (to destroy).[63] There is no requirement as to the actual destruction of all or part of the group.[64] As the Trial Chamber of the ICTR stated in the landmark *Akayesu* judgment:[65]

---

58 Apart from imprisonment (Rome Statute, article 77(1)), the Court may order the payment of fines (article 77(2)(a)) or '[a] forfeiture of proceeds, property and assets derived directly or indirectly from [the] crime' (article 77(2)(b)).

59 Mark A. Drumbl, 'Waging War against the World: The Need to Move from War Crimes to Environmental Crimes' (1998) 22 *Fordham International Law Journal* 122, 150.

60 Although article 75 of the Rome Statute provides for the possibility of the Court making an order for the award of reparations to 'victims', it is clear from the terms of that provision that this is intended to apply to 'human' victims, and not where the natural environment is the victim.

61 For a comprehensive discussion of possible structures for criminalization under environmental instruments, see Michael G. Faure and Marjolein Visser, 'How to Punish Environmental Pollution? Some Reflections on Various Models of Criminalization of Environmental Harm' (1995) 3 *European Journal of Crime, Criminal Law and Criminal Justice* 316.

62 As discussed below, due to the particular circumstances of the crime, the *mens rea* element of crimes against the environment is not just limited to a direct *dolus directus* intention, but also includes *dolus eventualis* and wilful blindness.

63 See, for example, Judgement and Sentence, *Prosecutor v. Georges Anderson Nderubumwe Rutaganda*, Case No. ICTR-96-3, Trial Chamber I, 6 December 1999, para 48.

64 Cassese's *International Criminal Law* (3rd ed, 2013) (revised by A. Cassese, P. Gaeta, L. Baig, M. Fan, C. Gosnell and A. Whiting), 45. See also Kittichaisaree, *supra* note 24, 71. Note, however, the possible inconsistency on this point raised by Gerhard Werle when considering the Elements of Crime for genocide: Gerhard Werle, *Principles of International Criminal Law* (2005), 204–5.

65 Judgement, *Prosecutor v. Jean-Paul Akayesu*, Case No. ICTR-96-4-T, Trial Chamber, 2 September 1998, para 497. In a 1996 report on the Draft Code of Offences against the Peace and Security of Mankind, the International Law Commission stated that 'it is not necessary to

> [c]ontrary to popular belief, the crime of genocide does not imply the actual extermination of [a] group in its entirety, but is understood as such once any one of the acts mentioned in Article 2(2)(a) through 2(2)(e) [of the ICTR Statute] is committed with the specific intent to destroy "in whole or in part" a national, ethnical, racial or religious group.

This is logical and appropriate – logical in the sense that one would otherwise have to wait until the actual destruction of a group *before* charging an alleged perpetrator, and appropriate in that the deterrence effect of the crime would be significantly diluted by such a requirement. Indeed, it would lead to the unacceptable consequence that if, for example, a perpetrator had killed many people with the requisite intention, but it could not be shown that he had *actually* destroyed a (part of a) group, then he could not be charged with the crime of genocide.[66] It is also appropriate because it emphasizes that the heinous nature of the crime arises by virtue of the *intent* to achieve such a result.

Although there are significant differences with respect to the two crimes, the crime of crimes against the environment is also defined as a crime of intent,[67] accompanied by specific acts. Like genocide, therefore, it is not necessary that the actual intended harm – 'widespread, long-term or severe damage to the natural environment' (as defined in article 8 *ter*(2) – see 5.3.2.1) – occurs before the crime is committed. Of course, any actual harm may be relevant as a material fact for other purposes, and may assist in a determination as to whether the alleged perpetrator did, in fact, intend to cause the threshold level of environmental damage.

It follows, therefore, that the general structural approach taken in this book to defining crimes against the environment is not completely 'new', or 'radical', or contrary to existing accepted practice. It is not intended here to completely overturn the existing approach to criminalizing those acts that meet a level of gravity such that they constitute an international crime. Rather, it is submitted that the proposed *sui generis* crime of crimes against the environment represents a logical, effective and consistent evolution of the codification of acts that both

---

achieve the final result of the destruction of a group in order for the crime of genocide to have been committed. It is enough to have committed any one of the acts listed in the article with the clear intention of bringing about the total or partial destruction of the group as such': Code of Offences against the Peace and Security of Mankind, Report of the International Law Commission on Its Forty-eighth session, United Nations General Assembly Official Records 51st Sess., Supp. No. 10, 9 UN Doc A/51/10 (1996), 126.

66 As discussed below, there has never been a prosecution initiated before the international criminal courts and tribunals for attempted genocide.

67 As noted, the 'intent' requirement for crimes against the environment incorporates a different (and lower) *mens rea* standard than the *dolus specialis* (special intent) requirement for the crime of genocide. The point here is the fact that, for both crimes, it is not necessary that the actual intended harm occurs.

'deeply shock the conscience of humanity'[68] and 'threaten the peace, security and well-being of the world'.[69]

That said, there is one significant difference. The focus of the crime of crimes against the environment relates to a threshold level of intended damage *to the environment*, and thus does differ from the other core international crimes in that it does not necessarily require or involve an anthropocentric element – for example, a concrete endangerment to human life or health – or a human rights element. However, once again these will be amongst those factors that may be relevant, in the particular circumstances, to determine the degree of seriousness of any such damage that does actually eventuate.

For all the reasons noted above, therefore, it would not be appropriate that the crime were to apply to *all* deliberate environmental damage, and certainly not to *de minimis* damage. In addition, given the nature of the international criminal enforcement mechanism that is to be utilized (the ICC), the crime does not specifically deal with acts constituting a 'mere' violation of the many MEAs that exist although, of course, those will often also have significant environmental consequences. As noted in chapter 2, the applicable sanctions under those instruments will, however, often be in the form of an obligation to make compensation, rather than any criminal responsibility.[70]

Nor is the international crime of crimes against the environment necessarily intended to apply to a breach of domestic legislation that regulates the environment in various jurisdictions, even where violations of this type may be classified as ordinary (non-international) environmental crimes, and may attract criminal sanctions under the relevant domestic law.[71] Having said this, acts

68 Rome Statute, preamble para 2.

69 *Ibid*, para 3.

70 There are, of course, other international instruments that are also relevant to conduct that might constitute this type of crime. For example, the Hague Convention for the Protection of Cultural Property in the Event of Armed Conflict, opened for signature 14 May 1954, 249 UNTS 240 (entered into force 7 August 1956) imposes obligations on States to refrain from any act of hostility directed against cultural property and to abstain from using this property for military purposes. This instrument does not, however, provide for international criminal responsibility for acts constituting a breach, although article 28 requires States Parties to 'take all necessary steps' to prosecute those who breach the treaty. See also the Second Protocol to the Hague Convention of 1954 for the Protection of Cultural Property in the Event of Armed Conflict, opened for signature 26 March 1999, 38 ILM 769 (entered into force 9 March 2004), article 15. For more detail, see chapter 2.

71 For a discussion on the position with regard to, for example, environmental crimes under the national law of the United States, see Rachel Glickman, Rose Standifer, Lory Stone and Jeremiah Sullivan, 'Environmental Crimes' (2003) 40 *American Criminal Law Review* 413.

that do constitute crimes against the environment cannot be justified on the basis of any national law.[72]

Instead, the proposed definition in 5.3.2.1 specifies a minimum threshold level of intentional damage for the crime (potentially) to be applicable. This is also in keeping with the understanding of what constitutes an international crime, since the definition should be restricted only to actions of such a magnitude as to be regarded as an affront to humanity. It is also appropriate to include a minimum threshold of damage as a constitutive element of the crime for the simple reason that most acts conducted during armed conflict will have some adverse environmental consequences. It is only where such acts are done with intent to cause some (defined) significant degree of damage that the crime might apply. The inclusion of a minimum threshold is also necessary given that the specified *mens rea* standard has been defined in relatively expansive terms and is lower than the 'default' standard set out in article 30 of the Rome Statute.

Moreover, as noted, the crime does not relate to activities occurring in peace time, but rather is applicable only in circumstances where there is a nexus to armed conflict.

Taking account of these various considerations, the approach to defining the crime begins with an acceptance that its terms cannot be overly broad. Nor can they be vague, ambiguous and uncertain.[73] Reference has been made in earlier chapters to the international criminal law principle of legality, which establishes an important basis for the codification of any crime. This principle is itself often also associated with the *lex certa* principle, which requires that the preconditions for criminal responsibility be expressed clearly and in sufficient detail, so that a person is in a position to be able to determine whether or not certain acts would be give rise to sanction under criminal law.[74]

---

72 See also Vienna Convention on the Law of Treaties, opened for signature 23 May 1969, 1155 UNTS 331; 8 ILM 679 (entered into force 27 January 1980) (VCLT), article 27. This also extends to other non-treaty international obligations: see International Law Commission, Draft Articles on Responsibility of States for International Wrongfully Acts contained in Report of the International Law Commission to the General Assembly on its work of the fifty-third session, [2001] 2:2 *Yearbook of the International Law Commission* 1, UN Doc A/CN.4/SER.A/2001/Add.1 (Part 2), article 32 and the commentary thereto.

73 Cherif Bassiouni described as 'overtly broad and ambiguous' the suggested crime of 'wilful and severe damage to the environment' as it appeared in the 1991 version of the Draft Code of Crimes Against the Peace and Security of Mankind prepared by the International Law Commission: M. Cherif Bassiouni, *Crimes Against Humanity in International Criminal Law* (2nd ed, 1999), 190.

74 Faure and Visser, *supra* note 61, 322. Those authors also refer (at footnote 24) to the XVth International Congress of Penal Law, which resolved that:
'Consistent with the principle of legality, there should be certainty in the definition of crimes against the environment'.

What is therefore required in the drafting of any crime, and perhaps even more so for a crime that will be codified in the statute of an international criminal tribunal for the first time – is the requisite degree of 'prescriptive specificity'.[75] The elements of the crime should be expressed as clearly as possible so as to satisfy these requirements.

This does not mean, however, that the definition should necessarily be too precise. If, for example, the terms of the crime are strictly encased within an exhaustive list of acts, then it may well be that an act that exhibits the characteristics of what would properly constitute an international crime might fall outside the scope of international criminal responsibility. This would also limit any deterrence effect that would otherwise be generated by the introduction of the crime within the jurisdiction of the Court.[76]

Of course, no amount of careful drafting will capture every possible scenario. Hence, for example, we have seen the necessary evolution of various international crimes to include acts that, in earlier codifications, would not have been included (perhaps largely because they were not contemplated by the drafters in the first place).[77] This demonstrates that, as reflected in chapter 1, international criminal law is typically 'reactive' – *responding* to acts that shock the conscience of the world rather than pre-empting them, as much as the latter would be the ideal approach.

It is therefore suggested that the terms of the crime of crimes against the environment to be included in the Rome Statute, whilst demonstrating sufficient clarity and detail to satisfy the *lex certa* principle, should also allow for some 'flexibility' as to a determination as to whether a particular act(s) warrants individual criminal responsibility in the particular circumstances.

This is also important in view of the continuing development of weapons technology giving rise to possible environmental damage in ways that we might not as yet be able to properly contemplate or comprehend, but which may be relevant in the future. This flexibility will, as is evident in the terms of the definition set out below, be inherent in the determination of the level of the (potential) damage to the environment, and also from the broad conception of what constitutes the 'natural environment' and 'damage to the natural environment' for the purposes of the crime.

---

75 Richards and Schmitt, *supra* note 57, 1090.

76 See, generally, Jakob von Holderstein Holtermann, 'A "Slice of Cheese" – a Deterrence-Based Argument for the International Criminal Court' (2010) 11 *Human Rights Review* 289.

77 See, for example, the instances referred to in chapter 4 relating to the inclusion of acts of sexual violence in later codifications of the crime of war crimes.

5.3.2. DEFINITION OF CRIMES AGAINST THE ENVIRONMENT

Taking into account the various considerations discussed above, it is proposed that a new article 8 *ter* be included in the text of the Rome Statute. As noted, the inclusion of this provision will also require some associated amendments to the Rome Statute (specified in Appendix I) and to the Elements of Crimes (5.3.2.3).

However, the most significant change is the proposed inclusion of article 8 *ter* in the terms as set out below.

5.3.2.1. Proposed article 8 ter of the Rome Statute

'Article 8 *ter*
Crimes against the environment

1. For the purposes of this Statute, "crimes against the environment" means employing, within the context of and associated with an armed conflict, a method or means of warfare with intent to cause widespread, long-term or severe damage to the natural environment.
2. For the purposes of paragraph 1:
   (a) a person has "intent" to cause widespread, long-term or severe damage to the natural environment where:
      (i) that person means to cause such damage; or
      (ii) that person is aware of the substantial likelihood that such damage will occur in the ordinary course of events; or
      (iii) that person consciously disregards information that clearly indicates a substantial likelihood that such damage will occur in the ordinary course of events;
   (b) "widespread" means an area on the scale of at least one hundred square kilometres;
   (c) "long-term" means lasting at least for a period of one or more seasons;
   (d) "severe" means serious disruption, degradation or harm;
   (e) "natural environment" includes those ecological, biological and resource systems necessary to sustain the continued existence of all forms of human, animal, or plant life;
   (f) "damage to the natural environment" includes but is not limited to circumstances that constitute a concrete endangerment to human life or health, and may include any of the following:
      (i) destruction or degradation of the marine environment, marine wildlife, or marine habitats;
      (ii) destruction or degradation of terrestrial fauna and flora, or their habitats;

(iii) pollution of the atmosphere;
(iv) destructive climate modification;
(v) any other form of environmental destruction, degradation or harm of comparable gravity.'

5.3.2.1.1. Discussion

Having set out the definition of the crime, the following discussion outlines the rationale underpinning the various elements specified.

5.3.2.1.1.1. Paragraph 1 (Article 8 *ter*(1))

(i) An armed conflict

As discussed in previous chapters, a distinction at law has historically been made between an 'international' armed conflict and an 'armed conflict not of an international character'. The ICTY was called upon to consider and clarify this distinction in *Tadic*,[78] the first case before the ICTY. This distinction is also maintained in the Rome Statute.[79]

The fact that there are considered to be two types of armed conflict under international law has given rise to significantly differing regulatory regimes within the *jus in bello*. As discussed in chapter 2, when the 1949 Geneva Conventions[80] were 'upgraded' by the conclusion of the two 1977 Additional Protocols,[81] what was striking was the enormous disparity in the range of rules specified in each of those instruments. Whereas the 1977 Additional Protocol I,

---

78 See Opinion and Judgment, *Prosecutor v. Duško Tadić*, Case No. IT-94-1-T, Trial Chamber, 7 May 1997; Judgement, *Prosecutor v Duško Tadić*, Case No. IT-94-1, Appeals Chamber, 15 July 1999.

79 For example, compare the respective *chapeau* of articles 8(2)(b) and 8(2)(c) of the Rome Statute.

80 Geneva Convention for the Amelioration of the Condition of the Wounded and Sick in Armed Forces in the Field, opened for signature 12 August 1949, 75 UNTS 31 (entered into force 21 October 1950); Geneva Convention for the Amelioration of the Condition of the Wounded, Sick and Shipwrecked Members of Armed Forces at Sea, opened for signature 12 August 1949, 75 UNTS 85 (entered into force 21 October 1950); Geneva Convention Relative to the Treatment of Prisoners of War, opened for signature 12 August 1949, 75 UNTS 135 (entered into force 21 October 1950); Geneva Convention Relative to the Protection of Civilian Persons in Time of War, opened for signature 12 August 1949, 75 UNTS 287 (entered into force 21 October 1950).

81 Protocol I Additional to the Geneva Conventions of August 12, 1949, and relating to the Protection of Victims of International Armed Conflicts, opened for signature 8 June 1977, 1125 UNTS 3, 16 ILM 1391 (entered into force 7 December 1978) (1977 Additional Protocol I); Protocol II Additional to the Geneva Conventions of August 12, 1949, and relating to the Protection of Victims of Non-International Armed Conflicts, opened for signature 8 June 1977, 1125 UNTS 609, 16 ILM 1442 (entered into force 7 December 1978) (1977 Additional Protocol II).

which relates to international armed conflicts, consists of over 100 articles, some of them quite detailed and 'radical' (at least for the time), the 1977 Additional Protocol II, which was the first international instrument to deal solely with non-international armed conflicts, was far more modest (28 articles, of which only 18 were substantive in content). While, of course, the absolute number of provisions is not necessarily indicative of the content of a treaty, it is clear from a comparison of both documents that there are many issues that were simply not addressed in relation to internal armed conflicts.

This regulatory gap is also reflected in the existing war crime in the Rome Statute dealing with damage to the environment. As noted, the acts specified in article 8(2)(b)(iv) are prohibited only within the context of an international armed conflict.[82] This is unacceptable for a number of reasons, even more so given that armed conflicts are (either primarily or partially) increasingly 'internal' in character as, with some obvious exceptions,[83] the nature of warfare has largely moved away from the traditional 'State versus State' conflict. Notwithstanding the general reluctance among States to agree to more rigorous binding legal standards regulating what have traditionally been regarded as internal matters, Antonio Cassese has noted 'the current trend to abolish the distinction', with the ultimate result of having 'one corpus of law applicable to all conflicts'.[84]

While such sentiments may perhaps be a little optimistic, at least in the relatively short term, there is no convincing reason – apart from this traditional political State-centered conservatism – to maintain this distinction when it comes to serious crimes that 'deeply shock the conscience of humanity'.[85] By contrast, there *are* compelling reasons to avoid this distinction with respect to such crimes. It is already not a relevant consideration with respect to the crimes of genocide and crimes against humanity under the Rome Statute. Whilst, as indicated above, the distinction is maintained in respect of war crimes, this largely reflects the bifurcated approach adopted when the 1977 Additional Protocols were negotiated and agreed.

Disproportionate damage to civilians or civilian objects is, under customary international law, 'also a crime in non-international armed conflict'.[86] In

82 See Rome Statute, *chapeau* article 8(2)(b).

83 For example, the armed conflict between the Russian Federation and Georgia that took place in 2008, although even that conflict also involved the separatist 'governments' of South Ossetia and Abkhazia. See, generally, *Case Concerning Application of the International Convention on the Elimination of all Forms of Racial Discrimination (Georgia v. Russian Federation)* (Preliminary Objections, Judgment) [2011] ICJ Rep 70, para 23–114.

84 Cassese *et al*, *supra* note 64, 82.

85 Rome Statute, preamble para 2.

86 Gerhard Werle and Florian Jessberger, *Principles of International Criminal Law* (3rd ed, 2014), 494.

this respect, therefore, article 8(2)(b)(iv) 'lags behind customary international law'.[87] The distinction is also neither necessary nor appropriate in the case of crimes against the environment. As noted in previous chapters, acts that might fall within the scope of this crime are just as likely to occur within an internal armed conflict as during an international one.[88] A required nexus only to an international armed conflict therefore means that such intentional damage would likely escape criminal accountability.

Moreover, certain forms of serious environmental damage or destruction (for example, the poisoning of waterways), even if perpetrated within an internal conflict, may well significantly impact upon the environment of countries outside of those specific borders, since environmental damage (and its effects) will not necessarily be confined to the area where the relevant method or means of warfare is initially employed.

Hence, for all of these reasons, whilst article 8 *ter* does require a nexus to an armed conflict, it does not distinguish between the *types* of armed conflict. In other words, the crime is applicable in either an international armed conflict, or an armed conflict not of an international character, or both.[89]

(ii) Method or means of warfare

Article 8(2)(b)(iv) of the Rome Statute can only apply in circumstances where *inter alia*: (i) there is an international armed conflict; and (ii) an 'attack' has been intentionally 'launched' in the knowledge that such attack will give rise to certain consequences. As discussed above, the crime of crimes against the environment applies in both international armed conflicts and armed conflicts not of an international character. In addition, it stems not from an 'attack', but rather from employing a particular 'method or means of warfare'.

The concept of an attack is different from that of an armed conflict,[90] and has been interpreted by the international criminal courts and tribunals in terms of both war crimes and crimes against humanity. In *Kordić and Čerkez*, the ICTY Appeals Chamber applied the definition of attack as specified in the 1977 Additional Protocol 1 – 'acts of violence against the adversary, whether

---

87 *Id.*

88 See Tara Weinstein, 'Prosecuting Attacks that Destroy the Environment: Environmental Crimes or Humanitarian Atrocities?' (2005) 17 *Georgetown International Environmental Law Review* 697, 722.

89 See further 5.3.2.3.2.1.

90 See, for example, Judgement, *Prosecutor v. Dragoljub Kunarac, Radomir Kovac and Zoran Vukovic*, Case No. IT-96-23 and IT-96-23/1-A, Appeals Chamber, 12 June 2002, para 86.

in offence or in defence' – in determining whether an unlawful attack against civilians occurred.[91]

The same definition was used by the Pre-Trial Chamber of the ICC in *Katanga and Chui* in the context of the war crime set out in article 8(2)(b)(i) of the Rome Statute.[92] It is therefore likely that this definition would also apply to an attack within the terms of article 8(2)(b)(iv), including with respect to the issue of damage to the natural environment.[93] This emphasizes the fact that article 8(2)(b)(iv) is principally concerned with acts that involve violence directed against humans rather than, as is the case with crimes against the environment as defined in 5.3.2.1, focusing on the intentional damage to the environment *per se*.

Thus, even though an 'attack' has been interpreted more broadly for the purposes of assessing whether a 'widespread or systematic attack directed against any civilian population' has been perpetrated within the context of a crime against humanity,[94] it is still appropriate that the crime of crimes against the environment instead relates to the employment of a method or means of warfare. The International Committee of the Red Cross has indicated the expansive scope of those concepts as follows:[95]

> International law limits the methods and means used to wage war. These restrictions apply to the type of weapons used, the way they are used and the general conduct of all those engaged in the armed conflict.

---

91 See also the other ICTY decisions referenced at footnote 622 in Werle and Jessberger, *supra* note 86, 477.

92 Decision on the Confirmation of Charges, *Prosecutor v. Germain Katanga and Mathieu Ngudjolo Chui*, Case No. ICC-01/04–01/07, 30 September 2008, para 266.

93 In the absence of any indication to the contrary, the word 'attack' would be interpreted in the same way for all the variants of the offence under article 8(2)(b)(iv). In other words, it would not be interpreted in one way for the first two variants but somehow mean something different when it comes to the third variant under the same provision – that is, relating to damage to the natural environment.

94 In Judgement and Sentence, *Prosecutor v. Laurent Semanza*, Case No. ICTR-97-20-T, Trial Chamber III, 15 May 2003, the Trial Chamber of the ICTR stated (at para 327) that:
'[a]n "attack" is generally defined as an unlawful act, event, or series of events of the kind listed in Article 3(a) through (i) of the Statute. An "attack" does not necessarily require the use of armed force, it could also involve other forms of inhumane mistreatment of the civilian population.'
Similarly, in Judgement, *Prosecutor v. Milomir Stakić*, Case No. IT-97-24-T, Trial Chamber II, 31 July 2003, the Trial Chamber of the ICTY, quoting from Judgement, *Prosecutor v. Dragoljub Kunarac, Radomir Kovac and Zoran Vukovic*, Case No. IT-96-23 and IT-96-23/1-A, Appeals Chamber, 12 June 2002, para 86, noted (at para 623) that:
'[a]n attack can "precede, outlast, or continue during the armed conflict, but it need not be part of it", and "is not limited to the use of armed force; it encompasses any mistreatment of the civilian population.".'

95 International Committee of the Red Cross, 'Methods and Means of Warfare' <www.icrc.org/eng/war-and-law/conduct-hostilities/methods-means-warfare/index.jsp> at 28 January 2015.

It is intended that the scope of crimes against the environment applies to *any* way in which the environment is targeted as an act of warfare – either by way of a 'method' (through employing a particular action) or a 'means' (for example, through employing a particular type of weapon), or both. It is not restricted to acts directed against humans – its application arises when the act is directed against the environment. The reference to methods and means of warfare does, however, limit the scope of the crime to acts (and certain 'inactions' – see below) that constitute a part of the *conduct* of the armed conflict. In other words, it does not extend to actions that are not a part of that conduct – for example, where natural resources are exploited for financial gain, or to actions that might contribute to climate change and/or global warming – even though, as noted in chapter 1, these may ultimately fuel (future) armed conflicts.

That said, the crime does extend to the (perhaps rarer) circumstance of deliberate inaction or an 'omission to act' (see (iii) below). This would, for example, include a failure either to limit the effects of a particular act that was part of the conduct of an armed conflict, or to take steps to stop such damage from occurring and/or escalating, even where it was not initially intended. Such inaction would, most likely, not constitute an 'attack', and would thus fall outside of the scope of the existing legal regime.

As a result, the crime of crimes against the environment may apply even where there has not necessarily been an act of violence as such by the perpetrator, contrary to the requirement of '[i]ntentionally launching an attack' in article 8(2)(b)(iv) of the Rome Statute.

(iii) Intent (also incorporating article 8 *ter*(2)(a))

As has been argued, the existing legal regime for addressing the intentional destruction of the environment during armed conflict is inadequate and must be strengthened. The rationale for including a *sui generis* crime of crimes against the environment in the Rome Statute is therefore to codify an appropriate accountability mechanism to deal with this issue. The definition of this crime must exhibit the appropriate level of specificity, clearly setting out both the acts (and/or omissions) that form the basis of the crime (*actus reus*), as well as any mental element related to its commission (*mens rea*). As discussed above, the *actus reus* of the crime consists of the employment, within the context of and associated with an armed conflict, of a method or means of warfare.

Turning to the *mens rea*, as noted, the crime of crimes against the environment is a crime of intent. The Rome Statute envisages that the different crimes might have specific *mens rea* thresholds appropriate to each crime, and that these can (and do) vary as between the crimes. There is therefore no express limitation as to the intent requirements that may be specified for crimes against the environment; rather they should encompass those circumstances that indicate an unacceptable level of disregard for the environment in the conduct of armed conflict. Accordingly, in order to comprehensively address the 'intentional' destruction of the environment during warfare, it is necessary to include a range of *mens rea* standards within the definition of crimes against the environment. These are reflected in article 8 *ter*(2)(a) and differ from the mental element standard specified in article 30 of the Rome Statute.

Before discussing the specific mental elements that apply to the crime of crimes against the environment, it is necessary to raise two important points. First, it is acknowledged that the *mens rea* threshold requirements proposed for this crime are, in certain respects, lower than the standards required for a number of the other core international crimes, even some that may address acts committed against humans. At first glance, therefore, some might criticize this disparity on the basis that article 8 *ter* may be seen as providing a greater degree of protection to the environment than other provisions in the Rome Statute do for civilians and civilian objects.

It is submitted that this argument misses the point about what is sought to be achieved by the terms of article 8 *ter*. It has been emphasized throughout this book that the intention is to provide an effective and workable mechanism to impose international criminal responsibility for acts directed against the environment that are unacceptable and, according to all relevant criteria, that constitute an international crime. It would be, in this author's view, illogical and counter-productive to limit the scope of this crime below what is necessary to achieve this purpose for the sole reason that other crimes have lower standards.

On the other hand, this observation does highlight a very important point. There is no doubt that the elements of some of the other core crimes should be reconsidered in light of experience and changing values. Reference was made (albeit briefly) in chapter 4, for example, as to the need to 'upgrade' the terms of the crime of genocide. This may well also apply to other core crimes. The shortcomings of article 8(2)(b)(iv) that have been highlighted in chapter 4 will still apply to civilians and civilian objects after the amendments proposed in this book. Naturally, this point raises complex questions that are beyond the scope of this book – which is focused on the intentional destruction of the environment – but are deserving of very careful consideration.

However, to unacceptably compromise the terms of the crime of crimes against the environment due to the shortcomings of other crimes does not represent a proper and appropriate approach to the fundamental question as to how to effectively address intentional destruction of the environment during armed conflict.

Secondly, the terms of article 8 *ter* as defined in 5.3.2.1 may be interpreted as imposing 'absolute' liability for the use of certain means of warfare, including, for example, nuclear weapons. It has been noted in chapter 2 that there are already in place restrictions with respect to certain weapons; however, there is currently no express comprehensive prohibition against the use of nuclear weapons. Of course, one also should bear in mind the (majority) opinion of the International Court of Justice in the *Legality of the Threat or Use of Nuclear Weapons* Advisory Opinion.

Nonetheless, once again the terms of article 8 *ter* must be viewed for what they seek to achieve. If they do, indeed, have the effect of limiting or even prohibiting the use of such weapons, it is submitted that is a necessary consequence of the problem that the inclusion of this *sui generis* crime seeks to effectively address.[96]

Having dealt with these preliminary issues, the various intent requirements in article 8 *ter* are described below, starting with a discussion of the relevance and effect of article 30 of the Rome Statute.

a) Article 30 of the Rome Statute

As noted, within the framework of the Rome Statute, article 30 sets out a what has been termed a 'default rule'[97] standard regarding the mental element for crimes within the jurisdiction of the Court. This codifying of a general definition of the mental element for individual criminal responsibility was unprecedented in international criminal law. Notwithstanding that there has been doubt expressed as to whether and, if so, to what extent it reflects customary international law,[98] article 30 is the starting point in terms of the intent standard for crimes under the Rome Statute.

Whilst, at first sight, the inclusion of this provision may appear to 'harmonise' the *mens rea* requirements for international crimes, and thus 'put an end to the long lasting debate regarding the *mens rea* enigma' that has emerged through

---

96 In drafting the terms of article 8 *ter*, this author had considered the inclusion of an additional paragraph (article 8 *ter*(3)) to address this issue. This draft paragraph is included in Appendix II, together with a discussion of its terms. However, it is emphasized here that, after due consideration, this author has concluded that pargaraph 3 does <u>not</u> form part of the proposal for a *sui generis* crime of crimes against the environment to be included as article 8 *ter* in the Rome Statute.

97 Gerhard Werle and Florian Jessberger, "Unless Otherwise Provided': Article 30 of the ICC Statute and the Mental Elements of Crimes under International Criminal Law' (2005) 3 *Journal of International Criminal Justice* 35, 35.

98 Cassese *et al*, *supra* note 64, 39.

the jurisprudence of the various tribunals,[99] this has, in fact, not been the case. This standard applies subject to any express rules for a specific crime,[100] as is indicated by the first three words of the article – '[u]nless otherwise provided'. In other words, the default rule standard operates *only to the extent* that specific express rules relating to the requisite mental element are not applicable. These rules for a particular crime may be found in other provisions of the Rome Statute, in the Elements of Crimes or, it has been suggested, in customary international law.[101] It is important therefore to determine the scope of article 30, the terms of any express rules relating to crimes against the environment, and the extent to which the article 30 standard consequently applies.

Article 30(1) specifies that, for a person to be criminally responsible and liable for punishment for a crime within the jurisdiction of the Court, the material elements (*actus reus*) must be committed 'with intent and knowledge'. The provision then defines what is meant by each of those concepts.[102]

With regard to the relevant 'conduct' in relation to the crime of crimes against the environment – employing a method or means of warfare – article 30(2)(a) continues to apply. The crime can only apply where a person 'means to engage in [that] conduct'. This is reflected in the proposed Elements of Crimes set out in 5.3.2.3.2.

As regards the relevant 'consequence', the situation is different. As discussed, the crime of crimes against the environment as defined in article 8 *ter* is characterized by a person's *intent* to target the environment during armed conflict so as to cause a specified level of damage. The damage represents the consequence of the conduct. In the absence of article 8 *ter*(2)(a), the existence (or not) of the requisite intent in relation to that consequence would be determined by the terms of article 30(2)(b) of the Rome Statute – that is, that a person 'means to cause that consequence' or, at least, in committing the act 'must be aware that it *will occur* in the ordinary course of events'.[103]

Antonio Cassese has suggested that article 30(2)(b) does not therefore purport to encompass the full continuum[104] of possible *mens rea* categories, which would

---

99 Mohamed Elewa Badar, *The Concept of Mens Rea in International Criminal Law: The Case for a Unified Approach* (2013), 382.

100 Werle, *supra* note 64, 104.

101 See Werle and Jessberger, *supra* note 97. See also Werle, *ibid*, 106–9.

102 Rome Statute, articles 30(2) and (3) respectively.

103 *Ibid*, article 30(2)(b) (emphasis added).

104 Amalie Frese, 'The Least Bad Option: A Pluralistic Theory of Discrimination', unpublished thesis, University of Copenhagen, 29 June 2012, 66 (copy with author).

include at least 'intent, recklessness [and] negligence',[105] although there have been mixed signals in the jurisprudence of the ICC as to whether the provision might in fact encompass recklessness.[106] Either way, it is clear that crimes against the environment includes *mens rea* elements that *do* differ from the article 30 standard.

b) *Dolus Directus* (article 8 *ter*(2)(a)(i))

It is not disputed that the standard set by article 30(2)(b) does cover direct intent to cause the requisite level of damage to the environment (*dolus directus*). Of course, the existence of this *mens rea* standard represents a clear case in which the crime of crimes against the environment is intended to apply. Given that the terms of article 8 *ter* encompass a different (broader) *mens rea* standard than article 30 in other respects (see below), the terms of the first part of article 30(2)(b) ('that person means to cause …') are reproduced in and encompassed by article 8 *ter*(2)(a)(i), so as to avoid any doubt that this standard applies to the crime.[107]

This intent standard will also capture a situation where, for example, a person intends to cause the requisite level of environmental damage, but employs a method or means of warfare that (unbeknown to him) is incapable of reaching the prescribed threshold. This is to be compared with article 8(2)(b)(iv), which applies only when the alleged perpetrator has actual or constructive knowledge that actions *will* cause widespread, long-term and severe damage to the environment.

c) *Dolus Eventualis* (article 8 *ter*(2)(a)(ii))

In many cases, of course, it may be difficult to prove that there exists a direct intent to cause the prescribed level of environmental damage. However, there will be other circumstances that fall outside of the *dolus directus* standard of intent, yet still indicate a complete disregard by a person for any adverse environmental consequences arising from acts undertaken during armed conflict. For these to be excluded from culpability, as they are under the existing legal regime specified

105 Cassese *et al*, *supra* note 64, 39.
106 See the contrasting views of two different Pre-Trial Chambers, in Decision on the Confirmation of Charges, *Prosecutor v. Thomas Lubanga Dyilo,* Case No. ICC-01/04–01/06, Pre-Trial Chamber 1, 29 January 2007, and Decision Pursuant to Article 61(7)(a) and (b) of the Rome Statute on the Charges of the Prosecutor Against Jean-Pierre Bemba Gombo, *Prosecutor v. Jean-Pierre Bemba Gombo*, Case No. ICC-01/05–01/08, Pre-Trial Chamber II, 15 June 2009. See also Werle and Jessberger, *supra* note 86, 180–2.
107 It is not necessary to incorporate into article 8 *ter*(2)(a)(i) the second part of article 30(2)(b) ('is aware that it will occur in the ordinary course of events'), since that mental element is subsumed by and covered by the broader standard specified in article 8 *ter*(2)(a)(ii).

in article 8(2)(b)(iv), significantly dilutes the scope of any crime relating to the infliction of damage to the environment. As noted, it is therefore necessary to augment the mental element for the crime to additionally allow for a 'lower'[108] standard than a *dolus directus* level, so as to also encompass the following circumstances;

(i) where that person, being aware of the *substantial likelihood* of such damage, decides to proceed with employing a method or means of warfare notwithstanding the risk (article 8 *ter*(2)(a)(ii));
(ii) where that person engages in 'wilful blindness' so as to avoid being made aware of the substantial likelihood of such damage (article 8 *ter*(2)(a)(iii)).

As a result, the wording of article 8 *ter*(2)(a), whilst incorporating the direct intent standard also specified article 30(2)(b), differs in two important respects from the remainder of that provision. First, article 8 *ter*(2)(a)(ii) incorporates a situation where a person is aware of the substantial likelihood of such damage, rather than that it 'will occur'. In other words, although the crime focuses on the fact that it is the intent to target the environment that reflects the heinous nature of this act, at the same time, it also includes circumstances involving *dolus eventualis* (which has also been referred to by the ICTY as 'advertent recklessness),[109] which is not otherwise contemplated by article 30(2).[110]

To ensure, therefore, that this aspect of the mental element is not excluded,[111] it is necessary to expressly provide for its incorporation within the meaning of 'intent' in article 8 *ter.* Its inclusion as article 8 *ter*(2)(a)(ii) also renders moot any uncertainty as to whether this standard of mental element is already encompassed by article 30.

---

108 Werle, *supra* note 64, 109.

109 See, for example, Judgement, *Prosecutor v. Duško Tadić*, Case No. IT-94-1, Appeals Chamber, 15 July 1999, para 220; Judgement, *Prosecutor v. Milomir Stakić*, Case No. IT-97-24-A, Appeals Chamber, 22 March 2006, para 99. The two expressions – recklessness and *dolus eventualis* – have been treated as more or less equivalent concepts by some authors; see, for example, Cassese *et al*, *supra* note 64, 41. Other commentators do point to differences, albeit that these are largely relatively minor in practice; Werle, *supra* note 64, 113 and the references in footnote 131. With a slightly contrary view, however, Kai Ambos suggests that recklessness is 'not encompassed' by article 30 of the Rome Statute, but that the question as to whether it does incorporate *dolus eventualis* 'is more complex': see generally, Kai Ambos, *Treatise on International Criminal Law: Volume 1: Foundations and General Part* (2013), chapter VII.

110 See, for example, Judgment pursuant to Article 74 of the Statute, *Prosecutor v. Thomas Lubanga Dyilo*, Case No. ICC-01/04–01/06, Trial Chamber I, 14 March 2012, para 1011, where Trial Chamber I of the ICC has concluded that:
'[t]he plain language of the Statute, and most particularly the use of the words "will occur" in Article 30(2)(b) as opposed to "may occur", excludes the concept of *dolus eventualis*'.
See also *ibid* Werle, 113–4.

111 Antonio Cassese, *International Criminal Law* (2nd ed, 2008), 137.

Dolus eventualis is, of course, different from *dolus directus*. That said, it is not an unknown concept in relation to international crimes and other relevant treaty instruments, or within the jurisprudence of the various mechanisms of international criminal justice. Indeed, recklessness represents a sufficient requisite *mens rea* element in a variety of different situations. For example, specific prohibited acts under the *jus in bello* require that certain conduct is committed 'wilfully'.[112] In its commentary to these provisions, the International Committee of the Red Cross (ICRC), when discussing the word 'wilfully', states that:[113]

> the accused must have acted consciously and with intent, i.e., with his mind on the act and its consequences ... ("criminal intent" or "malice aforethought"); this encompasses the concepts of "wrongful intent" or "*recklessness*", viz., the attitude of an agent who, without being certain of a particular result, accepts the possibility of it happening

Similarly, the requirement of 'wilfulness' is found with respect to a number of war crimes within the Rome Statute (although *not* in article 8(2)(b)(iv)), either expressly,[114] or impliedly through the underlying provisions of the 1977 Additional Protocol I,[115] which are regarded as 'embody[ing] customary international law'.[116] As a consequence, for those crimes 'reckless conduct is usually sufficient'.[117]

The jurisprudence of the various international criminal courts have confirmed that recklessness can represent a sufficient mental element with respect to a number of specific crimes, even if it is not expressly referred to in the definition of the crime.[118] In *Kupreskic et al*, for example, the ICTY Trial Chamber, citing the ICTR Trial Chamber in *Akayesu*,[119] concluded that the *mens rea* of the crime of murder as a crime against humanity was either an 'intent to kill' or

112 See, for example, 1977 Additional Protocol I, articles 85(3) and 85(4).

113 International Committee of the Red Cross, 'Commentary on the Additional Protocols of 8 June 1977 to the Geneva Conventions of 12 August 1949', para 3474 <www.icrc.org/applic/ihl/ihl.nsf/Comment.xsp?viewComments=LookUpCOMART&documentId=78C0DA9A7B459ACEC12563CD0042F649&action=openDocument> at 31 January 2015 (emphasis added). See also Cassese *et al*, *supra* note 64, 40–1.

114 See, for example, Rome Statute, articles 8(2)(a)(i), 8(2)(a)(iii), 8(2)(a)(vi).

115 See, for example, *ibid*, article 8(2)(a)(ii).

116 Werle, *supra* note 64, 310.

117 *Ibid*, 110. That author also concludes that the use of the descriptor 'wanton' in relation to conduct proscribed in other war crimes also means that recklessness is sufficient in those cases; see, for example, Rome Statute, article 8(2)(a)(iv).

118 Note, however, that this view is not universally supported; see, for example, the conflicting approaches taken by different ICTY Trial Chambers in relation to grave breaches of Common Article 3 of the 1949 Geneva Conventions in the various decisions noted in Werle, *supra* note 64, 298, footnote 211.

119 Judgement, *Prosecutor v. Jean-Paul Akayesu*, Case No. ICTR-96-4-T, Trial Chamber, 2 September 1998, para 589.

an 'intent to inflict serious injury in *reckless* disregard of human life'.[120] In the same decision, the ICTY Trial Chamber also found that the burning of property with '*recklessness* towards the lives of its inhabitants' may constitute the crime of persecution as a crime against humanity, if committed on discriminatory grounds.[121] In addition, notions of recklessness have been incorporated into the *mens rea* elements of crimes committed within the context of the third category of a Joint Criminal Enterprise (JCE).[122]

In *Stakic*, the ICTY Trial Chamber, referring both to civil law (German) and common law (United States) domestic legal principles, found that 'that both a *dolus directus* and a *dolus eventualis* are sufficient to establish the crime of murder under Article 3' of the ICTY Statute ('Violations of the laws or customs of war').[123] The Chamber went on to provide the following 'technical definition' of *dolus eventualis* in the context of that crime:[124]

> if the actor engages in life-endangering behaviour, his killing becomes intentional if he "reconciles himself" or "makes peace" with the likelihood of death.

This assessment of the requisite *mens rea* element for life-endangering behaviour leading to death is also appropriate in relation to the decision by a person to

---

120 Judgement, *Prosecutor v. Zoran Kupreskic, Mirjan Kupreskic, Vlatko Kupreskic, Drago Josipovic, Dragan Papic and Vladimir Santic*, Case No. IT-95-16-T, Trial Chamber, 14 January 2000, para 561 (emphasis added). See also Judgement, *Prosecutor v. Zejnil Delalic, Zdravko Mucic, Hazim Delic and Esad Landzo*, Case No. IT-96-21-T, Trial Chamber, 16 November 1998, para 439.

121 *Ibid, Kupreskic et al*, para 631 (emphasis added).

122 See, for example, Judgement, *Prosecutor v. Duško Tadić*, Case No. IT-94-1, Appeals Chamber, 15 July 1999, para 228, where the ICTY Appeals Chamber held that, with respect to the third category of Joint Criminal Enterprise, one *mens rea* element with respect to crimes that were not part of a 'common criminal purpose' was that the accused 'willingly took [the] risk' that a crime would be committed. See also Judgement, *Prosecutor v. Mitar Vasiljević*, Case No. IT-98-32-A, Appeals Chamber, 25 February 2004, para 101; Judgement, *Prosecutor v. Miroslav Kvočka, Mlado Radić, Zoran Žigić and Dragoljub Prcac*, Case No. IT-98-30/1-A, Appeals Chamber, 28 February 2005, para 83.

123 Judgement, *Prosecutor v. Milomir Stakić*, Case No. IT-97-24-T, Trial Chamber II, 31 July 2003, para 587.

124 *Id*. This definition has been expressly approved or referred to in subsequent decisions; see, for example, Decision on Motions for Acquittal Pursuant to Rule 98 *bis* of the Rules of Procedure and Evidence, *Prosecutor v. Enver Hadzihasanoinc and Amir Kubura*, Case No. IT-01-47-T, Trial Chamber, 27 September 2004, para 37; and also by the ICC in Decision on the Confirmation of Charges, *Prosecutor v. Thomas Lubanga Dyilo*, Case No. ICC-01/04-01/06, Pre-Trial Chamber 1, 29 January 2007, para 352, footnote 434. In addition, the concept of *dolus eventualis* is also referred to in international criminal law jurisprudence in various other contexts; see, for example, Judgement, *Prosecutor v. Radoslav Brdanin*, Case No. IT-99-36-A, Appeals Chamber, 3 April 2007, para 365; Decision Pursuant to Article 61(7)(a) and (b) of the Rome Statute on the Charges of the Prosecutor Against Jean-Pierre Bemba Gombo, *Prosecutor v. Jean-Pierre Bemba Gombo*, Case No. ICC-01/05-01/08, Pre-Trial Chamber II, 15 June 2009, para 363.

employ a method or means of warfare in circumstances where that person is aware of the substantial likelihood that it will cause significant damage to the environment in the ordinary course of events, but nonetheless 'reconcil[es] himself... with it or consent[s] to it.'[125] Conduct of this nature demonstrates a clear disregard for the likely (environmental) consequences of one's acts, even though they may be very serious.

The wording in article 8 *ter*(2)(a)(ii) thus allows for criminal responsibility even apart from circumstances of 'near inevitability or virtual certainty' of the damage occurring, as is represented by the article 30 standard ('will occur'),[126] so as to also incorporate a 'substantial likelihood' of damage. Although it may represent a lower *mens rea* standard than a direct intent to damage the environment, it is equally necessary to include this *dolus eventualis* mental element within the crime of crimes against the environment in order to cover such circumstances. On the other hand, the terms of the crime do *not* extend to a mere possibility that the damage 'may' occur. This distinction follows jurisprudence of the ICTY. In *Strugar*, for example, the Trial Chamber, citing the Appeals Chamber in *Blaskic*,[127] noted that:[128]

> it has been confirmed by the Appeals Chamber that the awareness of a mere possibility that a crime will occur is not sufficient in the context of ordering under Article 7(1) of the Statute. The knowledge of a higher degree of risk is required.

The *Strugar* Trial Chamber, when considering the crime of murder, then went on to conclude that:[129]

> knowledge by the accused that his act or omission might *possibly* cause death is not sufficient to establish the necessary *mens rea*. The necessary mental state exists when the accused knows that it is *probable* that his act or omission will cause death.

Similarly, in *Delić*, the ICTY Trial Chamber, confirmed that the *mens rea* for murder *inter alia* includes 'indirect intent', which comprises 'knowledge that

125 *Id, Lubanga.*

126 Decision Pursuant to Article 61(7)(a) and (b) of the Rome Statute on the Charges of the Prosecutor Against Jean-Pierre Bemba Gombo, *Prosecutor v. Jean-Pierre Bemba Gombo*, Case No. ICC-01/05-01/08, Pre-Trial Chamber II, 15 June 2009, para 363.

127 See Judgement, *Prosecutor v. Tihomir Blaskic*, Case No. IT-95-14-A, Appeals Chamber, 29 July 2004, where (at para 42), the Appeals Chamber concluded that:
'a person who orders an act or omission with the awareness of the substantial likelihood that a crime will be committed in the execution of that order, has the requisite *mens rea* for establishing liability under Article 7(1) pursuant to ordering.'

128 Judgement, *Prosecutor v. Pavel Strugar*, Case No. IT-01-42-T, Trial Chamber II, 31 January 2005, para 235.

129 *Ibid*, para 236.

the death of a victim was a "probable" or "likely" consequence of such act or omission.'[130]

The mental element standard specified in article 8 *ter*(2)(a)(ii) is consistent with this line of reasoning and, in any event, accords with the underlying rationale of the crime, which focuses on those persons whose behaviour is indicative of a disregard of the environmental consequences of their actions. In this regard, the circumstances must demonstrate an awareness of the *substantial* likelihood of the damage occurring in the ordinary course of events. The substantial likelihood criterion allows for the input of relevant, independent and objective scientific data to assist in a determination as to whether international criminal responsibility for acts that may lead to damage to the environment should apply in a particular circumstance.[131]

At the same time, it provides an appropriate balance to the lower damage threshold in article 8 *ter* as compared to article 8(2)(b)(iv) – through the use of the disjunctive ('or') in article 8 *ter*(1) rather than conjunctive ('and') (see below). In this regard, this awareness standard is stricter than the prohibition of environmental damage that is found in the 1977 Additional Protocol I, which provides that:[132]

> [i]t is prohibited to employ methods or means of warfare which are intended, *or may be expected*, to cause widespread, long-term *and* severe damage to the natural environment.

Moreover, this departure from the 'default' standard specified in article 30 is entirely consistent with the *schema* for *mens rea* established under the Rome Statute, which clearly indicates that, for a number of crimes, the 'requirements of the mental element are lowered in comparison to the standard of Article 30'.[133] Important also is the fact that this *dolus eventualis mens rea* element represents 'one of the genuine and independent pillars of criminal responsibility'[134] that

130 Judgement, *Prosecutor v. Rasim Delić*, Case No. IT-04-83-T, Trial Chamber I, 15 September 2008, para 48.

131 Robert McLaughlin, 'Improving Compliance: Making Non-State International Actors Responsible for Environmental Crimes' (2000) 11 *Colorado Journal of International Environmental Law and Policy* 377, 394.

132 1977 Additional Protocol I, article 35(3) (emphasis added). The same threshold is also used in the 1977 Additional Protocol I, article 55(1). See also Jean-Marie Henckaerts and Louise Doswald-Beck (eds), *Customary International Humanitarian Law* (Volume I: Rules, 2005) (ICRC Study) Rule 45, which is discussed in chapter 3.

133 Werle and Jessberger, *supra* note 86, 187.

134 Mohamed Elewa Badar, 'The *Mens Rea* Enigma in the Jurisprudence of the International Criminal Court', in Larissa van den Herik and Carsten Stahn (eds), *The Diversification and Fragmentation of International Criminal Law* (2012), 503, 504.

might legitimately be utilised in appropriate circumstances as a basis to support the existence of an international crime.[135]

d) Wilful blindness (article 8 *ter*(2)(a)(iii))

The second way in which 8 *ter*(2)(a) differs from article 30(2)(b) is reflected in article 8 *ter*(2)(a)(iii). This specific notion of *mens rea* is designed to capture the 'stick your head in the sand approach',[136] where a person consciously disregards information as to the likely consequences of an action (or inaction, which would still constitute a 'method' of warfare). This is a similar approach (albeit in a different context) to that pertaining to non-military superior responsibility pursuant to article 28(b)(i) of the Rome Statute – which specifies that individual criminal responsibility may arise *inter alia* where 'the superior … consciously disregarded information which clearly indicated' that subordinates under his effective authority or control were committing or about to commit crimes within the jurisdiction of the Court.

It is also important to also include this mental element standard within the crime of crimes against the environment. In a sense, it is a necessary extension of the *dolus eventualis* standard that arises under article 8 *ter*(2)(a)(ii). Given that, as argued above, it is appropriate to proscribe behaviour in circumstances where a person was aware of the substantial likelihood of the environmental consequences of certain action then, in the absence of 8 *ter*(2)(a)(iii), criminal culpability could still be circumvented if a person deliberately chooses not to make himself aware of that substantial likelihood, by engaging in what has been described as 'wilful blindness'.

How this standard sits with respect to *dolus eventualis* is the subject of some academic debate. There are those who believe that it 'stands *between knowledge and recklessness*',[137] while others equate it with recklessness.[138]

Irrespective of its 'place' in the spectrum of *mens rea* standards, it already represents an aspect of the mental element of some international crimes.[139] One of the more frequent applications of this standard is with respect to command/

---

135 Werle, *supra* note 64, 114.

136 William G. Eckhardt, 'Command Criminal Responsibility: A Plea for a Workable Standard' (1982) 97 *Military Law Review* 1, 14.

137 Ambos, *supra* note 109, 227 (emphasis added).

138 See *id*, footnote 441.

139 See, for example, *R. v. Finta* [1994] 1 SCR 701, where the Supreme Court of Canada concluded that:
'[t]he mental element required to be proven to constitute a crime against humanity is that the accused was aware of or wilfully blind to facts or circumstances which would bring his or her acts within crimes against humanity'.

superior responsibility under international criminal law.[140] In the *Celebici* case, a Trial Chamber of the ICTY, when considering the *mens rea* standard under the ICTY Statute for superior responsibility that includes 'had reason to know',[141] confirmed:[142]

> the principle that a superior is not permitted to remain wilfully blind to the acts of his subordinates. There can be no doubt that a superior who simply ignores information within his actual possession compelling the conclusion that criminal offences are being committed, or are about to be committed, by his subordinates commits a most serious dereliction of duty for which he may be held criminally responsible under the doctrine of superior responsibility.

Wilful blindness, sometimes referred to as 'deliberate ignorance',[143] has in certain circumstances been held to result in 'equal culpability' to that arising from actual knowledge, even without its express inclusion in the elements of a crime.[144] However, some regard such a conclusion with a degree of introspection, as noted by William Schabas when he quotes Glanville Williams as follows:[145]

> The rule that wilful blindness is equivalent to knowledge is essential, and is found throughout the criminal law. It is, at the same time, an unstable rule, because judges are apt to forget its very limited scope. A court can properly find wilful blindness only where it can almost be said that the defendant actually knew. He suspected the fact; he realised its probability; but he refrained from obtaining the final confirmation because he wanted in the event to be able to deny knowledge. This, and this alone, is wilful blindness.

What is clear, therefore, is that to ensure that the concept of wilful blindness is included as one mental element in the crime of crimes against the environment, it is necessary to make express reference to this standard in article 8 *ter*(2)(a)(iii), since it otherwise falls outside of the article 30(2))b) default standard. Secondly,

---

These observations were expressly noted by the ICTY Trial Chamber in Opinion and Judgment, *Prosecutor v. Duško Tadić*, Case No. IT-94-1-T, Trial Chamber, 7 May 1997, para 657.

140 See, generally, Greg R. Vetter, 'Command Responsibility of Non-Military Superiors in the International Criminal Court (ICC)' (2000) 25 *Yale Law Journal* 89, 123–4.

141 ICTY Statute, article 7(3). This can be compared with the 'knew ... or should have known' standard for military command responsibility specified in Rome Statute, article 28(a)(i). Note the comments of an ICTY Trial Chamber on this distinction in Judgement, *Prosecutor v. Zejnil Delalic, Zdravko Mucic, Hazim Delic and Esad Landzo*, Case No. IT-96-21-T, Trial Chamber, 16 November 1998, para 393.

142 *Ibid, Delalic et al*, para 387.

143 Judgement on Appeal by Anto Nobilo against Finding of Contempt, *Prosecutor v. Aleksovski*, Case No. IT-95-14/1-AR77, Appeals Chamber, 30 May 2001, para 43.

144 *Id.*

145 William A. Schabas, '*Mens Rea* and the International Criminal Tribunal for the Former Yugoslavia' (2003) 37:4 *New England Law Review* 1015, 1029–30, footnote 61.

the extent to which such deliberate ignorance would suffice as a *mens rea* element for a crime must be clearly expressed. The wording used – 'consciously disregarded information that clearly indicated' – is, as noted, identical to that used in article 28(b)(i) of the Rome Statute in respect of non-military superior responsibility, and appears to take into consideration the caveat offered by Glanville Williams. It assumes that the person who suspects the substantial likelihood of a particular consequence makes a deliberate effort to ignore the information that would confirm that likelihood. Once again, the inclusion of this mental element does not conflict with the scope for an expanded regime for *mens rea* as contemplated by the opening words of article 30 of the Rome Statute.

e) Knowledge

Article 30(1) of the Rome Statute specifies that criminal responsibility and liability for punishment will only arise if the 'material elements [of the crime] are committed with intent and knowledge'. Knowledge is defined as 'awareness that a circumstance exists or a consequence will occur in the ordinary course of events'.[146] A requirement in similarly unequivocal terms constitutes an element of article 8(2)(b)(iv) ('in the knowledge that such attack *will* cause ...').[147] As referred to in chapter 4, in the view of the 2000 Committee Report examining NATO's actions during Operation Allied Force, this requirement of actual or constructive knowledge of the consequences of certain acts (in that case an 'attack') represented a *mens rea* standard that 'would be difficult to establish for the purposes of prosecution'.[148]

Antonio Cassese has noted that, in civil law countries, knowledge 'is not regarded as an autonomous category of *mens rea*, being absorbed either by intent or recklessness' but that, on the other hand, it is quite common in countries like the United States.[149] Under the terms of article 30(3) of the Rome Statute, the requirement of knowledge relates to the circumstances and the consequences.[150]

In relation to the circumstances, the definition of crimes against the environment requires that a method or means of warfare is employed 'within the context of and associated with an armed conflict'. It is therefore necessary that the alleged perpetrator has knowledge – is aware – of the factual circumstances that established the existence of an armed conflict. This requirement is reflected

---

146 Rome Statute, article 30(3).

147 *Ibid*, article 8(2)(b)(iv) (emphasis added).

148 Final Report to the Prosecutor of the International Criminal Tribunal for the former Yugoslavia by the Committee Established to Review the NATO Bombing Campaign Against the Federal Republic of Yugoslavia, 13 June 2000, 39 ILM 1257, para 23.

149 Cassese *et al*, *supra* note 64, 49.

150 See also Werle and Jessberger, *supra* note 86, 178–9.

in the Elements of Crimes set out in 5.3.2.3.2. In this sense, the standard set by article 30(3) remains applicable.

However, it follows from the departure from the article 30 standard in relation to the awareness of the *consequences* of employing a method or means of warfare, as discussed above in the context of the 'intent' requirement, that the second leg of the knowledge standard of article 30(3) does not apply. As noted, articles 8 *ter*(2)(a)(ii) and (iii) are not limited to situations where there is an awareness of a virtual certainty that a consequence will flow; rather it also includes a lower threshold so as to apply where there is an awareness of a substantial likelihood that a consequence will occur. The inclusion of this threshold within the definition of 'intent' for the purposes of article 8 *ter*(1) is not contradictory, but rather expansive, thus making the crime more (rather than less) effective in addressing the underlying imperatives that necessitate its inclusion as a *sui generis* crime in the Rome Statute.

(iv) Other issues arising from the definition of intent

The expansive definition of 'intent' in article 8 *ter*(2)(a) gives rise to a number of other issues that are briefly discussed below.

a) Non-applicability of a negligence standard

The mental element of crimes against the environment does not extend to 'advertent (culpable)' or 'inadvertent' negligence,[151] which are different from, and not encompassed by either *dolus directus*, *dolus eventualis*,[152] or wilful blindness. It is considered by some commentators that 'mere' negligence would not be sufficient for individual criminal liability to arise[153] and, by extension therefore, to justify an act as an international crime within the Rome Statute.[154] However, differing views have emerged on this point. Other commentators believe that gross or culpable negligence may operate as a standard of liability under international criminal law in certain circumstances.[155] The Trial Chamber of the ICTR has concluded that acts or omissions amounting to the crime of extermination as a crime against humanity 'may be done with intention, recklessness or *gross negligence*.'[156]

151 Cassese *et al*, *supra* note 64, 42.
152 Judgement, *Prosecutor v. Milomir Stakić*, Case No. IT-97-24-T, Trial Chamber II, 31 July 2003, para. 587. See also Werle, *supra* note 64, 113.
153 Cassese *et al*, *supra* note 64, 53.
154 See Gray, *supra* note 21, 265 and the corresponding footnote.
155 See, for example, Cassese *et al*, *supra* note 64, 53–5.
156 Judgement, *Prosecutor v. Clément Kayishema and Obed Ruzindana*, Case No. ICTR-95-1-T, Trial Chamber II, 21 May 1999, para 146 (emphasis added).

On the other hand, the ICTY Trial Chamber, when confirming that the *mens rea* for murder includes both direct and indirect intent, concluded that '[n]egligence and gross negligence do not form part of indirect intent'.[157] In addition, the ICTY Appeals Chamber, when considering the mental elements necessary to substantiate a finding of command responsibility and citing the ICTR Appeals Chamber in *Bagilishema*,[158] noted that:[159]

> the ICTR Appeals Chamber has on a previous occasion rejected criminal negligence as a basis of liability in the context of command responsibility, and that it stated that "it would be both unnecessary and unfair to hold an accused responsible under a head of responsibility which has not clearly been defined in international criminal law." It expressed that "[r]eferences to 'negligence' in the context of superior responsibility are likely to lead to confusion of thought ..." The Appeals Chamber expressly endorses this view.

Overall, in most cases at least, when the mechanisms of international criminal justice have applied a *mens rea* standard that is lower than the default standard set out in article 30 of the Rome Statute, they have not extended it to negligence or gross negligence,[160] although, as noted, there are some exceptions to this.[161] In any event, it is submitted that a negligence *mens rea* standard would in general not encapsulate the type of heinous behaviour that warrants sanction under the crime of crimes against the environment. Moreover, to attempt to apply a regime incorporating negligence is, in most cases, fraught with difficulty. It introduces elements of fault, which may be difficult to prove in the context of armed conflict, and requires a standard of care that may vary depending upon how military regulations might be interpreted.[162]

b) Command / Superior responsibility

As discussed, the wording of article 8 *ter* draws from a number of elements that are found within the notion of command/superior responsibility as elaborated under article 28 of the Rome Statute. It has already been noted that the threshold in article 8 *ter*(2)(a)(iii) mirrors the wording of article 28(b)(i) ('consciously disregard[ing] information which clearly indicated ...'). The knowledge element

---

157 See Judgement, *Prosecutor v. Rasim Delić*, Case No. IT-04-83-T, Trial Chamber I, 15 September 2008, para 48 and the various decisions referred to in footnote 88.

158 Judgement (Reasons), *Prosecutor v. Ignace Bagilishema*, Case No. ICTR-95-1A-A, Appeals Chamber, 3 July 2002, para 34–5.

159 Judgement, *Prosecutor v. Tihomir Blaskic*, Case No. IT-95-14-A, Appeals Chamber, 29 July 2004, para 63.

160 Werle, *supra* note 64, 116.

161 See Werle and Jessberger, *supra* note 86, 187. Those authors (at page 188) also provide examples of provisions within the Elements of Crimes which, in their view, 'may be found to expand liability under Article 30 ... into the realm of negligent liability'.

162 Jabbari-Gharabagh, *supra* note 25, 104.

in article 8 *ter* does, however, differ from that relating to a military commander ('knew or … should have known'), since it incorporates an element of substantial likelihood rather than certainty, unlike article 28(a)(i), for the reasons outlined above.

Nonetheless, as well as potentially giving rise to individual criminal responsibility pursuant to article 25 of the Rome Statute – for example, by ordering the employment of a method or means of warfare – article 8 *ter* still allows for the possibility that a commander/superior may be criminally responsible by virtue of article 28. This may arise in circumstances where a person under his effective command or control (in the case of a military commander), or a subordinate under his effective authority or control (in the case of a non-military superior), was committing or was about to commit a crime against the environment as defined in article 8 *ter*(1). In the end, the applicable mode(s) of liability will depend on the circumstances,[163] to be determined on a case-by-case basis. The point to note here is that the inclusion of article 8 *ter* into the Rome Statute does not constrict the application of well-accepted principles relating to the criminal responsibility of commanders/superiors in appropriate situations.

c) Attempt

The Rome Statute contains a general provision relating to 'attempt' that applies to crimes within the jurisdiction of the Court (article 25(3)(f)). In practical terms, the concept of attempt would rarely be relevant to crimes against the environment. As noted, the crime is constituted when a method or means of warfare is employed during armed conflict by a person with the requisite intent (as defined in article 8 *ter*(2)(a)). For the reasons discussed below, it is worth reiterating that crimes against the environment is a crime of intent rather than result.

As noted, this is also the case with the crime of genocide – although, of course, unlike crimes against the environment, the *mens rea* requirement for genocide

163 In this regard, however, note should be taken of the Appeals Chamber's finding in Judgement, *Prosecutor v. Tihomir Blaskic*, Case No. IT-95-14-A, Appeals Chamber, 29 July 2004, where (at para 91) the Chamber stated that:
'The Appeals Chamber considers that the provisions of Article 7(1) and Article 7(3) of the Statute connote distinct categories of criminal responsibility. However, the Appeals Chamber considers that, in relation to a particular count, it is not appropriate to convict under both Article 7(1) and Article 7(3) of the Statute. Where both Article 7(1) and Article 7(3) responsibility are alleged under the same count, and where the legal requirements pertaining to both of these heads of responsibility are met, a Trial Chamber should enter a conviction on the basis of Article 7(1) only, and consider the accused's superior position as an aggravating factor in sentencing'.
See also Judgement, *Prosecutor v. Zejnil Delalic, Zdravko Mucic, Hazim Delic and Esad Landzo 'Celebici Case')*, Case No. IT-96-21-A, Appeals Chamber, 20 February 2001, para 745.

involves a *dolus specialis*. There has never been a prosecution for 'attempted genocide' within the *ad hoc* tribunals, even though both the ICTY and ICTR Statutes do include 'attempt to commit genocide' as a 'punishable act'.[164] This raises some interesting dilemmas, with Jens David Ohlin of the view that 'attempts ought to be punished because perpetrators of attempts share the same culpability as perpetrators of completed offences', although this is not a view that is universally shared.[165]

In any event, with respect to crimes against the environment, there may be circumstances where significant preparatory steps are undertaken by a person *prior to* employing a method or means of warfare (with the requisite intent) but, for reasons independent of that person's intentions, the relevant conduct is then not carried out. In these circumstances, there may perhaps be an argument to suggest that the person has 'attempted' to commit the crime, although it is probably less likely that the Prosecutor would decide to move forward in such a case.

As to the general circumstances contemplated in the second sentence of article 25(3)(f), these are factual issues that may be relevant as to a determination of the 'gravity' of the crime. For example, if a person employs a method or means of warfare with the requisite intent, the crime has *prima facie* been committed. However, should that person then have a 'change of mind' and take steps to remediate the situation and mitigate the actual consequences of his conduct, that will be a matter that the Prosecutor[166] and/or Court[167] would likely take into account when determining the gravity of the crime/case, or the appropriate sentence upon conviction.[168]

d) Territorial 'reach' of article 8 *ter*

It is important to note the territorial 'reach' of article 8 *ter*. As with all other crimes set out in the Rome Statute, the jurisdiction of the Court in respect of

---

164 ICTY Statute, article 4(3)(d); ICTR Statute, article 2(3)(d). As noted in Judgement, *Prosecutor v Radislav Krstić*, Case No. IT-98-33-T, Trial Chamber, 2 August 2001 (para 640), article 4(3) of the ICTY Statute:
'is taken *verbatim* from Article III of the Genocide Convention. Article 4(3) provides for a broad range of heads of criminal responsibility, including heads which are not included in Article 7(1) [of the ICTY Statute] such as "conspiracy to commit genocide" and "attempt to commit genocide".'
See also Judgement, *Prosecutor v. Jean-Paul Akayesu*, Case No. ICTR-96-4-T, Trial Chamber, 2 September 1998, para 473.

165 Jens David Ohlin, 'Attempt, Conspiracy and Incitement to Commit Genocide' (2009) *Cornell Law Faculty Publications*, paper 24, 182 <http://scholarship.law.cornell.edu/cgi/viewcontent.cgi?article=1023&context=facpub> at 28 January 2015.

166 See Rome Statute, articles 53(1)(c) (regarding the initiation of an investigation) and 53(2)(c) (regarding factors that point to a sufficient basis for a prosecution).

167 *Ibid*, article 17(1)(d) (regarding admissibility of a case).

168 *Ibid*, article 78(1).

crimes against the environment will be determined in accordance with, and limited by, the 'Preconditions to the exercise of jurisdiction' set out in article 12. Obviously, if the accused is a national of a State Party to the Rome Statute, the Court may exercise its jurisdiction (article 12(2)(b)). It is also clear that acts by a national of a non-State Party may fall within the jurisdiction of the Court, where:

(i) the conduct in question – employing the method or means of warfare – occurs on the territory of a State Party (article 12(2)(a));
(ii) that non-State Party makes a so-called 'self-declaration' (article 12(3)) with respect to the crime;[169] or
(iii) a situation in which crimes against the environment appears to have been committed is referred to the Prosecutor of the Court by the United Nations Security Council acting under its Chapter VII powers (article 13(b) 'Exercise of jurisdiction').[170]

Thus, the specific scope of the crime does not purport to 'extend' the jurisdiction of the Court beyond the limits already imposed by the terms of the Rome Statute. That said, like other crimes within the jurisdiction of the Court, crimes against the environment as defined in article 8 *ter* can in certain circumstances operate 'extraterritorially', in the sense that acts undertaken outside the express territorial coverage of the Rome Statute may be intended to have serious environmental consequences in locations that are within the Court's territorial jurisdiction. For example, a commander or superior having the requisite *mens rea* may be located in a different place (country) from the specific area in which the relevant conduct occurs (assuming that he is responsible for employing a method or means of warfare there). The crime would *prima facie* still be applicable provided that the general jurisdictional limits of the Court allow for this.

This extraterritoriality aspect of the crime is not in itself unusual in terms of the crimes currently within the Rome Statute[171] – although more generally, this

---

169 See Steven Freeland, 'How Open Should the Door Be? Declarations by non-States Parties under Article 12(3) of the Rome Statute of the International Criminal Court' (2006) 75(2) *Nordic Journal of International Law* 211.

170 See United Nations Security Council Resolution 1593 (31 March 2005) UN Doc S/RES/1593/2005 on the Reports of the Secretary-General on the Sudan (referral of the situation in Darfur to the Prosecutor of the ICC); United Nations Security Council Resolution 1970 (26 February 2011) UN Doc S/RES/1970/2011 on Peace and Security in Africa (referral of the situation in Libya to the Prosecutor of the ICC).

171 For example, individual criminal responsibility for a crime within the jurisdiction of the Court extends to a person who 'orders, solicits or induces the commission of the crime', notwithstanding the fact that that person might not be located at the specific place (or even country) in which the crime takes place. See, generally, Rome Statute, article 25(3).

is an issue regarding which the Statute could ideally have been more specific from the outset. However, the particular point to be made is that the inclusion of crimes against the environment within the Rome Statute does not operate to unduly broaden the jurisdictional reach of the Court – apart, of course, from codifying an additional crime – given that the crime is 'committed' by the employment of a method or means of warfare with the requisite intent, not by the actual result, even though that may be elsewhere (see (v) below). In this respect, therefore, the terms of the crime accord with accepted principles already encapsulated in the Rome Statute.

That said, none of the *mens rea* elements of the crime are limited territorially. The *dolus directus / dolus eventualis* / wilful blindness requirements specified in article 8 *ter*(2)(a) are all with reference to damage 'to the natural environment'. Even where, for example, an (non-State Party national) accused 'only' intends to cause such damage within the territory of that non-State Party (for example by massively polluting a river), or to an area beyond the territorial jurisdiction of *any* State (for example the high seas), the crime might still be applicable in circumstances where the conduct occurs in the territory of a State Party.

(v) 'Or' not 'and'

As noted in chapter 4, article 8(2)(b)(iv) of the Rome Statute can only apply where there has been 'widespread, long-term *and* severe damage to the natural environment' (emphasis added). This conjunctive requirement, that all three qualifications of damage be proven, sets the minimum threshold of damage at an extremely high level. On the other hand, the disjunctive 'or' is used in the definition of crimes against the environment in article 8 *ter*. It is sufficient that the intended damage meet any one of these thresholds. Obviously this removes the need to prove all three elements of damage, thus lowering the evidentiary burden of proof that may be required. This is consistent with the goal of not setting the minimum damage threshold at a level that would be impossibly difficult to 'achieve' (and prove), thus allowing for the crime to be of practical relevance and to apply to behaviour with respect to significant environmental damage that would otherwise escape culpability under the current legal regime.

(vi) Other features of paragraph 1

a) No requirement of proof of actual intended harm

As noted in 5.3.1, the terms of article 8 *ter* have been drafted such that there is no requirement that the *actual* intended harm – in this case, 'widespread, long-term or severe damage to the natural environment' – must eventuate before the crime can be considered as having been committed. What is required is that a method or means of warfare is employed, and that this is done with the requisite intent regarding such damage as defined in article 8 *ter*(2)(a). The crime of crimes against the environment is therefore not just applicable on an *ex post facto* basis – that is, only after the significant environmental damage has occurred.[172] Article 8 *ter* has been drafted such that it may apply even if the intended level of harm is not actually ever reached.

There are a number of reasons why this is both logical and appropriate. It avoids the need to wait until the actual damage has been shown to have occurred – which may take a considerable period of time, given the definition of the relevant threshold levels, and may in any event be difficult to prove, despite the (necessarily general) guidance provided by the definition of each of these thresholds in article 8 *ter*(2)(b)-(d).

It also avoids unacceptable results, perhaps best illustrated by reference to the example of the deliberate igniting of the oil wells in Kuwait by the retreating Iraqi forces. As noted in chapter 1, at the time, these acts were widely condemned by Governments and civil society alike and were viewed with absolute shock and dismay. Yet, imagine if, for whatever reason, and despite the (possible) intent of the perpetrators, it could not be proven over the following weeks and months that such an act had actually met any of the threshold levels of damage, not for want of trying, but perhaps due to a stroke of good fortune – say, a change in wind direction or weather conditions.

In such a case, the crime would not be applicable at all if it were to be structured as a crime of 'result', in the sense that a constituent element was that the actual intended harm had occurred, although it might fall within the purview of 'attempt' (see above). It would also not sit comfortably with increasingly clear expectations regarding culpability for such acts. In the case of the Kuwaiti oil wells, for example, the heinousness of the acts, committed with the intent to cause a specific type of harm, were not regarded as any less so simply because, in the end and after a period of time, it was ultimately concluded by scientists that the environmental consequences were not quite as dire as had originally been feared.[173] There was undoubted relief following this conclusion, but no less condemnation of the acts that had been perpetrated.

To suggest otherwise would serve to defeat the underlying rationale of the crime – to proscribe actions done with an *intent* to target the environment during

---

172 Drumbl, *supra* note 59, 150–1.

173 See, generally, Jessica E. Seacor, 'Environmental Terrorism: Lessons from the Oil Fires of Kuwait' (1994) 10 *American University Journal of International Law and Policy* 481, and specifically the references listed at footnote 7.

armed conflict. It is this intent, coupled with a specified conduct (employing a method or means of warfare), that gives rise to international criminal culpability.

Moreover, given the numerous variables that will be relevant in determining the actual environmental consequences of certain acts, it is unworkable to include an additional requirement that widespread, long-term or severe damage must actually occur before the crime is applicable. That may leave matters lingering, potentially for a lengthy period, and would significantly dilute any deterrence value that the inclusion of the crime might otherwise have.

By structuring the crime as one of intent and incorporating an expansive mental element with regard to the consequences of a method or means of warfare, the terms of article 8 *ter* will more likely put military and political leaders and decision makers clearly on notice that the environmental consequences of acts taken during armed conflict *are* important considerations. Their acts, based on the situation *at the time of perpetration*, will be judged in this regard, rather than having to wait until a particular end result may (or may not) be shown to have actually occurred. This is the approach also typically taken with respect to the *jus in bello* principles as they are applied to the actions of combatants – to judge the legality (or otherwise) of their acts at the time, and in the circumstances and context at that time.[174]

In addition, even though it is not an environmental protection provision but rather is focussed on criminal law, it is anticipated that, on the basis proposed in this book, article 8 *ter* can function in a far more effective manner than the existing war crime (article 8(2)(b)(iv)) in protecting the environment by reducing the risk of significant environmental destruction during armed conflict *before it actually occurs*.

That said, the reality will probably be that, in most circumstances, a prosecution for crimes against the environment might not be initiated unless and until *some* environmental damage is caused, albeit not necessarily to the threshold level of 'widespread, long-term or severe'. This is because the existence of some adverse environmental consequences may make it easier to demonstrate the intent (as defined) of the alleged perpetrator (to cause the threshold level of environmental damage). In *Milosevic*, the Trial Chamber of the ICTY, when discussing a

174 See, for example, the *Rendulic* case (*Trial of William List and Others* (The Hostages Trial) (Judgment), United States Military Tribunal, Nuremberg, (1949) VIII *Law Reports of Trials of War Criminals*, Case No. 47, 69), referred to in chapter 3. See also the Elements of Crimes in relation to article 8(2)(b)(iv) of the Rome Statute, in particular footnote 37.

similar point with respect to the inference of the requisite intent in relation to an allegation of genocide, stated that:[175]

> [s]ince the acts in Article 4(2) of the Statute are only required to be committed with an intent to destroy the protected group, it is clear that the actual destruction of the group need not take place. However, the extent of the actual destruction, if it does take place, will more often than not be a factor from which the inference may be drawn that the underlying acts were committed with the specific intent to destroy, in whole or in part, a specific group as such

Similarly, in most circumstances, the greater the actual environmental damage, the more apparent it might be that the alleged perpetrator possesses the requisite intent. In this regard, any actual damage, rather than being a material *element* of the crime, is rather a material *fact* that will, no doubt, be useful (but not legally necessary) for the Court to determine whether this intent is present, since one might anticipate that, in most cases, intent would be inferred from the relevant factual circumstances.[176] In addition, should there be actual damage, this would also be a pertinent factor when determining the level of seriousness or gravity of the act, both for the purposes of applying article 8 *ter* and, as noted, for determining the appropriate punishment in the case of a conviction.

In the end, it is the role of the Prosecutor of the ICC, after having 'evaluated the information made available' to him/her, to decide whether an investigation – potentially leading to a prosecution – should be initiated, based on specific criteria.[177] The 'gravity of the crime' is a relevant consideration in this regard,[178] which might suggest that it is more likely that, with respect to crimes against the environment, an investigation would be initiated only after some (significant) environmental damage has occurred. The structure of article 8 *ter* does not prevent the Prosecutor from so waiting, but likewise it does not restrict him/her from acting beforehand.

Either way, the provision allows for a degree of flexibility, which is important given the scientific uncertainties that may be associated with assessing the environmental impact of certain acts. The important point here is that the actual intended harm – that is, proven 'widespread, long-term or severe damage to the natural environment' – is not a pre-requisite for the crime to be applicable,

---

175 Decision on Motion for Judgement of Acquittal, *Prosecutor v. Slobodan Milosevic*, Case No. IT-02-54-T, Trial Chamber, 16 June 2004, para 125. See also Susanne Malmstrom, 'Genocide Case Law at the ICTY', in Roberto Bellelli (ed), *International Criminal Justice: Law and Practice from the Rome Statute to its Review* (2010), 267, 279.

176 Cassese, *supra* note 111, 141.

177 See Rome Statute, article 53.

178 *Ibid*, article 53(1)(c).

as long as the *mens rea* and *actus reus* elements specified in article 8 *ter*(1) are satisfied.

b) No consideration of military necessity / military advantage

As discussed in previous chapters, it has been common, when determining whether criminal responsibility arises for certain acts undertaken within the context of an armed conflict, that they are explicitly 'measured' against the (anticipated) military advantage or benefit to be gained by such acts. Article 8(2)(b)(iv) of the Rome Statute applies this qualification in respect of any widespread, long-term and severe damage to the natural environment. Indeed, as noted, that provision goes further by requiring that such damage be '*clearly excessive* in relation to the concrete and direct overall military advantage anticipated' (emphasis added).[179]

This express reference to the military advantage/benefit criterion represents a major obstacle to providing for effective criminal accountability for acts targeting the environment during armed conflict. Given the inherently destructive nature of warfare, it would be relatively easy for an alleged perpetrator to claim that such acts bring with them a distinct military advantage particularly, as is contemplated by article 8(2)(b)(iv), in a broad ('overall') context.

As a consequence, article 8 *ter* has been drafted without reference to any military considerations that may be associated with the intentional targeting of the natural environment. This also stems from the underlying rationale of the crime – to address acts that show a complete disregard for the natural environment. On this issue, the International Committee of the Red Cross has concluded that:[180]

> a deliberate attack on the environment, not required by military necessity, would also amount to a war crime because it would in effect be an attack on a civilian object.

Indeed, article 8(2)(b)(ii) of the Rome Statute specifies that, within the context of an international armed conflict, a war crime may be constituted by:[181]

---

179 The words 'clearly' and 'overall' were included at the suggestion of the United States, which managed to convince delegates at the 1998 Rome Conference that this was necessary in order to further limit potential liability for 'collateral damage': Bartram S. Brown, 'U.S. Objections to the Statute of the International Criminal Court: A Brief Response' (1999) 31 *International Law and Politics* 855, 866–7.

180 See International Committee of the Red Cross, 'Customary IHL: Rule 156. Definition of War Crimes' <www.icrc.org/customary-ihl/eng/docs/v1_cha_chapter44_rule156> at 27 January 2015.

181 Rome Statute, article 8(2)(b)(ii).

> [i]ntentionally directing attacks against civilian objects, that is, objects which are not military objectives

However, this provision is not an appropriate 'complement' to article 8(2)(b)(iv) in relation to acts done with intent to cause environmental damage. It incorporates a number of the same weaknesses found in article 8(2)(b)(iv) – a restriction only to international armed conflicts and the requirement of an attack. It also makes no specific reference to the natural environment. Given the presence of a provision that specifically deals with damage to the environment (article 8(2)(b)(iv)), it is unlikely that, in practice, a prosecution would be initiated under the more general provision (article 8(2)(b)(ii)) for such acts. Indeed, even if such a prosecution was initiated, it would be highly likely that an alleged perpetrator would raise as a defence the 'cloak' of military advantage, thus potentially bringing the much more restrictive terms of article 8(2)(b)(iv) into play.

It should be noted, however, that, as with the other crimes specified in the Rome Statute, the general 'defences' ('Grounds for excluding criminal responsibility') set out in article 31 may apply with respect to crimes against the environment in the specific circumstances specified in that provision. However, the point to be made here is that the absence of a direct reference to military advantage/benefit within article 8 *ter* itself limits the possibility that this additional legislative 'justification' might be available to an alleged perpetrator.

c) No relationship with article 124 'Transitional Provision'

For the reasons discussed earlier, the criminalization of the intentional destruction of the environment during armed conflict has been drafted as a separate and *sui generis* crime, rather than as a war crime such as article 8(2)(b)(iv). One consequence of this is that the potential does not exist for a new State Party to the Rome Statute to 'opt out' of accepting the jurisdiction of the Court (for a period of up to seven years) with respect to crimes against the environment.[182] Whilst, as noted in chapter 4, an article 124 declaration has been made only on very few occasions, the fact that it is not possible at all for crimes against the environment removes one layer of (potential) uncertainty, and also serves to highlight the importance that must be attached to environmental concerns during the conduct of armed conflict.

---

182 See Rome Statute, article 124. As noted in chapter 4, article 124 was considered at the Review Conference of the Rome Statute held in Kampala from 31 May to 11 June 2010; however it was decided at that time not to delete the provision, but rather to review it again five years later.

5.3.2.1.1.2. Paragraph 2 (article 8 *ter*(2))

(i) Intent (see 5.3.2.1.1.1 (iii))

(ii) Widespread, long-term or severe

No clarification is provided in article 8(2)(b)(iv) as to what is envisaged by each of the terms, 'widespread', 'long-term' and 'severe' as they appear in that provision. In the absence of guidance, this may lead to a significant degree of uncertainty and, on top of all of the other problems with that provision (see chapter 4), makes it even harder to determine whether the requisite threshold of environmental damage has been reached, given the broad range of technical factors that might be relevant in relation to the state of the environment. This uncertainty will make it even less likely that a prosecution would ever be initiated under article 8(2)(b)(iv), since it would be not be at all clear to the Prosecutor what relevant level of environmental damage is at issue.

Nor is there any other assistance in this regard to be gained as to the meaning of 'widespread' and 'severe' as they appear within the definition of crimes against humanity in the Rome Statute[183] ('long-term' only appears in article 8(2)(b)(iv)). Even in the absence of a definition, their meaning in the context of one crime cannot be assumed to be the same as when used to set the parameters of another crime – they operate in completely different contexts. For example, the ICTY has held that, in the context of crimes against humanity, the word widespread ('a widespread or systematic attack')[184] refers to 'the large scale nature of the attack and the number of targeted persons',[185] an interpretation that, as discussed in chapter 4, is not necessarily consistent with the context of 'widespread ... [environmental] damage' in article 8(2)(b)(iv).

Therefore, in drafting the terms of the crime of crimes against the environment, and in order to provide a level of clarity as to the application and scope of this

183 See article 7(1) ('widespread or systematic attack...'), and articles 7(1)(e) ('[i]mprisonment or severe deprivation of liberty...), 7(2)(e) ('... the intentional infliction of severe pain or suffering...') and 7(2)(g) ('the intentional and severe deprivation of fundamental rights ...).

184 Although, in contrast to both article 3 of the ICTR Statute and article 7(1) of the Rome Statute, the ICTY Statute does not make specific reference to this requirement when setting out the elements of crimes against humanity for the purposes of that instrument (see ICTY Statute, article 5), the jurisprudence of the ICTY has consistently confirmed that such an attack is a constitutive element in the crime: see, for example, Judgement, *Prosecutor v Duško Tadić*, Case No. IT-94-1, Appeals Chamber, 15 July 1999, para 248; Judgement, *Prosecutor v. Dragoljub Kunarac, Radomir Kovac and Zoran Vukovic*, Case No. IT-96-23 and IT-96-23/1-A, Appeals Chamber, 12 June 2002, para 85.

185 Judgement, *Prosecutor v. Tihomir Blaskic*, Case No. IT-95-14-A, Appeals Chamber, 29 July 2004, para 101.

crime, particularly in light of the fact that the minimum threshold of damage, as discussed above, has been lowered by the use of 'or' rather than 'and', it is important to specify as clearly as is appropriate in the circumstances what is envisaged by each of the terms, 'widespread', 'long-term' and 'severe'.

Paragraphs 2(b)-(d) of article 8 *ter* therefore set out parameters for each of those qualifications. These definitions expressly apply only with respect to article 8 *ter*(1). They are broadly based on the definitions in the Understanding[186] relating to the ENMOD Convention discussed in chapter 2,[187] but have been varied in order to provide some greater specificity,[188] and also to ensure that, as noted above, the crime applies in relation to the intent to cause a specific (and significant) degree of environmental damage.

Paragraph 2(d) has been drafted in relatively general terms so as to, on the one hand, emphasize that the crime only applies in circumstances where the intended damage is 'serious'[189] (and thus not in the case of minor or inconsequential environmental damage), but which still allows the Court some scope for analysing and characterizing the intended harm,[190] and thus to apply article 8 *ter* in circumstances that may be warranted in a particular case. In other words, it allows for the crime to apply without a specific restriction apart from the degree of intended damage (compared with the temporal or geographic requirements as per paragraphs 2(b) and (c)).

This more general definition of 'severe' in article 8(2)(d*)* *ter* is appropriate for a number of reasons. First, it may well be that certain instances of significant environmental destruction 'cannot necessarily be measured in time or space'.[191] Moreover, the harmful effects of such intended damage may not necessarily be felt or manifest themselves for a considerable period of time after the relevant method or means of warfare has been employed (for example, the effects of booby traps or land mines).[192] In addition, it is impossible to foresee all of the circumstances in which the environment might be intentionally targeted during

186 Understanding Relating to Article I of ENMOD, 31 United Nations General Assembly Official Records Supp. No. 27 (A/31/27), Annex I.

187 Convention on the Prohibition of Military or Any Other Hostile Use of Environmental Modification Techniques, opened for signature 10 December 1976, 1108 UNTS 151; 16 ILM 88 (entered into force 5 October 1978).

188 Note that article 1 of ENMOD specifies the term 'long-lasting' rather than 'long-term'; however, the latter term has subsequently been used in both the 1977 Additional Protocol I and article 8(2)(b)(iv), and has been included in article 8 *ter.*

189 This term is also used in the Rome Statute as a descriptor of the relevant threshold of damage (to humans) for particular acts that may constitute an act of genocide (article 6(b)), a crime against humanity (article 7(1)(k)), and a war crime (articles 8(2)(a)(iii) and 8(2)(b)(vii)).

190 McLaughlin, *supra* note 131, 398.

191 Weinstein, *supra* note 88, 722.

192 Jabbari-Gharabagh, *supra* note 25, 100.

armed conflict. The guiding principal for the Court in considering whether paragraph 2(d) is applicable will therefore be the extent (gravity) of the intended environmental damage.

(iii) Natural environment

It is not necessarily a simple task to clearly define the environment, which in some senses is an 'intangible concept'.[193] As noted, there is no widely accepted definition that will fit into every circumstance. Rather, the definition in any particular treaty regime will be contextual, based very much on the scope and object of that particular instrument. Nonetheless, again to provide clarity, it is important to define, within the terms of article 8 *ter*, the intended 'victim' of the crime of crimes against the environment – the natural environment. In this regard, as foreshadowed in chapter 1, an expansive definition has been included in article 8 *ter*(2)(e), due largely to the fact that it is important not to be excessively restrictive, given the (unknown) destructive capabilities of current and future weapons technology.

In addition, the environment itself is a broad and complex notion and, only as we continue to gather further scientific knowledge about the interdependence of the various natural elements that occur around us, will we become more cognizant of the increasingly destructive effects that significant damage to one (perhaps even on the face of it seemingly minor) element will have on other aspects of 'ecological, biological and resource systems necessary to sustain the continued existence of all forms of human, animal or plant life'.[194]

Of course, it would be unrealistic to expect that combatants in an armed conflict would be aware of all of the nuances of environmental and ecological interdependency, and in any event, environmental damage will almost always follow *any* acts taken during armed conflict. Hence, the crime has not been drafted as a 'strict liability' crime but rather, as discussed, one based on an intent to target the environment so as to cause a certain threshold of damage.

(iv) Damage to the natural environment

The essence of the crime of crimes against the environment stems from the fact that it is the environment itself that is targeted. As noted, unlike, in particular, the crimes of genocide, crimes against humanity and war crimes, which are each

193 Major Walter G. Sharp, Sr., 'The Effective Deterrence of Environmental Damage During Armed Conflict: A Case Analysis of the Persian Gulf War' (1992) 137 *Military Law Review* 1, 32.

194 Jensen, *supra* note 10, 183.

primarily directed towards acts that have 'human' effects, crimes against the environment focusses upon methods and means of warfare that are employed to have particular consequences on the natural environment. Those three core international crimes under the Rome Statute have an anthropocentric focus, as opposed to an 'ecological' focus for crimes against the environment.[195]

The concept of crimes against the environment was given very little attention in the negotiations leading to the conclusion of the Rome Statute.[196] Whilst earlier (pre-Rome Statute) suggestions for the proscription of acts that damage the environment did propose a direct relationship with aspects of human health,[197] it is argued that such an approach is not appropriate, given the underlying basis for the inclusion of this crime as a *sui generis* crime within the Rome Statute. In effect, what differentiates crimes against the *environment* from the core international crimes in the Rome Statute is that it emphasizes the heinous nature of acts that intentionally target the environment *per se*.

As a consequence, article 8 *ter*(2)(f) makes it clear that the relevant 'damage' to the natural environment need not ('includes but is not limited to') necessarily constitute a 'concrete endangerment to human life or health'.[198] To limit the threshold element in that way may have the effect, for example, of excluding from the scope of the crime such things as the deliberate burning of the Kuwaiti oil wells by Iraqi forces, even though that action was met with universal condemnation as a 'crime' at the time. Of course, should the damage in a particular case also constitute such an endangerment, this may be a relevant factor that the Court might take into account in determining the gravity of the acts of an alleged perpetrator. However, an actual threat to human life or health is not a necessary precondition to the commission of the crime.

Although the Rome Statute itself does not have as its primary goal the greater protection of the environment, it is anticipated that article 8 *ter* may have that

---

195 Faure and Visser, *supra* note 61, 344.

196 Joe Sills, Jerome C. Glenn, Elizabeth Florescu and Theodore J. Gordon, 'Environmental Crimes in Military Actions and the International Criminal Court (ICC) – United Nations Perspectives', April 2001, Executive Summary, para 3 <http://millennium-project.org/millennium/es-icc.html> at 27 January 2015.

197 See, for example, Draft Code of Offences against the Peace and Security of Mankind, Report of the International Law Commission on Its Forty-eighth session, United Nations General Assembly Official Records 51st Sess., Supp. No. 10, 9 UN Doc A/51/10 (1996), which (at page 54) included as a war crime (article 20(g)) the following (emphasis added):
'[i]n the case of armed conflict, using methods or means of warfare not justified by military necessity with the intent to cause widespread, long-term and severe damage to the natural environment and *thereby gravely prejudice the health or survival of the population* and such damage occurs'.

198 This criterion is often used to define the scope of various models of criminal environmental sanction under national law: see, generally, Faure and Visser, *supra* note 61, 341–63.

consequence by criminalizing acts that target the environment during armed conflict. In order to best achieve this, it should not be limited only to damage that effects human beings or their activities, but rather should focus on the natural environment.[199]

So as to avoid suggestions that, without an anthropocentric 'accent', a definition of crimes against the environment would be too 'vague' and possibly violate the *lex certa* principle,[200] examples of the types of environmental damage that may fall within its scope are set out in articles 8 *ter*(2)(f)(i)-(v).[201] This is a similar approach to that taken with the definition of each of the other crimes in the Rome Statute. However, it is important that the process of listing examples of the relevant acts associated with the crime should not produce a definition that is restricted only to a small number of very specific types of environmental damage, or unduly limits the purposes for which they must be undertaken, with the effect that culpable actions that should be covered would still fall outside the scope of the crime. In this respect, whilst the legality principle requires clarity, it does not demand total inflexibility in the definition of a crime.[202]

---

199 See also Susan F. Mandiberg and Michael G. Faure, 'A Graduated Punishment Approach to Environmental Crimes: Beyond Vindication of Administrative Authority in the United States and Europe' (2008) *Lewis and Clark Law School Legal Research Paper Series*, Paper No. 2008-21, 4.

200 Faure and Visser, *supra* note 61, 358–63.

201 See Ensign Florencio J. Yuzon, 'Deliberate Environmental Modification Through the Use of Chemical and Biological Weapons: "Greening" the Environmental Laws of Armed Conflict to Establish an Environmentally Protective Regime' (1996) 11 *American University Journal of International Law and Policy* 793, 843 and the corresponding footnote.

202 See, for example, Rome Statute, article 7(1), which lists various acts that may constitute a crime against humanity in particular circumstances. Article 7(1)(k) refers to '[o]ther inhumane acts of a similar character ...' without providing any additional guidance or explanation as to the meaning of 'inhumane' in this context, thus allowing for the Court to exercise a degree of flexibility in its application: see, for example, Judgement, *Prosecutor v. Zoran Kupreskic, Mirjan Kupreskic, Vlatko Kupreskic, Drago Josipovic, Dragan Papic and Vladimir Santic*, Case No. IT-95-16-T, Trial Chamber, 14 January 2000, par. 562–66, and the quote at footnote 825. Reference can also be made to the Dutch case of Frans van Anraat, who was convicted on several counts in relation to the supply of chemicals to the Saddam Hussein regime in Iraq used to make mustard gas and nerve gas that was subsequently utilized in the Al-Anfal campaign referred to in chapter 1. In its decision dismissing the original conviction for complicity in genocide, the Dutch Court of Appeal concluded that:
'... especially regarding the question which degree of intention is required for a conviction on account of complicity in genocide, international criminal law is still in a stage of development and does not seem to have crystallized out completely': *Re* Frans van Anraat, LJN: BA4676, Court of Appeal in The Hague, 2200050906 – 2, 9 May 2007.
The case subsequently went to the European Court of Human Rights (ECtHR), which quoted from *Kononov v. Latvia* No. 36376/04, 17 May 2010, where that Court (at para 185–7) had stated that:
'[a]s regards foreseeability in particular ... however clearly drafted a legal provision may be in any system of law including criminal law, there is an inevitable element of judicial interpretation. There will always be a need for elucidation of doubtful points and for adaptation to changing circumstances.'

As a result, the list is not exhaustive ('may include') and, in addition, a more general example in paragraph 2(f)(v) has been incorporated, so as to give the Court some flexibility to apply article 8 *ter* in circumstances that, for example, may not necessarily be specifically contemplated at this present time. This follows the structure of other crimes in the Rome Statute[203] and, in the case of crimes against the environment, is particularly important, given the continuous development of destructive weapons technology.

5.3.2.2. *Associated Amendments to the Rome Statute*

As noted above, the inclusion of a separate crime of crimes against the environment in the Rome Statute will necessitate some associated amendments in that instrument. These are set out in Appendix I.

5.3.2.3. *Associated Amendments to the Elements of Crimes*

The Elements of Crimes constitutes part of the first 'layer' of the applicable law to be applied by the Court, along with the Rome Statute and the Rules of Procedure and Evidence.[204] It is used to assist the Court in the interpretation of the various crimes set out in articles 6, 7 and 8, and its terms (and any amendments to them) are to be consistent with the Rome Statute.[205] As a consequence, the inclusion in the Rome Statute of article 8 *ter* also necessitates associated amendments to the Elements of Crimes. These are set out below (see also Appendix I).

5.3.2.3.1. Amendment to the Explanatory Note

The first sentence of the explanatory note to the Elements of Crimes should be amended to read as follows (amendments in **bold**):

In *van Anraat*, the ECtHR went on to note that:

'[t]he scope of the notion of foreseeability depends to a considerable degree on the content of the text in issue, the field it is designed to cover and the number and status of those to whom it is addressed. A law may still satisfy the requirement of foreseeability even if the person concerned has to take appropriate legal advice to assess, to a degree that is reasonable in the circumstances, the consequences which a given action may entail [and that] it is a logical consequence of the principle that laws must be of general application that the wording of statutes is not always precise. One of the standard techniques of regulation by rules is to use general categorisations as opposed to exhaustive lists': *van Anraat v. The Netherlands* No. 65389/09, 6 July 2010, para 81 and 83.

203 See Rome Statute, article 7(1)(k), as well as article 7(1)(g), which encompasses (emphasis added):

'Rape, sexual slavery, enforced prostitution, forced pregnancy, enforced sterilization, *or any other form of sexual violence of comparable gravity*'.

204 *Ibid*, article 21(1)(a).

205 *Ibid*, articles 9(1) and 9(3).

'The structure of the elements of the crimes of genocide, crimes against humanity, war crimes **and crimes against the environment** follows the structure of the corresponding provisions of articles 6, 7, 8 **and 8 *ter*** of the Rome Statute.'

5.3.2.3.1.1. DISCUSSION

Whilst the explanatory note is not a formal part of the Elements of Crimes, it should be amended in order to maintain the consistency of the document with the Rome Statute. This was not done following the inclusion of article 8 *bis*, which clarified the terms of the crime of aggression, most likely because the provision in the Elements of Crimes relating to aggression does not systematically deal with every paragraph of article 8 *bis*. An amendment to the explanatory note in relation to article 8 *ter* is probably warranted – although not necessarily vital – if only to clarify the structure of the respective provision in the Elements of Crimes.

5.3.2.3.2. Proposed Elements of Crimes for Crimes against the Environment

'Article 8 *ter*
Crimes against the environment

Elements

1. The perpetrator meant to employ a method or means of warfare.
2. The method or means of warfare was employed;
   (a) with intent to cause widespread, long-term or severe damage to the natural environment; or
   (b) with an awareness that, in the ordinary course of events, there is a substantial likelihood that it will cause widespread, long-term or severe damage to the natural environment; or
   (c) after consciously disregarding information that clearly indicates that, in the ordinary course of events, there is a substantial likelihood that it will cause widespread, long-term or severe damage to the natural environment.
3. The method or means of warfare was employed in the context of and was associated with either an international armed conflict, or an armed conflict not of an international character, or both.
4. The perpetrator was aware of factual circumstances that established the existence of an armed conflict.'

5.3.2.3.2.1. Discussion

The proposed elements of crimes against the environment systematically set out the components of the crime that need to be proven. These are structured in the order specified in the General Introduction to the Elements of Crimes.[206] They confirm the requisite conduct (paragraph 1), the relevant mental element (paragraph 2), and the contextual circumstances (paragraphs 3 and 4), the latter of which follow from the limitation, discussed in earlier sections of this chapter, that the crime is only applicable when the conduct takes place within the course of an armed conflict.

As noted above, however, for the purposes of this crime, no distinction is to be drawn between an international armed conflict and an armed conflict not of an international character.[207] In other words, the crime applies to conduct in any type of armed conflict. The words 'or both' are included in paragraph 3 since, as noted, there may be conflicts that display the characteristics of both forms of armed conflict.[208]

The connection with an armed conflict specified in both article 8 *ter* and the provision in the Elements of Crimes set out above – 'in the context of and was associated with' – mirrors the wording used in the provisions of the Elements of Crimes relating to war crimes. This is also the case with the terms of paragraph 4 of the proposed Elements of Crimes set out above, which has been included to emphasize that this crime applies only to acts undertaken in an armed conflict. This also further highlights that the focus of the crime is directed towards criminalizing the targeting of the environment as part of the conduct of warfare.

---

206 See General Introduction to the Elements of Crimes, para 7.

207 Because the crime of crimes against the environment is limited to specific actions 'within the context of and associated with an armed conflict', the qualifications as to what constitutes an armed conflict not of an international character for the purposes of article 8(2)(f) of the Rome Statute in relation to certain war crimes will apply *mutatis mutandis* to article 8 *ter* and the respective Elements of Crimes set out in 5.3.2.3.2.

208 See, for example, Judgement, *Prosecutor v Duško Tadić*, Case No. IT-94-1, Appeals Chamber, 15 July 1999, where the ICTY Appeals Chamber (at para 84) stated:
'It is indisputable that an armed conflict is international if it takes place between two or more States. In addition, in case of an internal armed conflict breaking out on the territory of a State, it may become international (or, depending upon the circumstances, *be international in character alongside an internal armed conflict*) if (i) another State intervenes in that conflict through its troops, or alternatively if (ii) some of the participants in the internal armed conflict act on behalf of that other State' (emphasis added).
See also Decision on the Confirmation of Charges, *The Prosecutor v. Thomas Lubanga Dyilo*, Case No. ICC-01/04–01/06, Pre-Trial Chamber 1, 29 January 2007, where (at para 209) the ICC Pre-Trial Chamber used similar wording in relation to the situation in the Democratic Republic of the Congo.

5.4. COMPARISON WITH ARTICLE 8(2)(B)(IV) – 'RIGHTING THE WRONGS'

The discussion in chapter 4 analysed the various elements of article 8(2)(b)(iv) of the Rome Statute – the only 'eco-centric'[209] provision in that instrument expressly referring to the environment and to environmental damage – and why its terms are insufficient and inappropriate to deal with the intentional targeting of the environment during an armed conflict, thus rendering it largely 'impotent'.[210] It is suggested that the terms of article 8 *ter* properly address and rectify each of those concerns so as to provide the basis for a more effective mechanism to deal with such acts. The primary differences between article 8(2)(b)(iv) and article 8 *ter* have been referred to in detail in the discussion above, but are briefly summarized again in the following table:

Article 8(2)(b)(iv)	Article 8 *ter*
'intentionally launching an attack'	'employing ... a method or means of warfare'
'knowledge that such attack will cause ... damage'	'intent to cause ... damage', which incorporates *dolus directus*, *dolus eventualis*, or wilful blindness
'international armed conflict'	'an armed conflict' (of any type)
'widespread, long-term and severe damage'	'widespread, long-term or severe damage'
no definition/clarification of relevant terms	definition/clarification of 'widespread', 'long-term', 'severe', 'natural environment', 'damage to the natural environment'
'clearly excessive' damage	no such requirement
relative to 'concrete and direct overall military advantage anticipated'	no such requirement
subject to article 124 Rome Statute 'Transitional Provision'	not subject to article 124 Rome Statute

In the end, the variances between these two provisions are not just a matter of semantics, but are tangible and would make a real difference in terms of regulating conduct during armed conflict. There is a real possibility that notorious historical examples of intentional environmental destruction during an armed conflict, such as the igniting of the Kuwaiti oil wells, the draining of the al-Hawizeh and al-Hammar marshes in Iraq, or the extensive use of Agent Orange during the Vietnam War, might *not* have fallen foul of the terms of article 8(2)(b)(iv), had it been operable at the relevant time.

---

209 Ryan Gilman, 'Expanding Environmental Justice after War: The Need for Universal Jurisdiction over Environmental War Crimes' (2011) *Colorado Journal of International Environmental Law & Policy* 447, 450.

210 *Ibid*, 456.

By contrast, if such acts were to be taken following the inclusion of article 8 *ter* in the Rome Statute, an effective and appropriate mechanism would then be available to the Court so as to properly address such clear-cut examples of the intentional targeting of the natural environment. It is submitted that this would also 'contribute to the prevention of' such acts in the future.[211]

In the end, therefore, the two provisions – the existing article 8(2)(b)(iv) and the proposed article 8 *ter* – represent very different, perhaps even diametrically opposed, approaches towards addressing the need to establish appropriate mechanisms of justice to deal with persons who target the environment during armed conflict. The standards they set are completely different, as are most of the relevant factors that constitute the elements of the respective crimes, even though they deal with a similar subject matter. Obviously, therefore, it is neither practicable nor logical to, on the one hand, include the crime of crimes against the environment, whilst at the same time retaining the relevant parts of the existing (flawed) provision. To do so may significantly compromise any attempt to prosecute an accused under article 8 *ter*. As a consequence, it is necessary to also amend article 8(2)(b)(iv) so as to delete the reference to 'widespread, long-term and severe damage to the natural environment'. This is dealt with by the amendment to the Rome Statute set out in 5.3.2.2.2 (see Appendix I).

5.5. PROCEDURE FOR AMENDING THE ROME STATUTE AND THE ELEMENTS OF CRIMES

The Rome Statute sets out a relatively clear procedure that would allow for amendment of the instrument so as to include the separate crime of crimes against the environment, as well as for the necessary associated changes. There is therefore no procedural impediment that would prevent this from (eventually) becoming a reality.

Any proposal to amend the Rome Statute is to come from a State Party. Article 121 permits any State Party to propose an amendment to the Rome Statute '[a]fter the expiry of seven years from the entry into force of this Statute';[212] that is, at any time after 1 July 2009. There is no restriction as to the type of amendment that may be proposed, and thus the inclusion of a *sui generis* crime within the jurisdiction of the Court is possible, subject to compliance with the relevant procedures set out in the remainder of article 121.

A proposal to amend the Rome Statute would be dealt with either by the Assembly of States Parties (ASP) directly or, if the ASP considers that 'the issue

211 Refer to Rome Statute, preamble para 5.
212 *Ibid*, article 121(1).

involved so warrants', by a Review Conference convened for that purpose.[213] The issue of crimes against the environment was not raised at the first Review Conference in Kampala in May-June 2010, but the Rome Statute allows for the convening of a Review Conference(s) in the future, to consider matters including 'the list of crimes contained in article 5'.[214]

Article 121(3) of the Rome Statute prescribes the voting procedure for the adoption of an amendment to the treaty – either by consensus or, if that is not possible, by at least a two-thirds majority of States Parties. As noted, the inclusion of the crime of crimes against the environment would involve the substantial (and substantive) amendment of articles 5 and 8 of the Rome Statute. As a consequence, in the event that the relevant amendments were adopted, they would enter into force:[215]

> ... for those States Parties which have accepted the amendment one year after the deposit of their instruments of ratification or acceptance. In respect of a State Party which has not accepted the amendment, the Court shall not exercise its jurisdiction regarding a crime covered by the amendment when committed by that State Party's nationals or on its territory.

Obviously, this leaves open the possibility that a State Party will not accept the application of the crime of crimes against the environment to actions perpetrated by its nationals or on its territory. This may somewhat limit the immediate impact of the inclusion of the crime, but it is to be hoped that, over time, an increasing number of State Parties would accept the amendment, so that its application would become universal, at least in respect of those States that have ratified the Rome Statute.

The amendment of the Element of Crimes is even more straightforward. The Rome Statute provides that an amendment may be proposed by:[216]

(a) Any State Party;
(b) The judges acting by an absolute majority;
(c) The Prosecutor.

Once an amendment has been proposed by any of these, it requires a two-thirds majority of the members of the ASP for adoption[217] and, as noted, is to be consistent with the terms of the Rome Statute.[218]

213 *Ibid*, article 121(2).
214 *Ibid*, articles 123(1) and 123(2).
215 *Ibid*, article 121(5).
216 *Ibid*, article 9(2).
217 *Ibid*, article 9(2).
218 *Ibid*, article 9(3).

5.6. FINAL REFLECTIONS – A 'WORK IN PROGRESS'

Over the ages there have been many types of victims of armed conflict but, where the environment has been specifically targeted, those responsible have largely gone unpunished.[219] Whatever the reasons for this in the past, the analysis of the key questions raised in chapter 1 has demonstrated that the current position under both treaty and customary international law is not sufficient or appropriate to properly deal with such acts. Moreover, while it is clear that the mechanisms of justice under international criminal law, in particular the ICC, should play an integral function in this regard, the existing regime under the Rome Statute is also flawed given the virtual impossibility of a successful prosecution. The conclusion is, therefore, that it is necessary to amend the Rome Statute in order to more appropriately deal with 'crimes against the environment', and this should be done by including a separate crime within that instrument.

The inclusion of a *sui generis* crime of crimes against the environment in the Rome Statute, and thus within the jurisdiction of the ICC will, however, not be an easy task. In the end, what is required is a significant degree of political will and courage, strong assertiveness by civil society, and an understanding by relevant stakeholders that the threats posed by the ever-increasing destructive capability of modern weapons technology should override military considerations in circumstances where the natural environment is intentionally targeted during armed conflict.

Yet, a sense of realism is necessary if any effective progress is to be made. It is a sad truism that warfare is inherently destructive by nature. The conduct of virtually all armed conflict will render damage to the environment in some way. Thus, to argue for an all-embracing definition for this crime would, at this stage at least, fall on deaf ears, and would run the risk of being unrealistic. To categorically protect the environment in all circumstances during armed conflict (let alone in peace time as well) is simply not possible, and presents a 'particularly vexing dilemma'[220] – the only way to achieve that goal would be the idealistic but unattainable (at least for the foreseeable future) 'solution' of prohibiting all armed conflict. Given the realities, such an aim, no matter how laudable, is simply a bridge too far.

Rather, this book, whilst arguing for the need for such a crime, offers, as a proposal intended to allow for far more effective regulation of the conduct of armed conflict, a definition that applies to acts done with intent (in an expansive sense) to target the environment during armed conflict. Other, broader definitions of what constitutes a crime against the environment may perhaps follow in succeeding

219 Weinstein, *supra* note 88, 721.
220 Sharp, *supra* note 193, 65.

discussions, with this author mindful that the definition offered in this book is, for the reasons set out in this chapter, deliberately limited in certain ways.

In this author's view at least, it is far better at this stage to put forward a specific crime that, on the one hand, is consistent with the general practice and legal approach already largely accepted by States and as reflected in the wording and structure of the Rome Statute, and which may be politically 'acceptable' in the circumstances, but which still addresses in a meaningful and workable way, heinous acts directed against the natural environment. Of course, this represents a compromise – similar to the circumstances that shaped the final version of the Rome Statute – and will perhaps not please those advocates who argue for an even greater range of accountability mechanisms directed towards (deliberate) acts that damage the environment. It may well be, of course, that other mechanisms may eventually be developed through other fora to deal with additional circumstances of deliberate environmental damage. The proposal in this book, however, represents the most appropriate and extensive 'crime' that is consistent with the aims of the Court, as reflected in the scope of the Rome Statute.

Indeed, to argue for something that is even broader than the express terms proposed in this book would 'dilute' the important focus of the Rome Statute and the work of the Court, not to mention the legal difficulties and political and diplomatic barriers that would likely follow from such an attempt. In reality, these would stall any such proposal from the outset. It is not the intention of this author to argue for something that does not have a realistic chance of being taken up and adapted at some stage. To argue too broadly too early may even harm the chances of having any suggestion on this issue *ever* being accepted.

So, in summary therefore, the central proposal of this book represents a 'work in progress' towards even greater protection of the environment, albeit one that, it is submitted, goes a considerable way towards achieving a much greater effectiveness in the regulation of acts that are intended to cause significant environmental damage during armed conflict. This proposal is in line with the standards required by contemporary international society and by the development of legal doctrine. It certainly does not, and does not purport to offer an all-encompassing 'solution' to every challenge posed by the intentional destruction of the natural environment in any circumstance.

Perhaps put more bluntly, to do nothing now is unacceptable and reckless in the extreme, but to attempt to do 'everything' at once will cast such a proposal into a vortex of political oblivion, thus achieving nothing at all.

The position therefore is, and should remain fluid, in response to technological and military developments, and changes in the way humanity prioritizes its natural surroundings. The proposed definition of the crime of crimes against the environment has attempted to take account of this need, whilst also both adhering to the requirements of clarity and paying due respect to the *lex certa* principle. It reflects the significance of the environment to our very future. It is submitted that this is a step that should and *must* be taken, and can be done without any significant or overwhelming legal or moral impediments. The imperative demands that such a step is taken now.

This book has been written against a background where, undoubtedly, our individual and collective consciousness continues to undergo a process of 'greening'. We are seeing a strong momentum towards the evolution of an 'environmental culture',[221] which also extends to renewed calls for the prohibition of many more aspects of environmental destruction.[222] There has been a gradual but perceptible 'hardening' of global attitudes with respect to those who intentionally (or even otherwise) damage the natural environment.[223] We have come much more to recognize the importance of respecting the environment, and this becomes increasingly more significant even in the face of traditional sovereignty and military considerations.

The proposal put forward in this book represents a logical step forward along this evolutionary path.

There is therefore some sense of optimism ahead, despite the legal and political challenges that are still to be faced. In the words of Ellen Goodman, writing almost a quarter of a century ago, shortly after one of the more notorious 'crimes against the environment' seen in modern times – the igniting of the oil wells in Kuwait by Saddam Hussein's military forces – it is to be hoped that, eventually:[224]

> … even in the devastating business of killing each other, there may be enough sanity for a consensus, enough sense of preservation to at least have a rule on the book and in our minds against waging war on the very environment we must share when the war is over.

---

221 Anthony Leibler, 'Deliberate Wartime Environmental Damage: New Challenges for International Law' (1992) 23 *California Western International Law Journal* 67, 136.

222 See, for example, 'Ecocide: The Missing 5th Crime Against Peace – End Ecocide in Europe' (undated) <http://eradicatingecocide.cmail1.com/t/ViewEmail/r/7EDF8501F7FB82C7/9E223A7C45EC01860CC2E75D3CF5869> at 6 February 2013.

223 Gray, *supra* note 21, 268.

224 Ellen Goodman, 'A crime against humanity and nature', *The Chicago Tribune*, 3 February 1991 <http://articles.chicagotribune.com/1991-02-03/features/9101100755_1_civil-war-sea-turtles-enemy territory> at 27 January 2015.

APPENDIX I

5.3.2.2. *Associated Amendments to the Rome Statute*

The inclusion in the Rome Statute of article 8 *ter* will require the following associated amendments to the Rome Statute. These amendments are largely self-explanatory and, apart from the proposed new article 15 *quater*, which deals with the exercise of jurisdiction by the Court over crimes against the environment, do not warrant additional discussion.

5.3.2.2.1. Amendment to article 5(1)

Article 5(1) of the Rome Statute should be amended to read as follows (amendment in **bold**):

'The jurisdiction of the Court shall be limited to the most serious crimes of concern to the international community as a whole. The Court has jurisdiction in accordance with this Statute with respect to the following crimes:

(a) The crime of genocide;
(b) Crimes against humanity;
(c) War crimes;
(d) The crime of aggression;
(e) **Crimes against the environment.**'

5.3.2.2.2. Amendment to article 8(2)(b)(iv) and the Associated Elements of Crimes

For the reasons discussed in chapters 4 and 5, following the inclusion of the separate crime of crimes against the environment as article 8 *ter*, article 8(2)(b)(iv) of the Rome Statute, as well as the associated Elements of Crimes, should be amended so as to delete any reference to damage to the natural environment. The amended article 8(2)(b)(iv) would therefore read:[1]

> '(iv) Intentionally launching an attack in the knowledge that such attack will cause incidental loss of life or injury to civilians or damage to civilian objects which would

[1] Of course, as noted, the inherent flaws in the provision will still apply in this amended version of article 8(2)(b)(iv) with respect to the other variants of that crime. It is hoped that others will continue to argue for further reconsideration of this provision in the future; however, this issue is beyond the scope of this book.

> be clearly excessive in relation to the concrete and direct overall military advantage anticipated;'

The associated Elements of Crimes for article 8(2)(b)(iv) would also be amended accordingly.

5.3.2.2.3. Amendment to article 9(1)

Article 9(1) of the Rome Statute should be amended to read as follows (amendment in **bold**):

> 'Elements of Crimes shall assist the Court in the interpretation and application of articles 6, 7, 8, 8 *bis* **and 8 *ter***. They shall be adopted by a two-thirds majority of the members of the Assembly of States Parties.'

5.3.2.2.4. Inclusion of article 15 *quater*

A new article 15 *quater* should be included in the Rome Statute as follows:

'Article 15 *quater*
Exercise of jurisdiction over crimes against the environment

1. Subject to paragraph 2, the Court may exercise jurisdiction over crimes against the environment in accordance with article 13.
2. The Court may exercise jurisdiction only with respect to crimes against the environment committed not earlier than one year after the ratification or acceptance of the amendments to articles 5, 8 and 9 that relate to that crime by [thirty] States Parties.
3. This article is without prejudice to the provisions relating to the exercise of jurisdiction with respect to other crimes referred to in article 5.'

5.3.2.2.4.1. Discussion

This new article proposes that the Court can exercise its jurisdiction with respect to crimes against the environment in the same ways that it can for the crimes of genocide, crimes against humanity and war crimes. These are differentiated from the unique jurisdictional limitations applicable to the crime of aggression, as agreed at the First Review Conference in Kampala in May-June 2010, which are specified in articles 15 *bis* and 15 *ter* of the Rome Statute and which stem largely from the 'political' and State-centered nature of that particular crime. There is no reason to apply those restrictions to crimes against the environment; rather the exercise of jurisdiction should be consistent with the core international crimes.

The only restriction that has been specifically included in article 15 *quater* is set out in paragraph 2. This specifies a minimum number of ratifications / acceptances by States Parties to the various relevant amendments to the Rome Statute before the Court can exercise jurisdiction with respect to crimes against the environment. This follows the 'formula' agreed by States Parties after they concluded an amendment to the Rome Statute that clarified the terms of the crime of aggression,[2] although the words 'not earlier than' have been included to make the provision even clearer.

Whilst this author would prefer that the jurisdiction of the Court over crimes against the environment was activated immediately following the adoption of all of the amendments to the Rome Statute proposed in this book, in accordance with already accepted practice it may be more appropriate to follow this minimum acceptance formula as a pre-condition. The minimum number (30) specified in square brackets is the same as that agreed with respect to the crime of aggression.[3] However, this need not necessarily be the case with respect to crimes against the environment, and the actual number specified will ultimately be the subject of further discussion and agreement.

Whatever the finally agreed number of required ratifications / acceptances, the inclusion of article 15 *quater*(2) will allow a State Party to consider its own position in relation to that of other States Parties, and also provide a period of transition for all relevant stakeholders before the Court is seized with jurisdiction in respect of the crime.

It should also be noted that additional limitations as to the applicability of articles 5(e) and 8 *ter* to a specific State – either a State Party or non-State Party – will arise by virtue of the terms of article 121(5) of the Rome Statute, which are referred to in 5.4.

5.3.2.2.5. Amendment to article 20(3)

The *chapeau* of article 20(3) should be amended to read as follows (amendment in **bold**):

'No person who has been tried by another court for conduct also proscribed under article 6, 7, 8, 8 *bis* **or 8 *ter*** shall be tried by the Court with respect to the same conduct unless the proceedings in the other court'.

---

2 See Rome Statute, article 15 *bis*(2). Note, however, the additional temporal limitation with respect to the crime of aggression specified in article 15 *bis*(3). A similar limitation is considered neither necessary nor appropriate in relation to crimes against the environment, and has therefore not been included in the amendments that are proposed in this book.

3 *Ibid*, article 15 *bis*(2).

APPENDIX II

5.3.3. PARAGRAPH 3 – NOT INCLUDED IN FINAL VERSION OF ARTICLE 8 *TER*

As noted in the discussion relating to the mental element applicable under article 8 *ter*, this provision might be interpreted in certain circumstances as applying 'absolute' criminal liability with respect to the use of nuclear weapons during the course of armed conflict. This author has considered this issue carefully, also taking into account the decision of the International Court of Justice in the *Legality of the Threat or Use of Nuclear Weapons* Advisory Opinion.[4] This author had initially contemplated the inclusion of an additional paragraph (3) in article 8 *ter* as set out below, but ultimately decided that this was not appropriate.

The terms of this rejected paragraph are set out below together with a discussion as to its effect. However, it is stressed that this paragraph is not included in the final version of article 8 *ter* proposed in this book, and is only included here to indicate that this issue was not ignored in the drafting of the final version of article 8 *ter*.

Paragraph 3 – not included in final version of article 8 *ter*

> **'(3) A person shall not be criminally responsible under this article where that person has deliberately employed a method or means of warfare as a result of an extreme circumstance of lawful self-defence on the part of the State of nationality of that person, in which the very survival of that State is at stake.'**

5.3.3.1. Discussion

As noted, in 1996, the International Court of Justice rendered a landmark advisory opinion, in response to a request transmitted to it under a resolution of the United Nations General Assembly,[5] which had asked the Court, pursuant to article 96(1) of the Charter of the United Nations, to 'urgently' render its advisory opinion on the following question:[6]

---

4 *Legality of the Threat or Use of Nuclear Weapons* (Advisory Opinion) [1996] ICJ Rep 226.

5 See United Nations General Assembly Resolution 49/75K (15 December 1994) UN Doc A/RES/49/75 on Request for an Advisory Opinion from the International Court of Justice on the Legality of the Threat or Use of Nuclear Weapons.

6 *Ibid*, preamble para 11.

> Is the threat or use of nuclear weapons in any circumstance permitted under international law?

Having concluded *inter alia* (by 11 votes to three) that there did not exist, either under treaty law or customary international law, 'any comprehensive and universal prohibition of the threat or use of nuclear weapons as such', the Court went on to hold (by seven votes to seven, with the President's casting vote determining the 'majority') that:[7]

> [i]t follows from the above-mentioned requirements that the threat or use of nuclear weapons would generally be contrary to the rules of international law applicable in armed conflict, and in particular the principles and rules of humanitarian law;
> However, in view of the current state of international law, and of the elements of fact at its disposal, *the Court cannot conclude definitively whether the threat or use of nuclear weapons would be lawful or unlawful in an extreme circumstance of self-defence, in which the very survival of a State would be at stake*

As noted, there were seven judges who disagreed with this finding, as reflected by a range of separate and dissenting opinions by individual judges as part of the Court record. The conclusions reached by the Court on this point have been the subject of considerable comment and criticism.[8] Nevertheless, in the years since that Advisory Opinion was rendered, no such 'comprehensive and universal prohibition' has since been agreed, or codified in a legally binding multilateral instrument.

In these circumstances, and noting the very significant environmental consequences that would arise from the use of a nuclear weapon, this author considered whether it would be appropriate to include in article 8 *ter* a caveat that, in effect, would render a person not criminally accountable for such damage if it was caused in the same extreme circumstances enunciated in the majority judgment in the *Legality of the Threat or Use of Nuclear Weapons* Advisory Opinion.[9]

Given that the ICC 'shall only have jurisdiction over natural persons',[10] paragraph 3 would need to be reworded so as to focus on the situation of the 'State of nationality

---

7 [1996] ICJ Rep 226, at para 105 (emphasis added).

8 See, for example, Timothy J. Heverin, 'Case Comment: Legality of the Threat or Use of Nuclear Weapons: Environmental and Humanitarian Limits on Self-Defense' (1997) 72 *Notre Dame Law Review* 1277; Michael J. Matheson, 'The Opinion of the International Court of Justice on the Threat or Use of Nuclear Weapons' (1997) 91 *American Journal of International Law* 417; Christyne J. Vachon, 'Sovereignty Versus Globalization: The International Court of Justice's Advisory Opinion on the Threat or Use of Nuclear Weapons' (1998) 26 *Denver Journal of International Law and Policy* 691; Laurence Boisson de Chazournes and Philippe Sands (eds), *International Law, The International Court of Justice and Nuclear Weapons* (1999); Erik V. Koppe, *The Use of Nuclear Weapons and the Protection of the Environment During International Armed Conflict* (2006).

9 See also Adam Roberts, 'Environmental Issues in International Armed Conflict: The Experience of the 1991 Gulf War', in Richard J. Grunawalt, John E. King and Ronald S. McClain (eds), *Protection of the Environment during Armed Conflict* (1996), 222, 268.

10 Rome Statute, article 25(1).

of [a] person' who has employed a method or means of warfare in such a way as might otherwise fall within the terms of article 8(1) *ter*.

It is noted however, that the inclusion of this paragraph would place the Court in the position where it may be required to determine whether acts were undertaken as part of the 'lawful self-defence' of a State. This is not a matter that would normally be required of a court tasked with determining individual criminal responsibility.

In the end, this author, having considered the possibility of addressing this issue in the express terms of article 8 *ter*, decided instead *not* to include any exception, in the interests of proposing a provision that achieves the goal of effectively addressing the intentional destruction of the environment during armed conflict. Paragraph 3 should therefore not be considered as part of the proposal for a new *sui generis* crime of crimes against the environment, and has only been included in this Appendix II to indicate that this issue was carefully considered by this author in formulating the final terms of article 8 *ter* set out in 5.3.2.1.

BIBLIOGRAPHY

I. INTERNATIONAL TREATIES AND AGREEMENTS

1863 Instructions for the Government of Armies of the United States in the Field, promulgated as General Orders No. 100, 24 April 1863

1864 Convention for the Amelioration of the Condition of the Wounded in Armies in the Field, opened for signature 22 August 1864, 18 Martens (1st) 607 (entered into force 22 June 1865)

1868 Additional Articles relating to the Condition of the Wounded in War, opened for signature 20 October 1868, 18 Martens (1st) 612 (not in force)

Declaration Renouncing the Use, in Time of War, of Explosive Projectiles under 400 Grammes Weight, opened for signature 29 November 1868, 18 Martens (1st) 474 (entered into force 11 December 1868)

1899 Hague Convention I for the Pacific Settlement of International Disputes, opened for signature 29 July 1899, 26 Martens (2nd) 720 (entered in force 4 September 1900)

Hague Convention II with Respect to the Laws and Customs of War on Land, opened for signature 29 July 1899, 26 Martens (2nd) 949 (entered into force 4 September 1900) chapter 2

Hague Convention III for the Adaptation to Maritime Warfare of Principles of the Geneva Convention of 1864 opened for signature 29 July 1899, 26 Martens (2nd) 979 (entered in force 4 September 1900)

Declaration on the Launching of Projectiles and Explosives from Balloons, 29 July 1899, 32 Stat. 1839 (entered in force 4 September 1900)

Declaration on the Use of Projectiles the Object of Which is the Diffusion of Asphyxiating or Deleterious Gases, date of signature 29 July 1899, 26 Martens (2nd) 998 (entered into force 4 September 1900)

Declaration on the Use of Bullets Which Expand or Flatten Easily in the Human Body, date of signature 29 July 1899, 26 Martens (2nd) 1002 (entered into force 4 September 1900)

1906 Convention for the Amelioration of the Condition of the Wounded and Sick in Armies in the Field, opened for signature 6 July 1906, 11 LNTS 440 (entered into force 9 August 1907)

1907 Hague Convention IV Respecting the Laws and Customs of War on Land, opened for signature 18 October 1907, 3 Martens (3rd) 461 (entered into force 26 January 1910)

Hague Convention IX Concerning Bombardment by Naval Forces in Time of War, opened for signature 18 October 1907, 3 Martens (3rd) 604 (entered into force 26 January 1910)

Hague Convention X for the Adaptation to Maritime Warfare of the Principles of the 1906 Geneva Convention, opened for signature 18 October 1907, 3 Martens (3rd) 360 (entered into force 26 January 1910)

Hague Convention XI Relative to Certain Restrictions with Regard to the Exercise of the Right of Capture in Naval War, opened for signature 18 October 1907, 3 Martens (3rd) 663 (entered into force 26 January 1910)

Hague Convention XIII Concerning the Rights and Duties of Neutral Powers in Naval War, opened for signature 18 October 1907, 3 Martens (3rd) 713 (entered into force 26 January 1910)

1919 Treaty of Peace between the Allied and Associated Powers and Germany, opened for signature 28 June 1919, 112 BFSP 317 (entered into force 10 January 1920)

1920 Covenant of the League of Nations, opened for signature 28 June 1919, 112 BFSP 13 (entered into force 10 January 1920)

Protocol of Signature relating to the Statute of the Permanent Court of International Justice provided for by Article 14 of the Covenant of the League of Nations, opened for signature 16 December 1920, 6 LNTS 379 (entered into force 20 August 1921)

Treaty of Sevres, opened for signature 10 August 1920, 1920 UKTS 11 (not in force)

1922 1922 Washington Treaty Relating to the Use of Submarines and Noxious Gases in Warfare (not in force)

1923 Rules concerning the Control of Wireless Telegraphy in Time of War and Air Warfare, (non-binding)

Treaty of Lausanne Between Principal Allied and Associated Powers and Turkey, opened for signature 24 July 1923, 28 LNTS 11 (entered into force 24 July 1923)

1925 Geneva Protocol for the Prohibition of the Use in War of Asphyxiating, Poisonous and Other Gases, and of Bacteriological Methods of Warfare, opened for signature 17 June 1925, 26 Martens (3rd) 643 (entered into force 8 February 1928)

1928 Treaty Providing for the Renunciation of War as an Instrument of National Policy, opened for signature 27 August 1928, 94 LNTS 57 (entered into force 24 July 1929)

1929 Convention for the Amelioration of the Condition of the Wounded and Sick in Armies in the Field, opened for signature 27 July 1929, 118 LNTS 303 (entered into force 19 June 1931)

Convention relative to the Treatment of Prisoners of War, opened for signature 27 July 1929, 118 LNTS 343 (entered into force 19 June 1931)

1937 Convention for the Creation of an International Criminal Court, LN Doc C.547(I).M.384(I).1937.V (not in force)

1945 Charter of the United Nations, opened for signature 26 June 1945, 1 UNTS 16; 59 Stat. 1031 (entered into force 24 October 1945)

Statute of the International Court of Justice, opened for signature 26 June 1945, 1 UNTS 16, 59 Stat. 1031 (entered into force 24 October 1945)

Charter of the Nuremberg International Military Tribunal, annexed to the 1945 London Agreement for the Establishment of an International Military Tribunal (8 August 1945) 82 UNTS 279

1946 Charter of the International Military Tribunal for the trial of the major war criminals in the Far East (19 January 1946) TIAS 1589; 4 Bevans 20

1948 Convention on the Prevention and Punishment of the Crime of Genocide, opened for signature 9 December 1948, 78 UNTS 277 (entered into force 12 January 1951)

1949 Geneva Convention for the Amelioration of the Condition of the Wounded and Sick in Armed Forces in the Field, opened for signature 12 August 1949, 75 UNTS 31 (entered into force 21 October 1950)

Geneva Convention for the Amelioration of the Condition of the Wounded, Sick and Shipwrecked Members of Armed Forces at Sea, opened for signature 12 August 1949, 75 UNTS 85 (entered into force 21 October 1950)

Geneva Convention Relative to the Treatment of Prisoners of War, opened for signature 12 August 1949, 75 UNTS 135 (entered into force 21 October 1950)

Geneva Convention Relative to the Protection of Civilian Persons in Time of War, opened for signature 12 August 1949, 75 UNTS 287 (entered into force 21 October 1950)

1950 European Convention for the Protection of Human Rights and Fundamental Freedoms, opened for signature 4 November 1950, 213 UNTS 221 (entered into force 3 September 1953)

1951 Treaty Establishing the European Coal and Steel Community, opened for signature 18 April 1951, 261 UNTS 140 (entered into force 24 July 1952, expired 23 July 2002)

1954 International Convention for the Prevention of Pollution of the Sea by Oil, opened for signature 12 May 1954, 327 UNTS 3 (entered into force 26 July 1958)

Hague Convention for the Protection of Cultural Property in the Event of Armed Conflict, opened for signature 14 May 1954, 249 UNTS 240 (entered into force 7 August 1956)

First Hague Protocol for the Protection of Cultural Property in the Event of Armed Conflict, opened for signature 14 May 1954, 249 UNTS 358 (entered into force 7 August 1956)

1957 Treaty Establishing the European Economic Community, opened for signature 25 March 1957, 298 UNTS 11 (entered into force 1 January 1958)

1959 Antarctic Treaty, opened for signature 1 December 1959, 402 UNTS 71 (entered into force 23 June 1961)

1960 Paris Convention on Third Party Liability in the Field of Nuclear Energy, opened for signature 29 July 1960, 956 UNTS 251 (entered into force 1 April 1968)

1962 Convention on the Liability of Operators of Nuclear Ships, opened for signature 25 May 1962, 57 AILJ 268 (not in force)

1963 Vienna Convention on Civil Liability for Nuclear Damage, opened for signature 29 May 1963, 1063 UNTS 255 (entered into force 12 November 1977)

Treaty Banning Nuclear Weapons Tests in the Atmosphere, in Outer Space and under Water, opened for signature 5 August 1963, 480 UNTS 43 (entered into force 10 October 1963)

1966 International Covenant on Economic, Social and Cultural Rights, opened for signature 16 December 1966, 993 UNTS 3 (entered into force 3 January 1976)

International Covenant on Civil and Political Rights, opened for signature 16 December 1966, 999 UNTS 171 (entered into force 23 March 1976)

1967 Treaty on Principles Governing the Activities of States in the Exploration and Use of Outer Space, including the Moon and Other Celestial Bodies, opened for signature 27 January 1967, 610 UNTS 205 (entered into force 10 October 1967)

1968 Convention on the Non-Applicability of Statutory Limitations to War Crimes and Crimes against Humanity, opened for signature 26 November 1968, 8 ILM 68 (entered into force 11 November 1970)

1969 Vienna Convention on the Law of Treaties, opened for signature 23 May 1969, 1155 UNTS 331; 8 ILM 679 (entered into force 27 January 1980)

American Convention on Human Rights, opened for signature 22 November 1969, 1144 UNTS 123; 9 ILM 673 (entered into force 18 July 1978)

1971 Convention Relating to Civil Liability in the Field of Maritime Carriage of Nuclear Material, opened for signature 17 December 1971, 974 UNTS 255 (entered into force 15 July 1975)

1972 Convention on the Prohibition of the Development, Production and Stockpiling of Bacteriological (Biological) and Toxic Weapons and on their Destruction, opened for signature 10 April 1972, 1015 UNTS 163 (entered into force 26 March 1975)

Convention for the Protection of the World Cultural and Natural Heritage, opened for signature 16 November 1972, 11 ILM 1358 (entered into force 17 December 1975)

1973 Convention on International Trade in Endangered Species of Wild Fauna and Flora, opened for signature 3 March 1973, 993 UNTS 243 (entered into force 1 July 1975)

International Convention on the Suppression and Punishment of the Crime of Apartheid, opened for signature 30 November 1973, 1015 UNTS 243 (entered into force 18 July 1976)

1976 Convention on the Prohibition of Military or Any Other Hostile Use of Environmental Modification Techniques, opened for signature 10 December 1976, 1108 UNTS 151; 16 ILM 88 (entered into force 5 October 1978)

Understanding Relating to Article I of ENMOD, 31 United Nations General Assembly Official Records Supp. No. 27 (A/31/27), Annex I

Understanding Relating to Article II of ENMOD, 31 United Nations General Assembly Official Records Supp. No. 27 (A/31/27), Annex II

1977 Protocol I Additional to the Geneva Conventions of August 12, 1949, and relating to the Protection of Victims of International Armed Conflicts, opened for signature 8 June 1977, 1125 UNTS 3, 16 ILM 1391 (entered into force 7 December 1978)

Protocol II Additional to the Geneva Conventions of August 12, 1949, and relating to the Protection of Victims of Non-International Armed Conflicts, opened for signature 8 June 1977, 1125 UNTS 609, 16 ILM 1442 (entered into force 7 December 1978)

1979 Convention on the Elimination of All Forms of Discrimination against Women, opened for signature 1 March 1980, 19 ILM 33 (entered into force 3 September 1981)

1981 African Charter on Human and People's Rights, adopted 27 June 1981, 1520 UNTS 217; 21 ILM 58 (entered into force 21 October 1986)

Convention on Prohibitions or Restrictions on the Use of Certain Conventional Weapons Which May be Deemed to be Excessively Injurious or to Have Indiscriminate Effects, opened for signature 10 April 1981, 1342 UNTS 137 (entered into force 2 December 1983)

Protocol II on Prohibitions or Restrictions on the Use of Mines, Booby-Traps and Other Devices to the Convention on Prohibitions or Restrictions on the Use of Certain Conventional Weapons Which May be Deemed to be Excessively Injurious or to Have Indiscriminate Effects, opened for signature 10 April 1981, 1342 UNTS 137 (entered into force 2 December 1983)

Protocol III on Prohibitions or Restrictions on the Use of Incendiary Weapons to the Convention on Prohibitions or Restrictions on the Use of Certain Conventional Weapons Which May be Deemed to be Excessively Injurious or to Have Indiscriminate Effects, opened for signature 10 April 1981, 1342 UNTS 137 (entered into force 2 December 1983)

1982 United Nations Convention on the Law of the Sea, opened for signature 10 December 1982, 1833 UNTS 3; 21 ILM 1261 (entered into force 16 November 1994)

1984 Convention against Torture and Other Cruel, Inhuman or Degrading Treatment or Punishment, opened for signature 10 December 1984, 23 ILM 1027 (entered into force 26 June 1987)

1985 Vienna Convention for the Protection of the Ozone Layer, opened for signature 22 March 1985, 26 ILM 1529 (entered into force 22 September 1988)

1986 Vienna Convention on the Law of Treaties between States and International Organizations or between International Organizations, opened for signature 21 March 1986, 25 ILM 543 (not in force)

1987 Montreal Protocol on Substances that Deplete the Ozone Layer, opened for signature 16 September 1987, 26 ILM 154 (entered into force 1 January 1989)

1988 Joint Protocol Relating to the Application of the Vienna Convention on Civil Liability for Nuclear Damage and the Paris Convention on Third Party Liability in the Field of Nuclear Energy, opened for signature 21 September 1988, 1672 UNTS 301 (entered into force 27 April 1992)

1989 Convention on the Rights of the Child, opened for signature 20 November 1989, 1577 UNTS 3 (entered into force 2 September 1990)

1992 United Nations Framework Convention on Climate Change, opened for signature 4 June 1992, 31 ILM 849 (entered into force 21 March 1994)

International Convention on Civil Liability for Oil Pollution Damage, opened for signature 27 November 1992, IMO LEG/CONF.9/15 (entered into force 30 May 1996)

Convention for the Protection of the Marine Environment of the North East Atlantic, opened for signature 22 September 1992, 32 ILM 1069, (entered into force 25 March 1998)

Convention on Biological Diversity, opened for signature 5 June 1992, 31 ILM 822 (entered into force 29 December 1993)

1993 Convention on the Prohibition of the Development, Production, Stockpiling and Use of Chemical Weapons and on Their Destruction, opened for signature 13 January 1993, 32 ILM 800 (entered into force 29 April 1997)

1994 Arab Charter on Human Rights, adopted by the Council of the League of Arab States on 15 September 1994, reprinted in (1997) 18 *Human Rights Law Journal* 151 (not in force)

1996 Amended Protocol II to the 1980 Convention on Certain Conventional Weapons on Prohibitions or Restrictions on the Use of Mines, Booby-Traps and Other Devices, date of adoption 3 May 1996, 35 ILM 1206 (entered into force 3 December 1998)

Comprehensive Nuclear Test Ban Treaty, opened for signature 24 September 1996, 35 ILM 1443 (not in force)

Protocol to the Convention on the Prevention of Marine Pollution by Dumping of Wastes and Other Matter, opened for signature 7 November 1996, 36 ILM 1 (entered into force 24 March 2006)

1997 Protocol to amend the 1963 Vienna Convention on Civil Liability for Nuclear Damage, opened for signature 12 September 1997, 36 ILM 1462 (entered into force 4 October 2003)

Ottawa Convention on the Prohibition of the Use, Stockpiling, Production and Transfer of Anti-Personnel Mines and on Their Destruction, opened for signature 3 December 1997, 36 ILM 1507 (entered into force 1 March 1999)

Kyoto Protocol to the United Nations Framework Convention on Climate Change, opened for signature 16 March 1998, 37 ILM 22 (entered into force 16 February 2005)

1998 Rome Statute of the International Criminal Court, opened for signature 17 July 1998, 2187 UNTS 3; 37 ILM 999 (entered into force 1 July 2002)

European Convention on the Protection of the Environment through Criminal Law, opened for signature 4 November 1998, 38 ILM 259 (not in force)

1999 Second Protocol to the Hague Convention of 1954 for the Protection of Cultural Property in the Event of Armed Conflict, opened for signature 26 March 1999, 38 ILM 769 (entered into force 9 March 2004)

2000 Optional Protocol to the Convention on the Rights of the Child on the Involvement of Children in Armed Conflict, opened for signature 25 May 2000, 2173 UNTS 222 (entered into force 12 February 2002)

2003 Protocol V on Explosive Remnants of War to the Convention on Certain Conventional Weapons on Prohibitions or Restrictions on the Use of Mines, Booby-Traps and Other Devices, date of adoption 28 November 2003, 2399 UNTC 100 (entered into force 12 November 2006)

2005 Protocol III Additional to the Geneva Conventions of August 12, 1949, and relating to the Adoption of an Additional Distinctive Emblem, date of adoption 8 December 2005, 45 ILM 558 (entered into force 14 January 2007)

2008 Convention on Cluster Munitions, opened for signature 3 December 2008, 48 ILM 357 (entered into force 1 August 2010)

II. UNITED NATIONS DOCUMENTS

A) GENERAL ASSEMBLY RESOLUTIONS

1946 United Nations General Assembly Resolution 95(I) (11 December 1946) UN Doc A/RES/95(I) on the Affirmation of the Principles of International Law Recognized by the Charter of the Nürnberg Tribunal

United Nations General Assembly Resolution 96(I) (11 December 1946) UN Doc A/RES/96/(I) on the Crime of Genocide

1947 United Nations General Assembly Resolution 174 (II) (21 November 1947) UN Doc A/519 on the Establishment of an International Law Commission

1948 United Nations General Assembly Resolution 217A (III) (10 December 1948) UN Doc A/810 on the Universal Declaration of Human Rights

United Nations General Assembly Resolution 260 (III) B (9 December 1948) UN Doc A/RES/260(III)B on the Prevention and Punishment of the Crime of Genocide: Study by the International Law Commission of the Question of an International Criminal Jurisdiction

1950 United Nations General Assembly Resolution 485 (V) (12 December 1950) UN Doc A/RES/485(V) on the Amendment to the Article 13 of the Statute of the International Law Commission

United Nations General Assembly Resolution 488 (V) (12 December 1950) UN Doc A/RES/488(V) on the Formulation of the Nürnberg Principles

United Nations General Assembly Resolution 489 (V) (12 December 1950) UN Doc A/RES/489(V) on International Criminal Jurisdiction

1954 United Nations General Assembly Resolution 897 (IX) (4 December 1954) UN Doc A/RES/897(IX) on the Draft Code of Offences against the Peace and Security of Mankind

United Nations General Assembly Resolution 898 (IX) (14 December 1954) UN Doc A/RES/898(IX) on the International Criminal Jurisdiction

1955 United Nations General Assembly Resolution 984 (X) (3 December 1955) UN Doc A/RES/984(X) on the Amendment to Article 12 of the Statute of the International Law Commission: Place of Meeting of the Commission

United Nations General Assembly Resolution 985 (X) (3 December 1955) UN Doc A/RES/985(X) on the Amendment to Article 10 of the Statute of the International Law Commission: Term of Office of Members of the Commission

1960 United Nations General Assembly Resolution 1514 (XV) (14 December 1960) UN Doc A/RES/1514(XV) on the Declaration on the Granting of Independence to Colonial Countries and Peoples

1968 United Nations General Assembly Resolution 2398 (XXIII) (3 December 1968) UN Doc A/RES/2398(XXIII) on the Problems of the Human Environment

United Nations General Assembly Resolution 2444 (XXIII) (19 December 1968) UN Doc A/RES/2444(XXIII) on the Respect for Human Rights in Armed Conflicts

1969 United Nations General Assembly Resolution 2603A (XXIV) (16 December 1969) UN Doc A/RES/2603(XXIV)A on the Question of Chemical and Bacteriological (Biological) Weapons

1970 United Nations General Assembly Resolution 2712 (XXV) (9 December 1970) UN Doc A/RES/2712(XXV) on the Question of the Punishment of War Criminals and of Persons who have Committed Crimes against Humanity

1972 United Nations General Assembly Resolution 2996 (XXVII) (15 December 1972) UN Doc A/RES/2996(XXVII) on the International Responsibility of States in Regard to the Environment

1974 United Nations General Assembly Resolution 3264 (XXIX) (9 December 1974) UN Doc A/RES/3264(XXIX) on the Prohibition of Action to Influence the Environment and Climate for Military and Other Purposes, Incompatible with the Maintenance of International Security, Human Well-Being and Health

United Nations General Assembly Resolution 3314 (XXIX) (14 December 1974) UN Doc A/RES/3314(XXIX) on the Definition of Aggression

1976 United Nations General Assembly Resolution 31/72 (10 December 1976) UN Doc A/RES/31/72 on the Convention on the Prohibition of Military or any Other Hostile Use of Environmental Modification Techniques

1981 United Nations General Assembly Resolution 36/39 (18 November 1981) UN Doc A/RES/36/39 on the Enlargement of the International Law Commission

United Nations General Assembly Resolution 36/106 (10 December 1981) UN Doc A/RES/36/106 on the Draft Code of Offences against the Peace and Security of Mankind

1982 United Nations General Assembly Resolution 37/7 (28 October 1982) UN Doc A/RES/37/7 on the World Charter for Nature

1989 United Nations General Assembly Resolution 44/39 (4 December 1989) UN Doc A/RES/44/39 on the International Criminal Responsibility of Individuals and Entities Engaged in Illicit Trafficking in Narcotic Drugs across National Frontiers and Other Transnational Criminal Activities

1992 United Nations General Assembly Resolution 47/37 (25 November 1992) UN Doc A/RES/47/37 on the Protection of the Environment in Times of Armed Conflict

United Nations General Assembly Resolution 47/39 (30 November 1992) UN Doc A/RES/47/39 on the Convention on the Prohibition of the Development, Production, Stockpiling and Use of Chemical Weapons and on Their Destruction

1994 United Nations General Assembly Resolution 49/50 (9 December 1994) UN Doc A/RES/49/50 on the United Nations Decade of International Law

United Nations General Assembly Resolution 49/75K (15 December 1994) UN Doc A/RES/49/75 on Request for an Advisory Opinion from the International Court of Justice on the Legality of the Threat or Use of Nuclear Weapons

1996 United Nations General Assembly Resolution 51/45 (10 December 1996) UN Doc A/RES/51/45 on General and Complete Disarmament

United Nations General Assembly Resolution 51/45(S) (10 December 1996) UN Doc A/RES/51/45 on an international agreement to ban anti-personnel landmines

1997 United Nations General Assembly Resolution 51/160 (30 January 1997) UN Doc A/RES/51/160 on the Report of the International Law Commission on the work of its 48th session

United Nations General Assembly Resolution 52/160 (15 December 1997) UN Doc A/RES/52/160 on the Establishment of an International Criminal Court

2001 United Nations General Assembly Resolution 56/83 (12 December 2001) UN Doc A/RES/56/83 on the Responsibility of States for Internationally Wrongful Acts

2003 United Nations General Assembly Resolution 57/228 B (13 May 2003) UN Doc A/RES/57/228 B on the Khmer Rouge Trials

2006 United Nations General Assembly Resolution 61/102 (6 December 2006) UN Doc A/RES/61/102 on the Convention on the Prohibition of the Development,

Production and Stockpiling of Bacteriological (Biological) and Toxic Weapons and on their Destruction

2013 United Nations General Assembly Resolution 67/234 B (2 April 2013) UN Doc A/RES/67/234 B on the Arms Trade Treaty

B) SECURITY COUNCIL RESOLUTIONS

1991 United Nations Security Council Resolution 687 (3 April 1991) UN Doc S/RES/687/1991 on Iraq-Kuwait

United Nations Security Council Resolution 692 (20 May 1991) UN Doc S/RES/692/1991 on Iraq-Kuwait

1993 United Nations Security Council Resolution 827 (25 May 1993) UN Doc S/RES/827/1993 on the Tribunal (Former Yugoslavia)

1994 United Nations Security Council Resolution 955 (8 November 1994) UN Doc S/RES/955/1994 on the Establishment of an International Tribunal and Adoption of the Statute of the Tribunal

1999 United Nations Security Council Resolution 1272 (25 October 1999) UN Doc S/RES/1272/1999 on the Situation in East Timor.

2000 United Nations Security Council Resolution 1315 (14 August 2000) UN Doc S/RES/1315/2000 on the Situation in Sierra Leone

2003 United Nations Security Council Resolution 1503 (28 August 2003) UN Doc S/Res/1503 on the International Criminal Tribunal for the former Yugoslavia and International Criminal Tribunal for Rwanda

2004 United Nations Security Council Resolution 1534 (26 March 2004) UN Doc S/Res/1534 on the International Criminal Tribunal for the former Yugoslavia and International Criminal Tribunal for Rwanda

United Nations Security Council Resolution 1556 (30 July 2004) UN Doc S/RES/1556/2004 on the Report of the Secretary-General on the Sudan

2005 United Nations Security Council Resolution 1593 (31 March 2005) UN Doc S/RES/1593/2005 on the Reports of the Secretary-General on the Sudan

United Nations Security Council Resolution 1644 (15 December 2005) UN Doc S/RES/1644/2005 on the Situation in the Middle East

2006 United Nations Security Council Resolution 1674 (28 April 2006) UN Doc S/RES/1674/2006 on the Protection of Civilians in Armed Conflict

United Nations Security Council Resolution 1738 (23 December 2006) UN Doc S/RES/1738/2006 on the Protection of Civilians in Armed Conflict

2007 United Nations Security Council Resolution 1757 (30 May 2007) UN Doc S/RES/1757/2007 on the Situation in the Middle East

2008 United Nations Security Council Resolution 1820 (19 June 2008) UN Doc S/RES/1820/2008 on Women, Peace and Security

2010 United Nations Security Council Resolution 1966 (22 December 2010) UN Doc S/Res/1966/2010 on the International Tribunal for the Prosecution of Persons Responsible for Serious Violations of International Humanitarian Law Committed in the Territory of the Former Yugoslavia since 1991 and the International Criminal Tribunal for the Prosecution of Persons Responsible for Genocide and Other Serious Violations of International Humanitarian Law Committed in the Territory of Rwanda and Rwandan Citizens Responsible for Genocide and Other Such Violations Committed in the Territory of Neighbouring States between 1 January 1994 and 31 December 1994

2011 United Nations Security Council Resolution 1970 (26 February 2011) UN Doc S/RES/1970/2011 on Peace and Security in Africa

C) GENERAL ASSEMBLY AND SECURITY COUNCIL DOCUMENTS AND REPORTS

1986 United Nations Secretary-General, 'Report of the Mission Dispatched by the Secretary-General to Investigate Allegations of the Use of Chemical Weapons in the Conflict between The Islamic Republic of Iran and Iraq' (12 March 1986) UN Doc S/17922

1992 United Nations Security Council, 'Note by the President of the Security Council' (31 January 1992) Presidential Statement S/23500

1993 United Nations Security Council, 'Report of the Secretary-General Pursuant to Paragraph 2 of Security Council Resolution 808' (3 May 1993) UN Doc S/25704

Statute of the International Tribunal for the Prosecution of Persons Responsible for Serious Violations of International Humanitarian Law Committed in the Territory of the Former Yugoslavia since 1991, 32 ILM 1159 (entered into force 25 May 1993)

United Nations Secretary-General, 'Report of the Secretary-General on the Protection of the Environment in Times of Armed Conflict' (29 July 1993) UN Doc A/48/269

1994 Statute of the International Criminal Tribunal for the Prosecution of Persons Responsible for Genocide and Other Serious Violations of International Humanitarian Law Committed in the Territory of Rwanda and Rwandan Citizens Responsible for Genocide and Other Such Violations Committed in the Territory of Neighbouring States, between 1 January 1994 and 31 December 1994, 33 ILM 1598 (entered into force 8 November 1994)

1998 Statement by the United Nations Secretary-General Kofi Annan at the Ceremony held at Campidoglio Celebrating the Adoption of the Statute of the International Criminal Court (18 July 1998) <http://untreaty.un.org/cod/icc/rome/proceedings/contents.htm>

2002 United Nations Security Council, 'Final Report of the Panel of Experts on the Illegal Exploitation of Natural Resources and Other Forms of Wealth of the Democratic Republic of the Congo' (16 October 2002) UN Doc: S/2002/1146

2005 International Commission of Inquiry on Darfur, 'Report of the International Commission of Inquiry on Darfur to the United Nations Secretary-General, Pursuant to Security Council Resolution 1564 of 18 September 2004' (25 January 2005)

2007 United Nations Security Council, 'Security Council Hold First-Ever Debate on Impact of Climate Change on Peace, Security, Hearing Over 50 Speakers' (17 April 2007) UN Doc SC/9000

2009 United Nations General Assembly, 'Implementing the responsibility to protect: Report of the Secretary-General' (12 January 2009) UN Doc A/63/677

D) OTHER UNITED NATIONS AND INTERGOVERNMENTAL AGENCIES / BODIES / DOCUMENTS

1919 Report of the Commission on the Responsibility of the Authors of the War and on Enforcement of Penalties, reprinted in (1920) 14 *American Journal of International Law* 95

1945 Statement by the Delegations of the Four Sponsoring Governments on Voting Procedure, 7 June 1945, 11 UNCIO, *Documents* 711

UNESCO Constitution (adopted 16 November 1945) <http://portal.unesco.org/en/ev.phpURL_ID=15244&URL_DO=DO_TOPIC&URL_SECTION=201.html>

1950 International Law Commission, 'Principles of International Law Recognized in the Charter of the Nürnberg Tribunal and in the Judgment of the Tribunal', [1950] 2 *Yearbook of the International Law Commission* 374

1951 Draft Code of Offences against the Peace and Security of Mankind, [1951] 2 *Yearbook of the International Law Commission* 58

1954 Draft Code of Offences against the Peace and Security of Mankind, [1954] 2 *Yearbook of the International Law Commission* 112

1972 Declaration of the United Nations Conference on the Human Environment (Stockholm Declaration) (16 June 1972) UN Doc A/CONF.48/14/Rev.1

1976 Fifth Report on State Responsibility, by Roberto Ago, Special Rapporteur [1976] 2 *Yearbook of the International Law Commission* Part I, Chapter III, sections 1–4, UN Doc A/CN.4/291 and Add. 1–2

Commentary by the International Law Commission on Article 19 'Articles on State Responsibility for International Wrongfully Acts' [1976] 2 *Yearbook of the International Law Commission* Part II, 96, UN Doc A/CN.4/SER.A/1976/Add.l (Part 2)

1980 Report of the International Law Commission to the General Assembly on its work of the thirty-second session, [1980] 2 *Yearbook of the International Law Commission* Part II, 32, UN Doc A/CN.4/SER.A/1980/Add.l (Part 2)

1981 Report of M.C. Bassiouni to the *Ad Hoc* Working Group of Experts on South Africa for the Commission of Human Rights, 'Study on ways and means of insuring the implementation of International instruments such as the International Convention on the Suppression and Punishment of the Crime of Apartheid, including the establishment of the international jurisdiction envisaged by the Convention', 19 January 1981, UN Doc E/CN/4/1426

1985 Third Report on the draft Code of Offences against the Peace and Security of Mankind, by Mr. Doudou Thiam, Special Rapporteur, International Law Commission, 8 April 1985, UN Doc. A/CN.4/387 and Corr.1 and Corr.2

1987 International Law Commission, 'Draft Code of Offences Against the Peace and Security of Mankind' [1987] 2:2 *Yearbook of the International Law Commission* 7, UN Doc A/CN.4/SER.A/1987/Add.1 (Part 2)

1991 International Law Commission, 'Draft Code of Crimes Against the Peace and Security of Mankind' [1991] 1 *Yearbook of the International Law Commission* 234, UN Doc A/CN.4/SER.A/1991

International Law Commission, 'Draft Code of Crimes Against the Peace and Security of Mankind' [1991] 2:2 *Yearbook of the International Law Commission* 107, UN Doc A/CN.4/SER.A/1991/Add.1 (Part 2)

1992 Commission on Human Rights, Report prepared by Special Rapporteur Walter Kälin in accordance with in accordance with Human Rights Commission Resolution 1991/67, 'Report on the situation of human rights in Kuwait under Iraqi occupation' (16 January 1992) UN Doc E/CN.4/1992/26

Declaration of the United Nations Conference on Environment and Development, (14 June 1992) UN Doc A/CONF.151/26 (Volume 1), 31 ILM 874

United Nations Conference on Environment and Development Report, 'Agenda 21' (14 June 1992) UN Doc A/CONF.151/26/Rev.1 (vol I) (1993)

Non-Legally Binding Authoritative Statement of Principles for a Global Consensus on the Management, Conservation and Sustainable Development of All Types of Forests (14 June 1992) UN Doc A/CONF.151/26 (Vol. III), 31 ILM 881

1994 Draft Statute for an International Criminal Court, Report of the International Law Commission on Its Forty-sixth session, United Nations General Assembly Official Records 49th Sess., Supp. No. 10, UN Doc A/49/10 (1994)

1996 Draft Code of Offences against the Peace and Security of Mankind, Report of the International Law Commission on Its Forty-eighth session, United Nations General Assembly Official Records 51st Sess., Supp. No. 10, 9 UN Doc A/51/10 (1996)

Final Declaration of the Fourth Review Conference of the Convention on the Prohibition of the Development, Production and Stockpiling of Bacteriological (Biological) and Toxic Weapons and on their Destruction, (6 December 1996) BWC/CONF.IV/9, part II

2000 United Nations Transitional Administration in East Timor (UNTAET) Regulation 2000/11 (6 March 2000)

United Nations Transitional Administration in East Timor (UNTAET) Regulation 2000/15 (6 June 2000)

2001 International Law Commission, Draft Articles on Responsibility of States for International Wrongfully Acts contained in Report of the International Law Commission to the General Assembly on its work of the fifty-third session, [2001] 2:2 *Yearbook of the International Law Commission* 1, UN Doc A/CN.4/SER.A/2001/Add.1 (Part 2)

International Law Commission, Articles on the Responsibility of States for Internationally Wrongful Acts, annex to United Nations General Assembly Resolution 56/83 (12 December 2001) UN Doc A/Res/56/83

2002 Elements of Crimes (9 September 2002) ICC Doc ICC-ASP/1/3 (part II-B)

United Nations Committee on Economic, Social and Cultural Rights, 'Substantive Issues arising in the Implementation of the International Covenant on Economic, Social and Cultural Rights: General Comment No. 15 (2002) – The right to water (arts. 11 and 12 of the International Covenant on Economic, Social and Cultural Rights)' (November 2002) UN Doc E/C.12/2002/11

United Nations, 'The International Criminal Court' (Fact Sheet, December 2002) <www.un.org/News/facts/iccfact.htm>

Report of the Secretary-General, Kofi A. Annan, 'Prevention of Armed Conflict'

2004 United Nations Commission on Human Rights, 'Report of the United Nations High Commissioner for Human Rights and Follow-Up to the World Conference on Human Rights: Situation of Human Rights in Darfur Region of the Sudan' (7 May 2004) E/CN.4/2005/3

Negotiated Relationship Agreement between the International Criminal Court and the United Nations (4 October 2004) ICC Doc ICC-ASP/3/Res.1

Report of the High-Level Panel on Threats, Challenges, and Change, 'A More Secure World: Our Shared Responsibility' (1 December 2004) UN Doc A/59/565 (2004)

2005 International Law Commission, 'The effect of armed conflict on treaties: an examination of practice and doctrine: Memorandum by the Secretariat' (1 February 2005) UN Doc A/CN.4/550

First Report on the Effects of Armed Conflicts on Treaties, by Ian Brownlie, Special Rapporteur, (21 April 2005) UN Doc A/CN.4/552

2006 United Nations Statistics Division, 'Environmental Glossary' <http://unstats.un.org/unsd/publication/SeriesF/SeriesF_67E.pdf>

Declaration on the Occasion of the Entry into Force of the Protocol on Explosive Remnants of War (Protocol V) Adopted by the Third Review Conference of the Convention on Prohibitions or Restrictions on the Use of Certain Conventional Weapons Which May be Deemed to be Excessively Injurious or to Have Indiscriminate Effects on 13 November 2006 <www.unog.ch/80256EDD006B8954/%28httpAssets%29/682877B24ABFF1BFC12572810063FB74/$file/Declaration+on+EIF+Protocol+V.pdf>

International Law Commission, Draft Principles on the Allocation of Loss in the Case of Transboundary Harm Arising out of Hazardous Activities: General Commentary, contained in Report of the International Law Commission to the General Assembly on its work of the fifty-eighth session, UN Doc A/61/10 (2006)

2007 Report of the Special Representative of the Secretary-General for Children and Armed Conflict, UN Doc A/62/228, 13 August 2007

2008 United Nations Compensation Commission, 'The Claims' <www.uncc.ch/theclaims.htm> (29 July 2008)

2011 Final Declaration of the Seventh Review Conference of the States Parties to the Convention on the Prohibition of the Development, Production and Stockpiling of Bacteriological (Biological) and Toxin Weapons and on Their Destruction, 'Final Document of the Seventh Review Conference' <http://daccess-dds-ny.un.org/doc/UNDOC/GEN/G12/600/60/PDF/G1260060.pdf?OpenElement>

2012 United Nations Environment Programme, 'About UNEP: The Organization' <www.unep.org/Documents.Multilingual/Default.asp?DocumentID=43>

United Nations High Commission for Refugees, '2011 Global Trends: A Year of Crisis' (2012) <www.unhcr.org/cgibin/texis/vtx/home/opendocPDFViewer. html?docid=4fd6f87f9&query=Global%20Trends>

2013 The United Nations Office in Geneva (UNOG) Press Release, 'Governing Council of United Nations Claims Commission Elects New President' (11 February 2013) <www.unog.ch/80256EDD006B9C2E/%28httpNewsByYear_en%29/EF37E1B2633EB758C1257B0F004162C9?OpenDocument>

UN News Centre, 'Head of UN probe into chemical weapons use in Syria says preparatory work has begun' (27 March 2013) <www.un.org/ apps/news/story.asp?NewsID=44507&Cr=syria&Cr1=chemical#.UVcaU1f-jdc>

International Law Commission, 'Protection of the environment in relation to armed conflicts' (5 June 2013) <www.un.org/law/ilc/index.html>

United Nations Treaty Collection, 'Convention on Cluster Munitions' (21 June 2103) <http://treaties.un.org/UNTC/Pages/ViewDetails.aspx?src=TREATY&mtdsg_no=XXVI6&chapter=26&lang=en>

III. INTERNATIONAL JUDICIAL DECISIONS, PROCEEDINGS AND OTHER COURT DOCUMENTS

A) PERMANENT COURT OF INTERNATIONAL JUSTICE / INTERNATIONAL COURT OF JUSTICE

1927 *Case Concerning the Factory at Chorzów (Germany v. Poland)* (Jurisdiction) [1927] PCIJ Rep, Series A, No. 9

The Case of the S.S. 'Lotus' (France v. Turkey) (Judgment) [1927] PCIJ Rep, Series A, No. 10

1949 *Corfu Channel Case (United Kingdom of Great Britain and Northern Ireland v. Albania)* (Merits) (Judgment) [1949] ICJ Rep 4

Reparation for Injuries Suffered in the Service of the United Nations (Advisory Opinion) [1949] ICJ Rep 174

1950 *International Status of South-West Africa* (Advisory Opinion) [1950] ICJ Rep 128

Interpretation of Peace Treaties Case (second phase) (Advisory Opinion) [1950] ICJ Rep 221

Colombian-Peruvian Asylum Case (Colombia v. Peru) (Judgment) [1950] ICJ Rep 266

1951 *Reservations to the Convention on the Prevention and Punishment of the Crime of Genocide* (Advisory Opinion) [1951] ICJ Rep 16

Fisheries Case (United Kingdom v. Norway) (Judgment) [1951] ICJ Rep 116

1969 *North Sea Continental Shelf Cases (Federal Republic of Germany v. Denmark and Federal Republic of Germany v. The Netherlands)* (Judgment) [1969] ICJ Rep 3

1970 *Barcelona Traction, Light and Power Company, Limited (Belgium v. Spain)* (Judgment) [1970] ICJ Rep 3

1971 *Legal Consequences for States of the Continued Presence of South Africa in Namibia (South West Africa) notwithstanding Security Council Resolution 276 (1970)* (Advisory Opinion) [1971] ICJ Rep 16

1974 *Nuclear Tests Case (Australia v. France)* (Judgment) [1974] ICJ Rep 253

Nuclear Tests Case (New Zealand v. France) (Judgment) [1974] ICJ Rep 457

1975 *Western Sahara Case* (Advisory Opinion) [1975] ICJ Rep 12

1982 *Case Concerning Delimitation of the Maritime Boundary in the Gulf of Maine Area (Canada v. United States of America)* (Constitution of Chamber) (Order of 20 January 1982) [1982] ICJ Rep 3

1985 *Continental Shelf (Libyan Arab Jamahiriya v. Malta)* (Judgment) [1985] ICJ Rep 13

1986 *Military and Paramilitary Activities in and against Nicaragua (Nicaragua v. United States of America)* (Merits) (Judgment) [1986] ICJ Rep 14

1994 *Territorial Dispute (Libyan Arab Jamahiriya v. Chad)* (Judgment) [1994] ICJ Rep 6

1995 *Maritime Delimitation and Territorial Questions (Qatar v. Bahrain)* (Judgment) [1995] ICJ Rep 6

East Timor (*Portugal v. Australia*) [1995] ICJ Rep 90

Verbatim Record, Public Sitting held on Monday 30 October 1995 in the case in *Legality of the Use by a State of Nuclear Weapons in Armed Conflict* (Request for Advisory Opinion Submitted by the World Health Organization) and in *Legality of the Threat or Use of Nuclear Weapons* (Request for Advisory Opinion Submitted by the General Assembly of the United Nations)

1996 *Legality of the Threat or Use of Nuclear Weapons* (Advisory Opinion) [1996] ICJ Rep 226

Application of the Convention on the Prevention and Punishment of the Crime of Genocide (Bosnia and Herzegovina v. Yugoslavia) (Preliminary Objections, Judgment) [1996] ICJ Rep 595

Case Concerning Oil Platforms (Islamic Republic of Iran v. United States of America) (Preliminary Objections, Judgment) [1996] ICJ Rep 803

1997 *Gabcíkovo-Nagymaros Project (Hungary v. Slovakia)* (Judgment) [1997] ICJ Rep 7

1999 *Legality of Use of Force (Yugoslavia v. France)* (Request for the Indication of Provisional Measures submitted by Applicant, 28 April 1999) para 3 <www.icj-cij.org/docket/files/105/10545.pdf>

Legality of Use of Force (Yugoslavia v. Spain) (Request for the Indication of Provisional Measures) [1999] ICJ Rep 761

Legality of Use of Force (Yugoslavia v. United States of America) (Request for Provisional Measures) [1999] ICJ Rep 916

2000 International Court of Justice, 'The judges of the International Court of Justice elect the members to the Court's Chambers and Committees', (Press Release, 10 February 2000) <www.icj-cij.org/presscom/index.php?pr=127&pt=&p1=6&p2=1>

2004 *Legal Consequences of the Construction of a Wall in the Occupied Palestinian Territory* (Advisory Opinion) [2004] ICJ Rep 136

Legality of Use of Force (Serbia and Montenegro v. Belgium) (Preliminary Objections, Judgment) [2004] ICJ Rep 328

2005 *Armed Activities on the Territory of the Congo (Democratic Republic of the Congo v Uganda)* (Separate Opinion of Judge Simma) [2005] ICJ Rep 168, 334

2006 Speech by H.E. Judge Rosalyn Higgins, President of the International Court of Justice, to the General Assembly of the United Nations, 26 October 2006 <www.icj-cij.org/court/index.php?pr=1874&pt=3&p1=1&p2=3&p3=1>

2007 *Case Concerning Pulp Mills on the River Uruguay (Argentina v. Uruguay)* (Provisional Measures, Order) [2007] ICJ Rep 3

Case Concerning the Application of the Convention on the Prevention and Punishment of the Crime of Genocide (Bosnia and Herzegovina v. Serbia and Montenegro) (Judgment) [2007] ICJ Rep 43

2008 *Aerial Herbicide Spraying (Ecuador v. Colombia)*, International Court of Justice, 'Ecuador institutes proceedings against Colombia with regard to a dispute concerning the alleged aerial spraying by Colombia of toxic herbicides over Ecuadorian territory', (Press Release, 1 April 2008) <www.icjcij.org/docket/files/138/14470>

2009 International Court of Justice, 'Belgium institutes proceedings against Senegal and requests the Court to indicate provisional measures', (Press Release, 19 February 2009) <www.icjcij.org/docket/files/144/15052.pdf?PHPSESSID=662d101b508c0ff67b1fcafb724d112f>

2010 *Case Concerning Pulp Mills on the River Uruguay (Argentina v. Uruguay)* (Judgment) [2010] ICJ Rep 14

International Court of Justice, 'Australia institutes Proceedings against Japan for alleged Breach of International Obligations concerning Whaling', (Press Release, 1 June 2010) <www.icj-cij.org/docket/files/148/15953.pdf>

2011 *Case Concerning Application of the International Convention on the Elimination of all Forms of Racial Discrimination (Georgia v. Russian Federation)* (Preliminary Objections, Judgment) [2011] ICJ Rep 70

2013 International Court of Justice, 'Whaling in the Antarctic (Australia v. Japan): The Court Authorizes New Zealand to Intervene in the Proceedings', (Press Release, 13 February 2013) <www.icj-cij.org/docket/files/148/17266.pdf>

2014 *Whaling in the Antarctic (Australia v. Japan: New Zealand intervening)* (Judgment) 31 March 2014 <www.icj-cij.org/docket/files/148/18136.pdf>

International Court of Justice, 'The Republic of the Marshall Islands files Applications against nine States for their alleged failure to fulfil their obligations with respect to the cessation of the nuclear arms race at an early date and to nuclear disarmament', (Press Release, 25 April 2014) <www.icj-cij.org/presscom/files/0/18300.pdf>

B) INTERNATIONAL CRIMINAL COURT

2006 Office of the Prosecutor, The International Criminal Court, 'Iraq Response', (9 February 2006) <www.iccnow.org/documents/OTP_letter_to_senders_re_Iraq_9_February_2006.pdf>

Office of the Prosecutor, The International Criminal Court, 'Third Report of the Prosecutor of the International Criminal Court to the UN Security Council Pursuant to UNSCR 1593 (2005)', 14 June 2006, <www.icc-cpi.int/library/cases/OTP_ReportUNSC_3-Darfur_English.pdf>

Judgment on the Prosecutor's Application for Extraordinary Review of Pre-Trial Chamber I's 31 March 2006 Decision Denying Leave to Appeal, *Situation in the Democratic Republic of the Congo*, Case No. ICC-01/04-168, Pre-Trial Chamber I, 13 July 2006

2007 Decision on the Confirmation of Charges, *Prosecutor v. Thomas Lubanga Dyilo*, Case No. ICC-01/04-01/06, Pre-Trial Chamber 1, 29 January 2007

Office of the Prosecutor, The International Criminal Court, 'ICC Prosecutor Presents Evidence on Darfur Crimes', (Press, Release, 27 February 2007) <www.icc-cpi.int/press/pressreleases/230.html>

Decision on the Prosecution Application under Article 58(7) of the Statute, *Prosecutor v. Ahmad Muhammad Harun ('Ahmad Harun') and Ali Muhammad Al Abd-Al-Rahman ('Ali Kushayb')*, Case No. ICC-02/05-01/07, Pre-Trial Chamber I, 27 April 2007

2008 Judgment on the appeal of Mr. Germain Katanga against the decision of Pre-Trial Chamber 1 entitled "Decision on the Defence Request Concerning Languages", *Prosecutor v. Germain Katanga*, Case No. ICC-01/04-01/07 (OA3), Appeals Chamber, 27 May 2008

Judgment on the Appeal Against the Decision on Joinder rendered on 10 March 2008 by the Pre-Trial Chamber in the Germain Katanga and Mathieu Ngudjolo Chui Cases, *Prosecutor v. Germain Katanga and Mathieu Ngudjolo Chui*, Case No. ICC-01/04-01/07, Appeals Chamber, 9 June 2008

Decision on the Confirmation of Charges, *Prosecutor v. Germain Katanga and Mathieu Ngudjolo Chui*, Case No. ICC-01/04-01/07, Pre-Trial Chamber, 30 September 2008

2009 Decision on the Prosecution's Application for a Warrant of Arrest against Omar Hassan Ahmad Al Bashir, *Prosecutor v. Omar Hassan Ahmad Al Bashir ('Omar Al Bashir')*, Case No. ICC-02/05-01/09, Pre-Trial Chamber I, 4 March 2009

Decision Pursuant to Article 61(7)(a) and (b) of the Rome Statute on the Charges of the Prosecutor Against Jean-Pierre Bemba Gombo, *Prosecutor v. Jean-Pierre Bemba Gombo*, Case No. ICC-01/05-01/08, Pre-Trial Chamber II, 15 June 2009

2011 Judgment on the appeal of the Republic of Kenya against the decision of Pre-Trial Chamber II of 30 May 2011 entitled "Decision on the Application by the Government of Kenya Challenging the Admissibility of the Case Pursuant to Article 19(2)(b) of the Statute", *Prosecutor v. William Samoei Ruto, Henry Kiprono Kosgey and Joshua Arap Sang*, Case No. ICC-01/09-01/11 OA, Appeals Chamber, 30 August 2011

Judgment on the appeal of the Republic of Kenya against the decision of Pre-Trial Chamber II of 30 May 2011 entitled "Decision on the Application by the Government of Kenya Challenging the Admissibility of the Case Pursuant to Article 19(2)(b) of the Statute", *Prosecutor v. Francis Kirimi Muthaura, Uhuru Muigai Kenyatta and Mohammed Hussein Ali*, Case No. ICC-01/09-02/11 OA, Appeals Chamber, 30 August 2011

2012 Judgment pursuant to Article 74 of the Statute, *Prosecutor v. Thomas Lubanga Dyilo*, Case No. ICC-01/04-01/06, Trial Chamber I, 14 March 2012

Decision on Sentence pursuant to Article 76 of the Statute, *Prosecutor v. Thomas Lubanga Dyilo*, Case No. ICC-01/04-01/06-2901, Trial Chamber I, 10 July 2012

C) INTERNATIONAL CRIMINAL TRIBUNAL FOR THE FORMER YUGOSLAVIA

1995 Decision on the Prosecutor's Motion Requesting Protective Measures for Victims and Witnesses, *Prosecutor v. Duško Tadić*, Case No. IT-94-1, Trial Chamber, 10 August 1995

Decision on Defence Motion for Interlocutory Appeal on Jurisdiction, *Prosecutor v. Duško Tadić*, Case No. IT-94-1, Appeals Chamber, 2 October 1995

1997 Opinion and Judgment, *Prosecutor v. Duško Tadić*, Case No. IT-94-1-T, Trial Chamber, 7 May 1997

Judgement on the Request of the Republic of Croatia for Review of the Decision of Trial Chamber II of 18 July 1997, *Prosecutor v. Tihomir Blaskic*, Case No. IT-95-14-AR, Appeals Chamber, 29 October 1997

1998 Judgement, *Prosecutor v. Zejnil Delalic, Zdravko Mucic, Hazim Delic and Esad Landzo*, Case No. IT-96-21-T, Trial Chamber, 16 November 1998

Judgement, *Prosecutor v. Furundzjia*, Case No. IT-95-17/1, Trial Chamber, 10 December 1998

1999 Judgement, *Prosecutor v. Duško Tadić*, Case No. IT-94-1, Appeals Chamber, 15 July 1999

Judgement, *Prosecutor v. Goran Jelesić*, Case No. IT-95-10-I, Trial Chamber, 14 December 1999

2000 Judgement, *Prosecutor v. Zoran Kupreskic, Mirjan Kupreskic, Vlatko Kupreskic, Drago Josipovic, Dragan Papic and Vladimir Santic*, Case No. IT-95-16-T, Trial Chamber, 14 January 2000

Judgement, *Prosecutor v. Tihomir Blaskic*, Case No. IT-95-14-T, Trial Chamber, 3 March 2000

Final Report to the Prosecutor of the International Criminal Tribunal for the former Yugoslavia by the Committee Established to Review the NATO Bombing

Campaign Against the Federal Republic of Yugoslavia, 13 June 2000, 39 ILM 1257

Office of the Prosecutor, Press Release, 'Prosecutor's Report on the NATO Bombing Campaign' (13 June 2000) ICTY Doc PR/P.I.S./510-e

2001 Judgement, *Prosecutor v. Zejnil Delalic, Zdravko Mucic, Hazim Delic and Esad Landzo 'Celebici Case')*, Case No. IT-96-21-A, Appeals Chamber, 20 February 2001

Judgement on Appeal by Anto Nobilo against Finding of Contempt, *Prosecutor v. Aleksovski*, Case No. IT-95-14/1-AR77, Appeals Chamber, 30 May 2001

Judgement, *Prosecutor v Radislav Krstić*, Case No. IT-98-33-T, Trial Chamber, 2 August 2001

Judgement on Defence Motions to Acquit, *Prosecutor v Dusko Sikirica, Damir Dosen and Dragan Kolundzjia*, Case No. IT-95-8-T, Trial Chamber, 3 September 2001

2002 Judgement, *Prosecutor v. Dragoljub Kunarac, Radomir Kovac and Zoran Vukovic*, Case No. IT-96-23 and IT-96-23/1-A, Appeals Chamber, 12 June 2002

2003 Decision on Dragoljub Ojdanić's Motion Challenging Jurisdiction – *Joint Criminal Enterprise, Prosecutor v. Milan Milutinović, Nikola Šainović and Dragoljub Ojdanić*, Case. No. IT-99-37-AR72, Appeals Chamber, 21 May 2003

Judgement, *Prosecutor v. Milomir Stakić*, Case No. IT-97-24-T, Trial Chamber II, 31 July 2003

Judgement and Opinion, *Prosecutor v. Stanislav Galic*, Case No. IT-98-29-T, Trial Chamber I, 5 December 2003

2004 Judgement, *Prosecutor v. Mitar Vasiljević*, Case No. IT-98-32-A, Appeals Chamber, 25 February 2004

Judgement, *Prosecutor v. Radislav Krstić*, Case No. IT-98-33-A, Appeals Chamber, 19 April 2004

Decision on Motion for Judgement of Acquittal, *Prosecutor v. Slobodan Milosevic*, Case No. IT-02-54-T, Trial Chamber, 16 June 2004

Judgement, *Prosecutor v. Tihomir Blaskic*, Case No. IT-95-14-A, Appeals Chamber, 29 July 2004

Decision on Motions for Acquittal Pursuant to Rule 98 *bis* of the Rules of Procedure and Evidence, *Prosecutor v. Enver Hadzihasanoinc and Amir Kubura*, Case No. IT-01-47-T, Trial Chamber, 27 September 2004

Judgement, *Prosecutor v. Dario Kordić and Mario Čerkez*, Case No. IT-95-14/2-A, Appeals Chamber, 17 December 2004

2005 Judgement, *Prosecutor v. Pavel Strugar*, Case No. IT-01-42-T, Trial Chamber II, 31 January 2005

Judgement, *Prosecutor v. Miroslav Kvočka, Mlado Radić, Zoran Žigić and Dragoljub Prcac*, Case No. IT-98-30/1-A, Appeals Chamber, 28 February 2005

Decision on Joint Defence Interlocutory Appeal of Trial Chamber Decision on Rule 98 *bis* Motions for Acquittal, *Prosecutor v. Enver Hadzihasanovic and Amir Kubura*, Case No. IT-01-47-AR73.3, Appeals Chamber, 11 March 2005

2006 Judgement, *Prosecutor v. Milomir Stakić*, Case No. IT-97-24-A, Appeals Chamber, 22 March 2006

2007 Judgement, *Prosecutor v. Radoslav Brdanin*, Case No. IT-99-36-A, Appeals Chamber, 3 April 2007

2008 Judgement, *Prosecutor v. Rasim Delić*, Case No. IT-04-83-T, Trial Chamber I, 15 September 2008

2014 Judgement, *Prosecutor v. Nikola Sainovic, Nebojsa Pavkovic, Vladimir Lazarevic and Sreten Lukic*, Case No. IT-05-87-A, Appeals Chamber, 23 January 2014

D) INTERNATIONAL CRIMINAL TRIBUNAL FOR RWANDA

1997 Decision on the Defence Motion on Jurisdiction, *Prosecutor v. Joseph Kanyabashi*, Case No. ICTR-96-15-T, Trial Chamber II, 18 June 1997

1998 Judgement, *Prosecutor v. Jean-Paul Akayesu*, Case No. ICTR-96-4-T, Trial Chamber, 2 September 1998

Judgement and Sentence, *Prosecutor v. Jean Kambanda*, Case No. ICTR-97-23-S, Trial Chamber, 4 September 1998

1999 Judgement, *Prosecutor v. Clément Kayishema and Obed Ruzindana*, Case No. ICTR-95-1-T, Trial Chamber II, 21 May 1999

Judgement and Sentence, *Prosecutor v. Georges Anderson Nderubumwe Rutaganda*, Case No. ICTR-96-3, Trial Chamber I, 6 December 1999

2001 Decision on the Kingdom Of Belgium's Application to File an *Amicus Curiae* Brief and on the Defence Application to Strike Out the Observations of the Kingdom of Belgium Concerning the Preliminary Response by the Defence, *Prosecutor v. Laurent Semanza*, Case No. ICTR-97-20-T, Trial Chamber, 9 February 2001

2002 Judgement (Reasons), *Prosecutor v. Ignace Bagilishema*, Case No. ICTR-95-1A-A, Appeals Chamber, 3 July 2002

2003 Judgement and Sentence, *Prosecutor v. Laurent Samanza*, Case No. ICTR-97-20-T, Trial Chamber III, 15 May 2003

Judgement and Sentence, *Prosecutor v. Ferdinand Nahimana, Jean-Bosco Barayagwiza and Hassan Ngeze*, Case No ICTR-99-52-T, Trial Chamber I, 3 December 2003

2007 Judgement, *Ferdinand Nahimana, Jean-Bosco Barayagwiza and Hassan Ngeze v. The Prosecutor*, Case No. ICTR-99-52-A, Appeals Chamber, 28 November 2007

E) SPECIAL TRIBUNAL FOR LEBANON

2011 Interlocutory Decision on the Applicable Law: Terrorism, Conspiracy, Homicide, Perpetration, Cumulative Charging, *Ayyash et al*, Case No. ST-11–01/I, Appeals Chamber, 16 February 2011

F) SPECIAL COURT FOR SIERRA LEONE

2012 Judgment, *Prosecutor v. Charles Gankay Taylor*, Case No. SCSL-03–01-T, Trial Chamber II, 18 May 2012

G) EUROPEAN COURT OF HUMAN RIGHTS

1975 *Golder v. United Kingdom* (1975) 1 EHRR 524

1995 *S.W.* v. *United Kingdom: C.R.* v. *United Kingdom* (1995) 21 EHRR 363

1996 *Cantoni v. France* (1996) REVS-1996 V, no. 20

1999 *Başkaya v. Turkey* (1999) 31 EHRR 10

2001 *Al-Adsani v. United Kingdom* (2001) 34 EHRR 273

2006 *Korchuganova v. Russia* No. 75039/01, 8 June 2006

2009 *Liivik v. Estonia* No. 12157/05, 25 June 2009

2010 *Kononov v. Latvia* No. 36376/04, 17 May 2010

van Anraat v. The Netherlands No. 65389/09, 6 July 2010

H) EUROPEAN COURT OF JUSTICE

2005 *Commission of the European Communities v Council of the European Union*, Case C-176/03, Grand Chamber, 13 September 2005, (2005) EC Rep I-07879

I) OTHER INTERNATIONAL COURTS AND TRIBUNALS

1893 *Bering Sea Fur Seals Fisheries Arbitration (Great Britain v. United States)* (1893) 1 Moore's International Arbitration Awards 755

1941 *Trail Smelter Arbitration (United States v. Canada)* (1941) 3 RIAA 1905

1947 *Judgment of the International Military Tribunal, Trial of the Major War Criminals* 1 October 1946, reprinted in (1947) 41 *American Journal of International Law* 172

1949 *Trial of William List and Others* (The Hostages Trial) (Judgment), United States Military Tribunal, Nuremberg, (1949) VIII *Law Reports of Trials of War Criminals*, Case No. 47

United States. v. Flick and Others (Judgment), United States Military Tribunal, Nuremberg, (1949) IX War Crimes Reports 1

United States. v. Krupp and Others (Judgment), United States Military Tribunal, Nuremberg, (1949) X War Crimes Reports 69

1998 *Beef Hormones, EC Measures Concerning Meat and Meat Products* (Decision), Panel of the World Trade Organisation, WT/DS26/AB/R, WT/DS48/AB/R, 16 January 1998

2006 *EC – Measures Affecting Biotech Products* (Decision), Panel of the World Trade Organisation, DS291, 292, 293, 26 September 2006

LG&E Energy Corp., LG&E Capital Corp., LG&E International Inc. v. Argentine Republic (Decision on Liability), International Centre for Settlement of Investment Disputes (ICSID), Case No. ARB/02/1, 3 October 2006

IV. DOMESTIC JUDICIAL DECISIONS

A) AUSTRALIA

1991 *Polyukhovich v Commonwealth and Another* (1991) 172 CLR 501

1999 *Nulyarimma v Thompson* (1999) 165 ALR 621

B) CANADA

1994 *R. v. Finta* [1994] 1 SCR 701, Supreme Court of Canada

C) FRANCE

1988 *Féderation National des Déportées et Internés Résistants et Patriots and Others v. Barbie* 100 ILR 330

D) ISRAEL

1961 *Attorney-General of the Government of Israel v Eichmann* (1961) 36 ILR 5, District Court of Jerusalem

1962 *Eichmann v Attorney-General of the Government of Israel* (1962) 36 ILR 277, Supreme Court of Israel

E) THE NETHERLANDS

2007 *Re* Frans van Anraat, LJN: BA4676, Court of Appeal in The Hague, 2200050906 – 2, 9 May 2007

F) UNITED KINGDOM

1999 *R v. Bow Street Metropolitan Magistrate and others, Ex Parte* Pinochet Ugarte (Amnesty International and others intervening) (No 3) (1999) 2 All ER 97 (HL)

G) UNITED STATES

1865 *United States v. Wirz* Am. St. Trials 657–874

1942 *Ex Parte Quirin*, United States Supreme Court, (1942) 317 US 1

2006 *Hamdan v. Rumsfeld, Secretary of Defense, et al* (29 June 2006), United States Supreme Court, 548 U.S. 557 (2006)

2007 *Commonwealth of Massachusetts v. EPA* (2 April 2007), United States Supreme Court, 549 U.S. 497 (2007)

2011 *American Electric Power v. Connecticut* (20 June 2011), United States Supreme Court, 131 S. CT. 2527 (2011)

V. DOMESTIC LEGISLATION AND REGULATIONS

A) AUSTRALIA

2002 *International Criminal Court Act* 2002

International Criminal Court (Consequential Amendments) Act 2002

B) BELGIUM

2006 Weapons Law of 8 June 2006

C) GERMANY

1945 Law No.10 of the Control Council of Germany, Punishment of Persons Guilty of War Crimes, Crimes Against Peace and Against Humanity (20 December 1945) 36 ILR 31

D) IRAQ

2003 Statute of the Iraqi Special Tribunal 2003 (10 December 2003), English version reproduced in Al-Waqai Al-Iraqiya, Official Gazette of Iraq, Volume 44 <www.loc.gov/law/help/hussein/docs/CoalitionProvisionalAuthorityOrder48en.pdf>

E) ISRAEL

1951 Nazi and Nazi Collaborators (Punishment) Law

F) NEW ZEALAND

1999 *International Crimes and International Criminal Court Act* 2000

G) UNITED STATES

1963 Clean Air Act

VI. LITERATURE

A) BOOKS AND MONOGRAPHS / BOOK CHAPTERS

Hans Addink, Bas Arts and Arthur Mol, 'Climate change policy in changing contexts: globalization, political modernization and legal innovation', in Ekko C. van Ierland, Joyeeta Gupta and Marcel T.J. Kok (eds), *Issues in International Climate Policy: Theory and Policy* (2003)

Yusuf Aksar, *Implementing Humanitarian Law: From the Ad Hoc Tribunals to a Permanent International Criminal Court* (2004)

Kai Ambos, *Treatise on International Criminal Law: Volume 1: Foundations and General Part* (2013)

David Armstrong, Lorna Lloyd and John Redmond, *From Versailles to Maastricht: International Organisation in the Twentieth Century* (1996)

Kelly Askin, 'Women's Issues in International Criminal Law: Recent Developments and the Potential Contribution of the ICC', in Dinah Shelton (ed), *International Crimes, Peace, and Human Rights: The Role of the International Criminal Court* (2000), 47

Anthony Aust, *Handbook of International Law* (2005)

Anthony Aust, *Modern Treaty Law and Practice* (2nd ed, 2007)

Jay E. Austin and Carl E. Bruch (eds), *The Environmental Consequences of War: Legal, Economic, and Scientific Perspectives* (2000)

Mohamed Elewa Badar, 'The *Mens Rea* Enigma in the Jurisprudence of the International Criminal Court', in Larissa van den Herik and Carsten Stahn (eds), *The Diversification and Fragmentation of International Criminal Law* (2012), 503

Mohamed Elewa Badar, *The Concept of Mens Rea in International Criminal Law: The Case for a Unified Approach* (2013)

Ilias Bantekas and Susan Nash, *International Criminal Law* (3rd ed, 2007)

Gary Bass, *Stay the Hand of Vengeance: The Politics of War Crimes Tribunals* (2000)

M. Cherif Bassiouni (ed), *A Draft International Criminal Code and Draft Statute for an International Criminal Tribunal* (1987)

M. Cherif Bassiouni, *Crimes Against Humanity in International Criminal Law* (2nd ed, 1999)
M. Cherif Bassiouni (ed), *A Manual on International Humanitarian Law and Arms Control Agreements* (2000)
M. Cherif Bassiouni, 'The Normative Framework of International Humanitarian Law: Overlaps, Gaps, and Ambiguities', in Michael J.N. Schmitt (ed), *International Law Studies Volume 75: International Law Across the Spectrum of Conflict* (2000), 1
M. Cherif Bassiouni, *Introduction to International Criminal Law* (2003)
Yves Beigbeder, *Judging War Criminals: The Politics of International Justice* (1999)
Geoffrey Best, *Humanity in Warfare: The Modern History of the International Law of Armed Conflicts* (1983)
Geoffrey Best, 'The restraint of war in historical and philosophical perspective', in A.J.M. Delissen and G.J. Tanja (eds), *Humanitarian Law of Armed Conflict: Challenges Ahead, Essays in Honour of Frits Kalshoven* (1991), 3
Geoffrey Best, *War and Law since 1945* (1997)
Daniel Bethlehem, 'The methodological framework of the Study', in Elizabeth Wilmshurst and Susan Breau (eds), *Perspectives on the ICRC Study on Customary International Humanitarian Law* (2007), 3
Robert Bevan, *The Destruction of Memory: Architecture at War* (2006)
Patricia Birnie and Alan Boyle, *International Law and the Environment* (1st ed, 1992)
Patricia Birnie and Alan Boyle, *International Law and the Environment* (2nd ed, 2002)
Gilbert Bitti, 'Two Bones of Contention Between Civil and Common Law: The Record of the Proceedings and the Treatment of a Concursus Delictorum', in Horst Fischer, Claus Kress and Sascha Rolf Lüder (eds), *International and National Prosecution of Crimes Under International Law: Current Developments* (2001), 273
Gideon Boas, *The Milošević Trial: Lessons for the Conduct of Complex International Criminal Proceedings* (2007)
Daniel Bodansky, Jutta Brunnée and Ellen Hey (eds), *The Oxford Handbook of International Environmental Law* (2007)
Ben Boer, Ross Ramsay and Donald R. Rothwell, *International Environmental Law in the Asia Pacific* (1998)
Shelia Bone (ed), *Osborn's Concise Law Dictionary* (9th ed, 2001)
Cherie Booth, 'Prospects and issues for the International Criminal Court: lessons from Yugoslavia and Rwanda', in Philippe Sands (ed), *From Nuremberg to The Hague: The Future of International Criminal Justice* (2003), 157
Michael Bothe, Karl Josef Partsch and Waldemar A. Solf, *New Rules for Victims of Armed Conflicts: Commentary on the Two 1977 Protocols Additional to the Geneva Conventions of 1949* (1982)
Bill Bowring, *The Degradation of the International Legal Order? The Rehabilitation of Law and the Possibility of Politics* (2008)
Patrick J. Boylan, *Review of the Convention for the Protection of Cultural Property in the Event of Armed Conflict: (The Hague Convention of 1954)* (1993)
Alan E. Boyle and Michael R. Anderson (eds), *Human Rights Approaches to Environmental Protection* (1998)
Ian Brownlie, *International Law and the Use of Force by States* (1963)
Michael Byers, *War Law: International Law and Armed Conflict* (2005)

Michael Byers and Simon Chesterman, 'Changing the rules about rules? Unilateral humanitarian intervention and the future of international law', in J.L. Holzgrefe and Robert O. Keohane (eds), *Humanitarian Intervention: Ethical, Legal, and Political Dilemmas* (2003), 177

Neil Carter, *The Politics of the Environment: Ideas, Activism, Policy* (2nd ed, 2007)

Antonio Cassese, 'Means of Warfare: the traditional and the new law', in A. Cassese (ed), *The New Humanitarian Law of Armed Conflict* (1979)

Antonio Cassese, Albin Eser, Giorgio Gaja, Philip Kirsch, Alain Pellet and Bert Swart (eds), *The Rome Statute of the International Criminal Court: A Commentary* (2002)

Antonio Cassese, *International Criminal Law* (2003)

Antonio Cassese, 'The Role of Internationalized Courts and Tribunals in the Fight Against International Criminality', in Cesare P.R. Romano, André Nollkaemper and Jann K. Kleffner (eds), *Internationalized Criminal Courts: Sierra Leone, East Timor, Kosovo, and Cambodia* (2004), 3

Antonio Cassese, *International Law* (2nd ed, 2005)

Antonio Cassese, *International Criminal Law* (2nd ed, 2008)

Antonio Cassese, Paoloa Gaeta and John R.W.D. Jones (eds), *The Rome Statute of the International Criminal Court: A Commentary* (2002)

Cassese's *International Criminal Law* (3rd ed, 2013) (revised by A. Cassese, P. Gaeta, L. Baig, M. Fan, C. Gosnell and A. Whiting)

Laurence Boisson de Chazournes and Philippe Sands (eds), *International Law, The International Court of Justice and Nuclear Weapons* (1999)

Simon Chesterman, Thomas M. Franck and David M. Malone, *Law and Practice of the United Nations: Documents and Commentary* (2008)

Marie Isabelle Chevrier, Krzysztof Chomiczewski, Henri Garrigue, György Granasztói, Malcolm R. Dando and Graham S. Pearson (eds), *The Implementation of Legally Binding Measures to Strengthen the Biological and Toxin Weapons Convention* (2004)

R.R. Churchill and A.V. Lowe, *The Law of the Sea* (2nd ed, 1988)

Alex Conte, *Security in the 21st Century: The United Nations, Afghanistan and Iraq* (2005)

Michael Cottier, 'Did NATO Forces Commit War Crimes During the Kosovo Conflict? Reflections on the Prosecutor's Report of 13 June 2000', in Horst Fischer, Claus Kress and Sascha Rolf Lüder (eds), *International and National Prosecution of Crimes Under International Law: Current Developments* (2001), 505

James Crawford, 'The drafting of the Rome Statute', in Philippe Sands (ed), *From Nuremberg to The Hague: The Future of International Criminal Justice* (2003), 109

John H. Currie, Craig Forcese and Valerie Oosterveld, *International Law: Doctrine, Practice, and Theory* (2007)

Ciara Damgaard, *Individual Criminal Responsibility for Core International Crimes: Selected Pertinent Issues* (unpublished doctoral thesis submitted to the Faculty of Law, University of Copenhagen, Denmark 2007)

Gerard DeGroot, *The Bomb: A Life* (2004)

Ingrid Detter, *The Law of War* (2nd ed, 2000)

Yoram Dinstein, *War, Aggression and Self-Defence* (3rd ed, 2001)

Yoram Dinstein, *The Conduct of Hostilities under the Law of International Armed Conflict* (2004)

Knut Dörmann, 'War Crimes in the Elements of Crimes', in Horst Fischer, Claus Kress and Sascha Rolf Lüder (eds), *International and National Prosecution of Crimes Under International Law: Current Developments* (2001), 95

Knut Dörmann, *Elements of War Crimes under the Rome Statute of the International Criminal Court: Sources and Commentary* (2004)

Mark A. Drumbl, 'Waging War against the World: The Need to Move from War Crimes to Environmental Crimes', in Jay E. Austin and Carl E. Bruch (eds), *The Environmental Consequences of War: Legal, Economic, and Scientific Perspectives* (2000), 620

Mark A. Drumbl, *Reimagining Child Soldiers in International Law and Policy* (2012)

Henry Dunant, *A Memory of Solferino* (reprinted 1986)

Treasa Dunworth, 'The International Crimes and International Criminal Court Act 2000 (NZ): A Model for the Region?', in Neil Boister and Alberto Costi (eds), *Regionalising International Criminal Law in the Pacific* (2006), 145

Dan Fahey, 'Environmental and Health Consequences of the Use of Depleted Uranium Weapons', in Avril McDonald, Jann K. Kleffner and Brigit Toebes (eds), *Depleted Uranium Weapons and International Law: A Precautionary Approach* (2008), 29

Richard A. Falk, *Law, Morality, and War in the Contemporary World* (1984)

Richard Falk, 'The Environmental Law of War: An Introduction', in Glen Plant (ed), *Environmental Protection and the Law of War: A "Fifth Geneva" Convention on the Environment in Time of Armed Conflict* (1992), 78

William J. Fenrick, 'Article 8: War Crimes – article 8(2)(b)(iv)', in Otto Triffterer (ed), *Commentary on the Rome Statute of the International Criminal Court: Observers' Notes, Article by Article* (1999), 197

Mark Findlay, *Governing through Globalised Crime: Futures for international criminal justice* (2008)

Oonagh E. Fitzgerald, 'Implementation of International Humanitarian and Related International Law in Canada', in Oonagh E. Fitzgerald (ed), *The Globalized Rule of Law: Relationships between International and Domestic Law* (2006), 625

Malgosia Fitzmaurice, 'International Responsibility and Liability', in Daniel Bodansky, Jutta Brunnée and Ellen Hey (eds), *The Oxford Handbook of International Environmental Law* (2007), 1010

Dieter Fleck (ed), *The Handbook of Humanitarian Law in Armed Conflicts* (1999)

Gary W. Florkowski, *Managing Global Legal Systems* (2006)

Steven Freeland, 'Commentary on 'Decision on Prosecutor's Appeal on Admissibility of Evidence, *Prosecutor v Zlatko Aleksovski*', in André Klip and Göran Sluiter (eds), *Annotated Leading Cases of International Criminal Tribunals, Volume III, The International Criminal Tribunal for the Former Yugoslavia, 1997–1999* (2001), 260

Steven Freeland, 'Commentary on the Findings of the Prosecutor of the International Criminal Tribunal for the Former Yugoslavia into the actions of NATO in Serbia and Kosovo during 1999', in André Klip and Göran Sluiter (eds), *Annotated Leading Cases of International Criminal Tribunals, Volume V, The International Criminal Tribunal for the Former Yugoslavia, 1999–2000* (2003), 38

Steven Freeland, 'Crimes against the Environment – A Role for the International Criminal Court?', in Alberto Costi and Yves-Louis Sage (eds), *Droit de l'Environnement dans le Pacifique: Problématiques et Perspectives Croisées /*

Environmental Law in the Pacific: International and Comparative Perspectives (2005), 335
Steven Freeland, 'International Criminal Justice and the Death Penalty', in Jon Yorke (ed), *The Right to Life and the Value of Life: Orientations in Law, Politics and Ethics* (2010), 193
Steven Freeland, 'International Criminal Governance: Will the International Criminal Court be an 'Effective' Mechanism for Justice?', in Michael Head, Scott Mann and Simon Kozlina (eds), *Transnational Governance: Emerging Models of Global Legal Regulation* (2012), 213
Steven Freeland, 'The Future Regulation of Outer Space – Technology, Warfare and International Law' (2014) *Yearbook of the European Space Policy Institute* (forthcoming)
Mark Freeman and Gibran van Ert, *International Human Rights Law* (2004)
Judith Gardam (ed), *Humanitarian* Law (1999)
Elizabeth Glass Geltman, *Modern Environmental Law: Policy and Practice* (1997)
Alexis Goh and Steven Freeland, 'Australia and the International Criminal Court', in Gabriel A. Moens and Rudolphe Biffot (eds), *The Convergence of Legal Systems in the 21st Century – An Australian Approach* (2002), 285
Jozef Goldblat, 'The Mitigation of Environmental Disruption by War: Legal Approaches', in Arthur Westing (ed), *Environmental Hazards of War: Releasing Dangerous Forces in an Industrialized World* (1990) 48
John P. Grant and J. Craig Barker, *International Criminal Law Deskbook* (2006)
Leslie C. Green, *The contemporary law of armed conflict* (2nd ed, 2000)
Hugo Grotius, *De Jure Belli ac Pacis* (On the Law of War and Peace) (1625)
Michael Haas, *International Human Rights: A Comprehensive Introduction* (2008)
John Hagan, *Justice in the Balkans: Prosecuting War Crimes in The Hague Tribunal* (2003)
Françoise Hampson, 'Other areas of customary law in relation to the Study', in Elizabeth Wilmshurst and Susan Breau (eds), *Perspectives on the ICRC Study on Customary International Humanitarian Law* (2007), 50
Xue Hanqin, *Transboundary Damage in International Law* (2003)
D.J. Harris, *Cases and Materials on International Law* (5th ed, 1998)
D.J. Harris, *Cases and Materials on International Law* (6th ed, 2004)
James Harrison, 'The role of international conventions in solving transboundary pollution disputes', in Michael Faure and Song Ying (eds), *China and International Environmental Liability: Legal Remedies for Transboundary Pollution* (2008), 38
Martin Hedemann-Robinson, *Enforcement of European Union Environmental Law: Legal Issues and Challenges* (2007)
Wolff Heintschel von Heinegg, 'The Law of Armed Conflict at Sea', in Dieter Fleck (ed), *The Handbook of Humanitarian Law in Armed Conflicts* (1999), 405
Jean-Marie Henckaerts and Louise Doswald-Beck (eds), *Customary International Humanitarian Law* (Volume I: Rules, 2005)
Jean-Marie Henckaerts and Louise Doswald-Beck (eds), *Customary International Humanitarian Law* (Volume II: Practice, Part 1, 2005)
Tim Hillier, *Sourcebook on Public International Law* (1998)
Jane Holder and Maria Lee, *Environmental Protection, Law and Policy: Text and Materials* (2nd ed, 2007)

Karen Hulme, *War Torn Environment: Interpreting the Legal Threshold* (2004)
Karen Hulme, 'Natural Environment', in Elizabeth Wilmshurst and Susan Breau (eds), *Perspectives on the ICRC Study on Customary International Humanitarian Law* (2007), 204
David Hunter, James Salzman and Durwood Zaelke, *International Environmental Law and Policy* (2nd ed, 2002)
Nina H.B. Jørgensen, *The Responsibility of States for International Crimes* (2000)
Barry Kellman, 'The Chemical Weapons Convention: A Verification and Enforcement Model for Determining Legal Responsibility for Environmental Harm Caused by War', in Jay E. Austin and Carl E. Bruch (eds), *The Environmental Consequences of War: Legal, Economic, and Scientific Perspectives* (2000), 579
Kriangsak Kittichaisaree, *International Criminal Law* (2001)
André Klip, *European Criminal Law* (2nd ed, 2012)
André Klip and Göran Sluiter (eds), *Annotated Leading Cases of International Criminal Tribunals* (first published 1999)
Dr. Abdul G. Koroma, 'Foreword', in Jean-Marie Henckaerts and Louise Doswald-Beck (eds), *Customary International Humanitarian Law* (Volume I: Rules, 2005), xii
Erik V. Koppe, *The Use of Nuclear Weapons and the Protection of the Environment During International Armed Conflict* (2006)
Helle Krunke, 'Introduction: European Law as a Process between Constitutionalization and Democratization', in Hanne Petersen, Anne Lise Kjaer, Helle Krunke and Mikael Rask Madsen (eds), *Paradoxes of European Legal Integration* (2008), 9
Flavia Lattanzi, 'Crimes Against Humanity in the Jurisprudence of the International Criminal Tribunals for the Former Yugoslavia and Rwanda', in Horst Fischer, Claus Kress and Sascha Rolf Lüder (eds), *International and National Prosecution of Crimes Under International Law: Current Developments* (2001), 473
Andreas Laursen, *Changing International Law to Meet New Challenges: Interpretation, Modification and the Use of Force* (2006)
Raphael Lemkin, *Axis Rule in Occupied Europe, Laws of Occupation, Analysis of Government, Proposals for Redress* (1944)
Elli Louka, *International Environmental Law: Fairness, Effectiveness, and World Order* (2006)
Susanne Malmstrom, 'Genocide Case Law at the ICTY', in Roberto Bellelli (ed), *International Criminal Justice: Law and Practice from the Rome Statute to its Review* (2010), 267
Michael Matheson, *Council Unbound: The Growth of UN Decision Making on Conflict and Postconflict Issues after the Cold War* (2006)
Hilaire McCoubrey, *International Humanitarian Law: Modern Developments in the Limitation of Warfare* (2nd ed, 1998)
Lindsay Moir, *The Law of Internal Armed Conflict* (2002)
Frederic de Mulinen, *Handbook on the Law of War for the Armed Forces* (1987)
Claud Mullins, *The Leipzig Trials: An Account of the War Criminals' Trials and a Study of German Mentality* (1921)
Henry Kibet Mutai, 'Lessons from the Recent Decisions of the East African Court of Justice (EACJ) for Southern African Trade Blocs', in Trudi Hartzenberg (ed), *WTO Dispute Settlement: An African Perspective* (2008), 295

Eric P.J. Myjer (ed), *Issues of Arms Control Law and the Chemical Weapons Convention* (2001)
Karma Nabulsi, *Traditions of War: Occupation, Resistance, and the Law* (2004)
John O'Brien, *International Law* (2002)
Mary Ellen O'Connell, *International Law and the Use of Force: Cases and Materials* (2nd ed, 2009)
Phoebe Okowa, 'Environmental Justice in Situations of Armed Conflict', in Jonas Ebbesson and Phoebe Okawa (eds), *Environmental Law and Justice in Context* (2009), 231
Richard Overy, 'The Nuremberg trials: international law in the making', in Philippe Sands (ed), *From Nuremberg to The Hague: The Future of International Criminal Justice* (2003), 1
Jacqueline Peel, *The Precautionary Principle in Practice: Environmental Decision-Making and Scientific Uncertainty* (2005)
Glen Plant (ed), *Environmental Protection and the Law of War: A "Fifth Geneva" Convention on the Environment in Time of Armed Conflict* (1992)
René Provost, *International Human Rights and Humanitarian Law* (2002)
Thomas Richter, 'Transboundary environmental crimes: an analysis of Chinese and European Law', in Michael Faure and Song Ying (eds), *China and International Environmental Liability: Legal Remedies for Transboundary Pollution* (2008), 253
Adam Roberts, 'Failures in protecting the environment in the 1990–91 Gulf War', in Peter Rowe (ed), *The Gulf War 1990–91 in International and English Law* (1993), 111
Adam Roberts, 'Environmental Issues in International Armed Conflict: The Experience of the 1991 Gulf War', in Richard J. Grunawalt, John E. King and Ronald S. McClain (eds), *Protection of the Environment during Armed Conflict* (1996), 222
Adam Roberts, 'The Law of War and Environmental Damage', in Jay E. Austin and Carl E. Bruch (eds), *The Environmental Consequences of War: Legal, Economic, and Scientific Perspectives* (2000), 47
Adam Roberts and Richard Guelff, *Documents on the Laws of War* (3rd ed, 2005)
Naomi Roht-Arriaza and Javier Mariezcurrena (eds), *Transitional Justice in the Twenty-First Century: Beyond Truth versus Justice* (2006)
Steven D. Roper and Lilian A. Barria, *Designing Criminal Tribunals – Sovereignty and International Concerns in the Protection of Human Rights* (2006)
Gerard C. Rowe, 'Environmental Justice as an Ethical, Economic and Legal Principle', in Benjamin J. Richardson and Klaus Bosselmann (eds), *Environmental Justice and Market Mechanisms* (1999), 58
Wiebke Rückert and Georg Witschel, 'Genocide and Crimes Against Humanity in the Elements of Crimes', in Horst Fischer, Claus Kress and Sascha Rolf Lüder (eds), *International and National Prosecution of Crimes Under International Law: Current Developments* (2001), 59
Philippe Sands, 'After Pinochet: the role of national courts', in Philippe Sands (ed), *From Nuremberg to The Hague: The Future of International Criminal Justice* (2003), 68
Philippe Sands, *Principles of International Environmental Law* (2nd ed, 2005)
Philippe Sands and Paolo Galizzi (eds), *Documents in International Environmental Law* (2nd ed, 2004)

William Schabas, 'Commentary on Judgement, *Prosecutor v Furundžija*', in André Klip and Göran Sluiter (eds), *Annotated Leading Cases of International Criminal Tribunals, Volume III, The International Criminal Tribunal for the Former Yugoslavia, 1997–1999* (2001), 753

William A. Schabas, *Genocide in International Law* (2000)

William A. Schabas, *An Introduction to the International Criminal Court* (2001)

William A. Schabas, 'The Crime of Genocide in the Jurisprudence of the International Criminal Tribunals for the Former Yugoslavia and Rwanda', in Horst Fischer, Claus Kress and Sascha Rolf Lüder (eds), *International and National Prosecution of Crimes Under International Law: Current Developments* (2001), 447

Dietrich Schindler and Jiri Toman (eds), *The Law of Armed Conflicts* (4th ed, 2004)

Michael N. Schmitt, Charles H.B. Garraway and Yoram Dinstein, *The Manual on the Law of Non-International Armed Conflict* (2006)

Georg Schwarzenberger, *International Law* (3rd ed, Vol 1, 1957)

Iain Scobbie, 'The approach to customary international law in the Study', in Elizabeth Wilmshurst and Susan Breau (eds), *Perspectives on the ICRC Study on Customary International Humanitarian Law* (2007), 15

Malcolm N. Shaw, 'Genocide and International Law', in Yoram Dinstein (ed), *International Law at a Time of Perplexity* (1989), 797

Malcolm Shaw, *International Law* (5th ed, 2003)

Dinah Shelton, *Commitment and Compliance; The Role of Non-Binding Norms in the International Legal System* (2000)

Carsten Stahn, 'The Future of *Jus Post Bellum*', in Carsten Stahn and Jann K. Kleffner (eds), *Jus Post Bellum: Towards a Law of Transition From Conflict to Peace* (2008) 231

Henry J. Steiner and Philip Alston, *International Human Rights in Context: Law, Politics, Morals* (2nd ed, 2000)

Christopher D. Stone, 'The Environment in Wartime: An Overview', in Jay E. Austin and Carl E. Bruch (eds), *The Environmental Consequences of War: Legal, Economic, and Scientific Perspectives* (2000), 16

Lyal S. Sunga, *The Emerging System of International Criminal Law: Developments in Codification and Implementation* (1997)

Phan Nguyen Toan, 'Vietnam', in Terri Mottershead (ed), *Environmental Law and Enforcement in the Asia-Pacific Rim* (2002), 547

Samuel Totten and Paul R. Bartrop (eds), *The Genocide Studies Reader* (2009)

Gillian D. Triggs, *International Law: Contemporary Principles and Practices* (2006)

Pietro Verri, *Dictionary of the International Law of Armed Conflict* (translated from French into English by Edward Markee and Susan Mutti) (1992)

Lotta Viikari, *The Environmental Element in Space Law: Assessing the Present and Charting the Future* (2008)

L.C. Vohrah, Kelly Dawn Askin and Daryl A. Mundis, 'Contemporary Law Regulating Armed Conflict at Sea', in Nisuke Ando, Edward McWhinney and Rüdiger Wolfrum (eds), *Liber Amicorum: Judge Shigeru Oda* (Volume 2, 2002),1523

Silja Vöneky, 'Peacetime Environmental Law as a Basis of State Responsibility for Environmental Damage Caused by War', in Jay E. Austin and Carl E. Bruch (eds), *The Environmental Consequences of War: Legal, Economic, and Scientific Perspectives* (2000), 190

Weapons of Mass Destruction Commission, *Weapons of Terror: Freeing the World of Nuclear, Biological and Chemical Arms* (2006)

Ruth G. Wedgwood, 'Proportionality, Cyberwar, and the Law of War', in Michael N. Schmitt and Brian T. O'Donnell (eds), *Computer Network Attack and International Law* (2002), 219

Edith Brown Weiss (ed), *Environmental Change and International Law: New Challenges and Dimensions* (1992)

Edith Brown Weiss, 'Opening the Door to the Environment and to Future Generations', in Laurence Boisson de Chazournes and Philippe Sands (eds), *International Law, The International Court of Justice and Nuclear Weapons* (1999), 338

Gerhard Werle, *Principles of International Criminal Law* (2005)

Gerhard Werle and Florian Jessberger, *Principles of International Criminal Law* (3rd ed, 2014)

Mark Wheelis, 'A Short History of Biological Warfare and Weapons', in Marie Isabelle Chevrier, Krzysztof Chomiczewski, Henri Garrigue, György Granasztói, Malcolm R. Dando and Graham S. Pearson (eds), *The Implementation of Legally Binding Measures to Strengthen the Biological and Toxin Weapons Convention* (2004), 15

Sharon A. Williams, 'Article 17: Issues of Admissibility', in Otto Triffterer (ed), *Commentary on the Rome Statute of the International Criminal Court: Observers' Notes, Article by Article* (1999), 383

Sharon A. Williams, 'The Rome Statute on the International Criminal Court - Universal Jurisdiction or State Consent - To Make or Break the Package Deal', in Michael J.N. Schmitt (ed), *International Law Studies Volume 75: International Law Across the Spectrum of Conflict* (2000), 539

B) JOURNAL ARTICLES

Rosemary Abi-Saaf, 'The 'general principles' of humanitarian law according to the International Court of Justice' (1987) 766 *International Review of the Red Cross* 381

George Aldrich, 'Yugoslavia's Television Studios as Military Objectives' (1999) 1:3 *International Law Forum* 149

Nicholas G. Alexander, 'Notes and Comments: Airstrikes and Environmental Damage: Can the United States be Held Liable for Operation Allied Force?' (2000) 11 *Colorado Journal of International Environmental Law and Policy* 471

Kai Ambos, 'Current Issues in International Criminal Law' (2003) 14 *Criminal Law Forum* 225

Kai Ambos, 'Judicial Creativity at the Special Tribunal for Lebanon: Is There a Crime of Terrorism under International Law?' (2011) 24:3 *Leiden Journal of International Law* 655

Mohammed Amin Al-Midani and Mathilde Cabanettes (with revisions by Susan M. Akram), 'Arab Charter on Human Rights 2004' (2006) 24 *Boston University International Law Journal* 147

Betsy Baker, 'Legal Protections for the Environment in Times of Armed Conflict' (1993) 33 *Virginia Journal of International Law* 351

Stephanie H. Bald, 'Searching for a Lost Childhood: Will the Special Court of Sierra Leone Find Justice for Its Children?' (2002) 18 *American University International Law Review* 537

M. Cherif Bassiouni, 'Introduction to the Symposium: The Normative Framework of International Humanitarian Law: Overlaps, Gaps and Ambiguities' (1998) 8 *Transnational Law and Contemporary Problems* 199

Jack M. Beard, 'The Shortcomings of Indeterminacy in Arms Control Regimes: The Case of the Biological Weapons Convention' (2007) 101(2) *American Journal of International Law* 271

Karima Bennoune, 'Do We Need International Law to Protect Women in Armed Conflict?' (2006–2007) 38 *Case Western Reserve Journal of International Law* 363

Lynn Berat, 'Defending the Right to a Healthy Environment: Toward a Crime of Genocide in International Law' (1993) 11 *Boston University International Law Journal* 327

Louis Rene Beres, 'Iraqi Crimes and International Law: The Imperative to Punish' (1993) 21 *Denver Journal of International Law and Policy* 335

Nancye L. Bethurem, 'Environmental Destruction in the Name of National Security: Will the Old Paradigm Return in the Wake of September 11?' (2002) 8 *Hastings West-Northwest Journal of Environmental Law and Policy* 109

Laurence Boisson de Chazournes, Book Review and Note – Richard J. Grunawalt, John E. King and Ronald S. McClain (eds), 'Protection of the Environment During Armed Conflict' (1998) 92 *American Journal of International Law* 595

Michael Bothe, 'Customary International Humanitarian Law: Some Reflections on the ICRC Study' (2005) 8 *Yearbook of International Humanitarian Law* 143

Michel Bourbonièrre, 'The Ambit of the Law of Neutrality and Space Security' (2006) 36 *Israel Yearbook on Human Rights* 205

Kathleen F. Brickey, 'Environmental Crime at the Crossroads: The Intersection of Environmental and Criminal Law Theory' (1996) 71 *Tulane Law Review* 487

Avi Brisman, 'Toward a More Elaborate Typology of Environmental Values: Liberalizing Criminal Disenfranchisement Laws and Policies' (2007) 33 *New England Journal on Criminal and Civil Confinement* 283

Bartram S. Brown, 'Nationality and Internationality in International Humanitarian Law' (1998) 34 *Stanford Journal of International Law* 348

Bartram S. Brown, 'Primacy or Complementarity: Reconciling the Jurisdiction of National Courts and International Criminal Tribunals' (1998) 23 *The Yale Journal of International Law* 391

Bartram S. Brown, 'U.S. Objections to the Statute of the International Criminal Court: A Brief Response' (1999) 31 *International Law and Politics* 855

Carl E. Bruch, 'All's Not Fair in (Civil) War: Criminal Liability for Environmental Damage in Internal Armed Conflict' (2001) 25 *Vermont Law Review* 695

Jutta Brunneé, 'Environmental Security in the Twenty-First Century: New Momentum for the Development of International Environmental Law' (1995) 18 *Fordham International Law Journal* 1742

François Bugnion, 'Law of Geneva and Law of The Hague' (2001) 844 *International Review of the Red Cross* 901
Mark J.T. Caggiano, 'The Legitimacy of Environmental Destruction in Modern Warfare: Customary Substance over Conventional Form' (1993) 20 *Boston College Environmental Affairs Law Review* 479
James Cameroon and Juli Abouchar, 'The Precautionary Principle: A Fundamental Principle of Law and Policy for the Protection of the Global Environment' (1991) 14:1 *Boston College International and Comparative Law Review* 1
Juliana V. Campagna, 'War or Peace: It is Time for the United States to Ratify the 1954 Hague Convention for the Protection of Cultural Property in the Event of Armed Conflicts' (2005) 17 *Florida Journal of International Law* 271
Burrus M. Carnahan, 'Lincoln, Lieber and the Laws of War: The Origins and Limits of the Principle of Military Necessity' (1998) 92 *American Journal of International Law* 213
Heather Cash, 'Security Council Resolution 1593 and Conflicting Principles of International Law: How the Future of the International Criminal Court is at Stake' (2007) 45 *Brandeis Law Journal* 573
Anthony E. Cassimatis, 'International Humanitarian Law, International Human Rights Law, and Fragmentation of International Law' (2007) 56 *International and Comparative Law Quarterly* 623
Jonathan I. Charney, 'Universal International Law' (1993) 87 *American Journal of International Law* 529
Vincent Chetail, 'The Contribution of the International Court of Justice to International Humanitarian Law' (2003) 850 *International Review of the Red Cross* 235
Christine Chinkin, 'The Challenge of Soft Law: Development and Change in International Law' (1989) 38 *International and Comparative Law Quarterly* 850
Christine Chinkin, 'Editorial Comments: NATO's Kosovo Intervention: Kosovo: A "Good" or "Bad" War?' (1999) 93 *American Journal of International Law* 841
Byung-Sun Cho, 'Emergence of an International Environmental Criminal Law?' (2001) 19 *UCLA Journal of Environmental Law and Policy* 11
Roger S. Clark, 'International Court of Justice: Advisory Proceedings on the Legality of the Threat or Use of Nuclear Weapons (Question Posed by the General Assembly): The Laws of Armed Conflict and the Use or Threat of Use of Nuclear Weapons' (1996) 7 *Criminal Law Forum* 265
John Alan Cohan, 'Modes of Warfare and Evolving Standards of Environmental Protection Under the International Law of War' (2003) 15 *Florida Journal of International Law* 481
Alberto Costi, 'Hybrid Tribunals as a Viable Transitional Justice Mechanism to Combat Impunity in Post-Conflict Situations' (2006) 22 *New Zealand Universities Law Review* 213
Robin Coupland and Dominique Loye, 'The 1899 Hague Declaration concerning Expanding Bullets: A treaty effective for more than 100 years faces complex contemporary issues' (2003) 85 *International Review of the Red Cross* 135
Michael J. Dennis, 'Current Development: Newly Adopted Protocols to the Convention on the Rights of the Child' (2000) 96 *American Journal of International Law* 789

Daniela de Vito, Aisha Gill and Damien Short, 'Rape characterized as genocide' (2009) 10 *SUR International Journal on Human Rights* 29
Kristin Dorman, 'Proportionality and Distinction in the International Criminal Tribunal for the Former Yugoslavia' (2005) 12 *Australian International Law Journal* 83
Knut Dörmann, 'The legal situation of "unlawful / unprivileged combatants"' (2003) 85:849 *International Review of the Red Cross* 45
Mark A. Drumbl, 'Waging War against the World: The Need to Move from War Crimes to Environmental Crimes' (1998) 22 *Fordham International Law Journal* 122
Mark A. Drumbl, 'International Human Rights, International Humanitarian Law, and Environmental Security: Can the International Criminal Court Bridge the Gaps?' (2000) 6 *ILSA Journal of International and Comparative Law* 305
Pierre-Marie Dupuy, 'Soft Law and the International Law of the Environment' (1991) 12 *Michigan Journal of International Law* 420
William G. Eckhardt, 'Command Criminal Responsibility: A Plea for a Workable Standard' (1982) 97 *Military Law Review* 1
Mark Ellis, 'Breaking the Silence: Rape as an International Crime' (2006–2007) 38 *Case Western Reserve Journal of International Law* 225
Joel F. England, 'The Response of the United States to the International Criminal Court: Rejection, Ratification or Something Else?' (2001) 18:3 *Arizona Journal of International and Comparative Law* 941
Aaron Ezekiel, 'The Application of International Criminal Law to Resource Exploitation: Ituri, Democratic Republic of the Congo' (2007) 47 *Natural Resources Journal* 225
Dana Zartner Falstrom, 'Can International Law Survive the 21st Century? Yes, with Patience, Persistence and a Peek at the Past' (2007) 8 *San Diego International Law Journal* 291
Michael G. Faure and Marjolein Visser, 'How to Punish Environmental Pollution? Some Reflections on Various Models of Criminalization of Environmental Harm' (1995) 3 *European Journal of Crime, Criminal Law and Criminal Justice* 31
Paul Fautex and Lothar Gündling, 'The Role of War' (1991) 22(3) *IUCN Bulletin* 25
W. Jason Fisher, 'Targeted Killing, Norms, and International Law' (2007) 45 *Columbia Journal of Transnational Law* 711
Craig J.S. Forrest, 'The Doctrine of Military Necessity and the Protection of Cultural Property During Armed Conflicts' (2007) 37 *California Western International Law Journal* 177
Steven Freeland, 'The Bombing of Kosovo and the Milosevic Trial: Reflections on Some Legal Issues' [2002] *Australian International Law Journal* 150
Steven Freeland, 'Child Soldiers and International Crimes – How Should International Law be Applied?' (2005) 3 *New Zealand Journal of Public and International Law* 303
Steven Freeland, 'Human Rights, the Environment and Conflict – Addressing Crimes against the Environment' (2005) 2 *United Nations SUR International Journal on Human Rights* 112
Steven Freeland, 'How Open Should the Door Be? Declarations by non-States Parties under Article 12(3) of the Rome Statute of the International Criminal Court' (2006) 75(2) *Nordic Journal of International Law* 211

Steven Freeland, 'The Applicability of the *Jus in Bello* Rules of International Humanitarian Law to the Use of Outer Space' (2006) 49 *Proceedings of the Colloquium on the Law of Outer Space* 338

Steven Freeland, 'Towards Universal Justice – Why Countries in the Asia / Pacific Region should Embrace the International Criminal Court' (2007) 5 *New Zealand Journal of Public and International Law* 49

Steven Freeland, 'The Internationalization of Justice – A Case for the Universal Application of International Criminal Norms' (2007) 4 *New Zealand Yearbook of International Law* 45

Steven Freeland, 'Mere Children or Weapons of War – Child Soldiers and International Law' (2008) 29 *University of La Verne Law Review* 19

Steven Freeland, 'For Better or For Worse? The Use of 'Soft Law' within the International Legal Regulation of Outer Space' (2011) XXXVI *Annals of Air and Space Law* 409

Steven Freeland, 'International Criminal Justice in the Asia-Pacific Region – The Role of the International Criminal Court Treaty Regime' (2013) 11(5) *Journal of International Criminal Justice* 1029

Steven Freeland, 'Judicial Decision-Making in International Criminal Courts: 'Effective' Justice?' (2014) 3(1) *Griffith Journal of Law and Human Dignity* (forthcoming)

Jochen A. Frowein, Book Review – Theodor Meron, 'The Humanization of International Law' (2007) 101 *American Journal of International Law* 680

James D. Fry, 'Contextualized Legal Reviews for the Methods and Means of Warfare: Cave Combat and International Humanitarian Law' (2006) 44 *Columbia Journal of Transnational Law* 453

Giorgio Gaja, 'Should All References to International Crimes Disappear from the ILC Draft Articles on State Responsibility?' (1999) *European Journal of International Law* 365

Kenneth S. Gallant, 'Jurisdiction to Adjudicate and Jurisdiction to Prescribe in International Criminal Courts' (2003) 48 *Villanova Law Review* 764

Hans-Peter Gasser, 'For Better Protection of the Natural Environment in Armed Conflict: A Proposal for Action' (1995) 89 *American Journal of International Law* 637

Hans-Peter Gasser, 'Guidelines for Military Manuals and Instructions on the Protection of the Environment in Times of Armed Conflict' (1996) 311 *International Review of the Red Cross* 230

Ryan Gilman, 'Expanding Environmental Justice after War: The Need for Universal Jurisdiction over Environmental War Crimes' (2011) *Colorado Journal of International Environmental Law & Policy* 447

Rachel Glickman, Rose Standifer, Lory Stone and Jeremiah Sullivan, 'Environmental Crimes' (2003) 40 *American Criminal Law Review* 413

Mark Allan Gray, 'The International Crime of Ecocide' (1996) 26 *California Western International Law Journal* 215

Lothar Gündling, 'The Status in International Law of the Principle of Precautionary Action' (1990) 5 *International Journal of Estuarine and Coastal Law* 23

Mark Halsey, 'Environmental Crime: Towards an Eco-Human Rights Approach' (1996) 8 *Current Issues in Criminal Justice* 217

Günther Handl, Book Review – Glen Plant, 'Environmental Protection and the Law of War: A "Fifth Geneva" Convention on the Environment in Time of Armed Conflict' (1993) 42:4 *The International and Comparative Law Quarterly* 976

Jean-Marie Henckaerts, 'Study on customary international humanitarian law: A contribution to the understanding and respect for the rule of law in armed conflict' (2005) 87:857 *International Review of the Red Cross* 175

Jean-Marie Henckaerts, 'Assessing the Laws and Customs of War: The Publication of Customary International Humanitarian Law' (2006) 13 *Human Rights Brief* 8

Major Thomas J. Herthel, 'On the Chopping Block: Cluster Weapons and the Law of War' (2001) 51 *The Air Force Law Review* 229

Timothy J. Heverin, 'Case Comment: Legality of the Threat or Use of Nuclear Weapons: Environmental and Humanitarian Limits on Self-Defense' (1997) 72 *Notre Dame Law Review* 1277

Michael H. Hoffman, 'The Legal Status And Responsibilities of Private Internet Users under the Law of Armed Conflict: A Primer for the Unwary on the Shape of Law to Come' (2003) 2 *Washington University Global Studies Law Review* 415

Jakob von Holderstein Holtermann, 'A "Slice of Cheese" – a Deterrence-Based Argument for the International Criminal Court' (2010) 11 *Human Rights Review* 289

Laura Horn and Steven Freeland, 'More than Hot Air: Reflections on the Relationship between Climate Change and Human Rights' (2009) 13 *University of Western Sydney Law Review* 101

Laurent R. Hourcle, 'Environmental Law of War' (2001) *Vanderbilt Law Review* 653

J.H. Huebert and Walter Block, 'Space Environmentalism, Property Rights, and the Law' (2007) 37 *University of Memphis Law Review* 281

International Committee of the Red Cross, 'Response of Jean-Marie Henckaerts to the Bellinger/Haynes Comments on Customary International Law Study' (2007) 46 *International Legal Materials* 959

International Institute of Humanitarian Law, 'San Remo Manual on International Law Applicable to Armed Conflicts at Sea' (1995) 309 *International Review of the Red Cross* 595

Mansour Jabbari-Gharabagh, 'Type of State Responsibility for Environmental Matters in International Law' (1999) 33 *Revue Juridique Themis* 59

Eric Talbot Jensen, 'The International Law of Environmental Warfare: Active and Passive Damage During Armed Conflict' (2005) 38 *Vanderbilt Journal of Transnational Law* 145

Eric Talbot Jensen and James J. Teixeira, Jr., 'Prosecuting Members of the U.S. Military for Wartime Environmental Crimes' (2005) 17 *Georgetown International Environmental Law Review* 651

Florian Jessberger and Julia Geneuss, 'Down the Drain or Down to Earth? International Criminal Justice under Pressure' (2013) 11 *Journal of International Criminal Justice* 501

Joshua H. Joseph, 'Rethinking Yamashita: Holding Military Leaders Accountable for Wartime Rape' (2007) 28 *Women's Rights Law Reporter* 107

Peggy Rodgers Kalas, 'International Environmental Dispute Resolution and the Need for Access by Non-State Entities' (2001) 12 *Colorado Journal of International Environmental Law and Policy* 191

Elaina I. Kalivretakis, 'Are Nuclear Weapons Above the Law? A Look at the International Criminal Court and the Prohibited Weapons Category' (2001) 15 *Emory International Law Review* 683

Donald Kaniaru, Rajendra Shende, Scott Stone and Durwood Zaelke, 'Strengthening the Montreal Protocol: Insurance Against Abrupt Climate Change' (2007) 7(2) *Sustainable Development Law and Policy Journal* 3

Alexandra Knight, 'Global Environmental Threats: Can the Security Council Protect Our Earth?' (2005) 80:5 *New York University Law Review* 1549

Andreas Laursen, 'NATO, the War over Kosovo, and the ICTY Investigation' (2002) 17 *American University International Law Review* 765

Anthony Leibler, 'Deliberate Wartime Environmental Damage: New Challenges for International Law' (1992) 23 *California Western International Law Journal* 67

Raphael Lemkin, 'Genocide: A Modern Crime' (1945) 4 *Free World* 39

Raphael Lemkin, 'Genocide as a Crime Under International Law' (1947) 44 *American Journal of International Law* 145

Aurelie Lopez, 'Criminal Liability for Environmental Damage Occurring in Times of Non-International Armed Conflict: Rights and Remedies' (2007) 18 *Fordham Environmental Law Review* 231

Luan Low and David Hodgkinson, 'Compensation for Wartime Environmental Damage: Challenges to International Law After the Gulf War' (1995) 35 *Virginia Journal of International Law* 405

Susan F. Mandiberg and Michael G. Faure, 'A Graduated Punishment Approach to Environmental Crimes: Beyond Vindication of Administrative Authority in the United States and Europe' (2008) *Lewis and Clark Law School Legal Research Paper Series*, Paper No. 2008–21

Dennis Mandsager, 'Introductory Note to Response of Jean-Marie Henckaerts to the U.S. Joint Letter from John Bellinger III, Legal Advisor, U.S. Department of State, and William J. Haynes, General Counsel, U.S. Department of Defense to Dr. Jakob Kellenberger, President, International Committee of the Red Cross, Regarding Customary International Law Study' (2007) 46 *International Legal Materials* 957

Jackson Maogoto and Steven Freeland, 'The Final Frontier: The Laws of Armed Conflict and Space Warfare' (2007) 23:1 *Connecticut Journal of International Law* 165

Louis Maresca, 'A New Protocol on Explosive Remnants of War: The History and Negotiation of Protocol V to the 1980 Convention on Certain Conventional Weapons' (2004) 86:856 *International Review of the Red Cross* 815

Michael J. Matheson, 'The Opinion of the International Court of Justice on the Threat or Use of Nuclear Weapons' (1997) 91 *American Journal of International Law* 417

Michael J. Matheson, 'Current Development – The Fifty-Eighth Session of the International Law Commission' (2007) 101 *American Journal of International Law* 407

Stephen C. McCaffery, 'Crimes Against the Environment' (1986) 1 *International Criminal Law* 541

Kenneth F. McCallion, 'International Environmental Justice: Rights and Remedies' (2003) 26 *Hastings International and Comparative Law Review* 427

Kenneth F. McCallion and H. Rajan Sharma, 'Environmental Justice Without Borders: The Need for an International Court of the Environment to Protect Fundamental

Environmental Rights' (2000) 32 *George Washington Journal of International Law and Economics* 351

Tim McCormack and Sue Robertson, 'Jurisdictional Aspects of the Rome Statute for the New International Criminal Court' (1999) 23 *Melbourne University Law Review* 635

Paul V. McDade, 'The Effect of Article 4 of the Vienna Convention on the Law of Treaties 1969' (1986) 35 *International and Comparative Law Quarterly* 499

John O. McGinnis and Ilya Somin, 'Symposium: Global Constitutionalism: Global Influence on U.S. Jurisprudence: Should International Law be Part of our Law?' (2007) 59 *Stanford Law Review* 1175

Robert McLaughlin, 'Improving Compliance: Making Non-State International Actors Responsible for Environmental Crimes' (2000) 11 *Colorado Journal of International Environmental Law and Policy* 377

Keith P. McManus, 'Civil Liberty for Wartime Environmental Damage: Adapting the United Nations Compensation Commission for the Iraq War' (2006) 33 *Boston College Environmental Affairs Law Review* 417

Frédéric Mégret, 'The Problem of an International Criminal Law of the Environment' (2011) 36:2 *Columbia Journal of Environmental Law* 195

Theodor Meron, 'The Humanization of Humanitarian Law' (2000) 94 *American Journal of International Law* 239

Michael K. Murphy, 'Note: Achieving Economic Security with Swords as Ploughshares: The Modern Use of Force to Combat Environmental Degradation' (1999) 39 *Virginia Journal of International Law Association* 1181

Michael A. Newton, Symposium, 'Modern Military Necessity: The Role and Relevance of Military Lawyers' (2007) 12 *Roger Williams University Law Review* 877

Myron N. Nordquist, 'Panel Discussion on International Environmental Crimes: Problems of Enforceable Norms and Accountability' (1997) 3 *ILSA Journal of International Law* 697

Christiana Ochoa, 'The Individual and Customary International Law Formation' (2007) 48 *Virginia Journal of International Law* 119

Mary Ellen O'Connell, Book Review – Judith Gardam, 'Necessity, Proportionality and the Use of Force by States' (2006) 100 *American Journal of International Law* 973

Jens David Ohlin, 'Attempt, Conspiracy and Incitement to Commit Genocide' (2009) *Cornell Law Faculty Publications*, paper 24 <http://scholarship.law.cornell.edu/cgi/viewcontent.cgi?article=1023&context=facpub>

Marcos A. Orellana, 'Criminal Punishment for Environmental Damage: Individual and State Responsibility at a Crossroad' (2005) 17 *Georgetown International Environmental Law Review* 673

W. Hays Parks, 'Conventional Weapons and Weapons Reviews' (2005) 8 *Yearbook of International Humanitarian Law* 55

Rymn James Parsons, 'The Fight to Save the Planet: U.S. Armed Forces, "Greenkeeping," and Enforcement of the Law Pertaining to Environmental Protection During Armed Conflict' (1998) 10 *Georgetown International Environmental Law Review* 441

Jordan J. Paust, 'ILSA Panel Oct. 18, 2003, at Loyola University New Orleans – Panel on History of International Tribunals Prior to Nuremberg: Selective History of

International Tribunals and Efforts Prior to Nuremberg' (2003) 10 *ILSA Journal of International and Comparative Law* 207

Alain Pellet, 'Can a State Commit a Crime? Definitely, Yes!' (1999) *European Journal of International Law* 425

James Podgers, 'The World Cries for Justice' (1996) *American Bar Association Journal* 52

Rosemary Rayfuse, 'The Draft Code of Crimes Against the Peace and Security of Mankind: Eating Disorders at the International Law Commission' (1997) 8 *Criminal Law Forum* 43

Jesse S Reeves, 'The Neutralization of Belgium and the Doctrine of Kriegsraison' (1915) 13:3 *Michigan Law Review* 179

W. Michael Reisman, 'Editorial Comment: Holding the Center of the Law of Armed Conflict' (2006) 100 *American Journal of International Law* 852

Alfred Rest, 'Enhanced Implementation of International Environmental Treaties by Judiciary – Access to Justice in International Environmental Law for Individuals and NGOs: Efficacious Enforcement by the Permanent Court of Arbitration' (2004) 1 *Macquarie Journal of International and Comparative Environmental Law* 1

Roman Reyhani, 'Comment: Protection of the Environment During Armed Conflict' (2007) 14 *Missouri Environmental Law and Policy Review* 323

Jefferson D. Reynolds, 'Collateral Damage on the 21st Century Battlefield: Enemy Exploitation of the Law of Armed Conflict, and the Struggle for a Moral High Ground' (2005) 56 *Air Force Law Review* 1

Andy Rich, 'The Environment: Adequacy of Protection in Times of War' (2004) 12 *Penn State Environmental Law Review* 445

Peter J. Richards and Michael N. Schmitt, 'Mars Meets Mother Nature: Protecting the Environment During Armed Conflict' (1999) 28 *Stetson Law Review* 1047

Captain J. Ashley Roach, 'Certain Conventional Weapons Convention: Arms Control or Humanitarian Law?' (1984) 105 *Military Law Review* 3

Adam Roberts, 'The Laws of War: Problems of Implementation in Contemporary Conflicts' (1995) 6 *Duke Journal of Comparative and International Law* 11

Thomas G. Robisch, 'General William T. Sherman: Would the Georgia Campaigns of the First Commander of the Modern Era Comply with Current Law of War Standards?' (1995) 9 *Emory International Law Review* 459

Rudy S. Salo, 'When the Logs Roll Over: The Need for an International Convention Criminalizing Involvement in the Global Illegal Timber Trade' (2003) 16 *Georgetown International Environmental Law Review* 127

Danesh Sarooshi, 'The Statute of the International Criminal Court' (1999) 48 *International and Comparative Law Quarterly* 387

Ben Saul, 'Legislating from a Radical Hague: The United Nations Special Tribunal for Lebanon Invents an International Crime of Transnational Terrorism' (2011) 24:3 *Leiden Journal of International Law* 677

Jaume Saura, 'Lawful Peacekeeping: Applicability of International Humanitarian Law to United Nations Peacekeeping Operations' (2007) 58 *Hastings Law Journal* 479

William A. Schabas, '*Mens Rea* and the International Criminal Tribunal for the Former Yugoslavia' (2003) 37:4 *New England Law Review* 1015

Michael P. Scharf, 'Clear and Present Danger: Enforcing the International Ban on Biological and Chemical Weapons Through Sanctions, Use of Force, and Criminalization' (1999) 20 *Michigan Journal of International Law* 477

Michael N. Schmitt, 'Humanitarian Law and the Environment' (2000) 28 *Denver Journal of International Law and Policy* 265

Aaron Schwabach, 'Environmental Damage Resulting From the NATO Military Action Against Yugoslavia' (2000) 25 *Columbia Journal of Environmental Law* 117

Aaron Schwabach, 'NATO'S War in Kosovo and the Final Report to the Prosecutor of the International Criminal Tribunal for the Former Yugoslavia' (2001) 9 *Tulane Journal of International and Comparative Law* 167

Aaron Schwabach, 'Ecocide and Genocide in Iraq: International Law, the Marsh Arabs, and Environmental Damage in Non-International Conflicts' (2004) 15 *Colorado Journal of International Law and Policy* 1

Tullio Scovazzi, 'State Responsibility for Environmental Harm' (2001) 12 *Yearbook of International Environmental Law* 43

Jessica E. Seacor, 'Environmental Terrorism: Lessons from the Oil Fires of Kuwait' (1994) 10 *American University Journal of International Law and Policy* 481

Peter Sharp, 'Prospects for Environmental Liability in the International Criminal Court' (1999) 18 *Virginia Environmental Law Journal* 217

Major Walter G. Sharp, Sr., 'The Effective Deterrence of Environmental Damage During Armed Conflict: A Case Analysis of the Persian Gulf War' (1992) 137 *Military Law Review* 1

Dinah Shelton, 'Centennial Essay: In Honor of the 100th Anniversary of the AIJL and the ASIL: Normative Hierarchy in International Law' (2006) 100 *American Journal of International Law* 291

Daphna Shraga, 'UN Peacekeeping Operations: Applicability of International Humanitarian Law and Responsibility for Operations-Related Damage' (2000) 94 *American Journal of International Law* 406

Elies van Sliedregt and Larissa van den Herik, 'Introduction: The STL Interlocutory Decision on the Definition of Terrorism – Judicial Ingenuity or Radicalism?' (2011) 24:3 *Leiden Journal of International Law* 651

Jonathan Somer, 'Jungle justice: passing sentence on the equality of belligerents in non-international armed conflict' (2007) 89:867 *International Review of the Red Cross* 655

Carsten Stahn, 'Why some Doors may be Closed Already: Second Thoughts on a 'Case-by-Case' Treatment of Article 12(3) Declarations' (2006) 75 *Nordic Journal of International Law* 243

Symposium, 'The International Responses to the Environmental Impacts of War' (2005) 17 *Georgetown International Environmental Law Review* 565

Ludwik A. Teclaff, 'Beyond Restoration – the Case of Ecocide' (1994) 34 *Natural Resources Journal* 933

Colonel James P. Terry, 'The Environment and the Laws of War: The Impact of Desert Storm' (1992) 45 *Naval War College Review* 61

Colonel James P. Terry, 'The Vietnam War in Perspective: Lessons Learned in the Law of War as Applied in Subsequent Conflict' (2007) 54 *Naval Law Review* 79

Catherine Tinker, "Environmental Security' in The United Nations: Not a Matter for the Security Council' (1992) 59 *Tennessee Law Revue* 787

Lisa C. Toohey, 'Borders and Boundaries in International Law: Compensation for Agent Orange Damage in Vietnam' (2005) 13 *Williamette Journal of International Law and Dispute Resolution* 287

Nsongurua J. Udombana, 'War is Not Child's Play! International Law and the Prohibition of Children's Involvement in Armed Conflicts' (2006) 20 *Temple International and Comparative Law Journal* 57

Christyne J. Vachon, 'Sovereignty Versus Globalization: The International Court of Justice's Advisory Opinion on the Threat or Use of Nuclear Weapons' (1998) 26 *Denver Journal of International Law and Policy* 691

Barend van Niekerk, 'Environmental Pollution – The New International Crime' (1976) 93 *South African Law Journal* 68

Greg R. Vetter, 'Command Responsibility of Non-Military Superiors in the International Criminal Court (ICC)' (2000) 25 *Yale Law Journal* 89

Francisco Orrego Vicuńa, Rapporteur on the Subject of Environmental Responsibility and Liability, Final Report Prepared for the Eighth Committee of the *Institut de Droit International*, 'Responsibility and Liability for Environmental Damage Under International Law: Issues and Trends' (1998) 10 *Georgetown International Environmental Law Review* 279

Kenneth Watkin, 'Controlling the Use of Force: A Role for Human Rights Norms in Contemporary Armed Conflict' (2004) 98 *American Journal of International Law* 1

Kenneth Watkin, 'Assessing Proportionality: Moral Complexity and Legal Rules' (2005) 8 *Yearbook of International Humanitarian Law* 3

Michael Watson, 'Environmental Offences: The Reality of Environmental Crime' (2005) 7 *Environmental Law Review* 190

Mohammed Saif-Alden Wattad, 'The Rome Statute & Captain Planet: What lies between 'Crimes Against Humanity' and the 'Natural Environment'' (2009) XIX *Fordham Environmental Law Review* 265

Tara Weinstein, 'Prosecuting Attacks that Destroy the Environment: Environmental Crimes or Humanitarian Atrocities?' (2005) 17 *Georgetown International Environmental Law Review* 697

Gerhard Werle and Florian Jessberger, "Unless Otherwise Provided': Article 30 of the ICC Statute and the Mental Elements of Crimes under International Criminal Law' (2005) 3 *Journal of International Criminal Justice* 35

Rob White, 'Environmental Crime in Global Context: Exploring the Theoretical and Empirical Complexities' (2004) 16 *Current Issues in Criminal Justice* 271

Major Richard M. Whitaker, 'Environmental Aspects of Overseas Operations' (1995) 4 *Army Lawyer* 27

Ensign Florencio J. Yuzon, 'Deliberate Environmental Modification Through the Use of Chemical and Biological Weapons: "Greening" the Environmental Laws of Armed Conflict to Establish an Environmentally Protective Regime' (1996) 11 *American University Journal of International Law and Policy* 793

C) OTHER ATTRIBUTED DOCUMENTS

African Union, 'Draft Protocol on Amendments to the Protocol on the Statute of the African Court of Justice and Human Rights', 15 May 2012, AU Doc Exp/Min/IV/Rev.7 <http://africlaw.files.wordpress.com/2012/05/au-final-court-protocol-as-adopted-by-the-ministers-17-may.pdf>

Agence France-Presse, 'Clean-burning fools: BP fined $413', *The Sydney Morning Herald* (Sydney), 27–28 October 2007, 44

Alliance des Démocrates et des Libéraux pour l'Europe, 'Charter of Brussels for the creation of European and International Criminal Court of the Environment and Health', 30 January 2014

Al-Waqai Al-Iraqiya, Official Gazette of Iraq, Volume 44 <www.loc.gov/law/help/hussein/docs/CoalitionProvisionalAuthorityOrder48en.pd>

The Asia Pacific Forum of National Human Rights Institutions and the Advisory Council of Jurists, 'Human Rights and the Environment: Final Report and Recommendations', (September 2007) <www.ohchr.org/Documents/issues/ClimateChange/Submissions/Asia_Pacific_Forum_of_NHRIs_1_HR_and_Environment_ACJ_Report_Recommendations.pdf>

Baker & McKenzie, 'Supreme Court Rules on Key Global Warming Case', *Global Climate Change Newsletter* (Chicago), April 2007 <www.bakernet.com/NR/rdonlyres/2DA7D733-A828-4A48-B06C-755F08970AC1/42059/ClimateChangeAlert.pdf>

Robert Bevan, 'Heritage caught in the crossfire', *The Australian* (Sydney), 24 March 2006, 15

Ian Black, 'Iraq calls for debt relief from neighbours' *The Guardian* (London) 29 May 2008 <www.guardian.co.uk/world/2008/may/29/iraq.middleeast1?gusrc=rss&feed=networkfront>

David Blair, 'Toll of Infamy: Regime Killed 2 Million', *The Sydney Morning Herald* (Sydney), 16 December 2003, 8

Julian Borger, 'Conflicts fuelled by climate change causing new refugee crisis, warns UN', *The Guardian* (London), 17 June 2008, 15

Andrew Bounds, 'Brussels in push to lay down 'green crime' laws', *Financial Times* (London), 7 February 2007, 1

George H.W. Bush, Address Before a Joint Session of Congress, 11 September 1990

James Carroll, 'Proving nuclear realists wrong', *The International Herald Tribune* (Paris), 23 June 2008 <www.iht.com/articles/2008/06/23/opinion/edcarroll.php>

Center for Clean Air Policy, 'Letter from the President' (April 2007) <www.ccap.org/newsletter.htm#Pres>

Louis Charbonneau, 'Update 5 – Iran, North Korea, Syria block UN arms trade treaty' *Reuters News* (28 March 2013) <www.reuters.com/article/2013/03/29/arms-treaty-un-idUSL2N0CK24K20130329>

Kamari Maxine Clarke, 'Treat Greed in Africa as a War Crime', *The New York Times* (New York), 29 January 2013 <www.nytimes.com/2013/01/30/opinion/treat-greed-in-africa-as-a-war-crime.html?emc=eta1>

The CNA Corporation, 'National Security and the Threat of Climate Change' (April 2007) <www.cna.org/sites/default/files/news/FlipBooks/Climate%20Change%20 web/flipviewerxpress.html>

Coalition for the International Criminal Court, 'Delivering on the Promise of a Fair, Effective and Independent Court – Review Conference of the Rome Statute: Article 124', October 2010 <www.iccnow.org/?mod=article124>

Council of Europe – Treaty Office, 'Convention of the Protection of the Environment through Criminal Law' Status as at 26/1/2014 <http://conventions.coe.int/Treaty/Commun/ChercheSig.asp?NT=172&CM=1&DF=&CL=ENG>

Council of the European Union, 'Presidency Conclusions' Doc No 7224/07 CONCLU 1, 9 March 2007 <www.consilium.europa.eu/ueDocs/cms_Data/docs/pressData/en/ec/93135.pdf>

Alister Doyle, 'UN Aims to Study Link Between Environment, Wars' *Planet Ark World Environment News*, reproduced in United Nations Environment Programme, 'The Environment in the News', 14 January 2004

Phillip Elmer-Dewitt, 'A Man-Made Hell on Earth', *Time* (London), 18 March 1999, 36

Europa, 'Commission strengthens environmental protection through criminal law, abolishing "safe havens" of environmental crime' (Press Release, 9 February 2007, Doc IP/07/166) <http://europa.eu/rapid/press-release_IP-07-166_en.htm>

European Council, 'Directive 92/43/EEC of the European Council on the conservation of natural habitats and of wild fauna and flora', 21 May 1992

European Parliament and Council of the European Union, 'Directive 2008/99/EC of the European Parliament and of the Council on the protection of the environment through criminal law', 19 November 2008

David R. Francis, 'Fueling the Fire: "Resource Wars" Spurred by Assets of Developing Nations', *Christian Science Monitor* (Boston), 6 December 2002, A3

Steven Freeland, 'Saddam trial underlines need to take tough stand on genocide', *The Canberra Times* (Canberra) 29 August 2006, 11

Steven Freeland, 'Is It Iraqi Justice – Or Revenge?', *The Canberra Times* (Canberra), 7 November 2006, 9

Steven Freeland, 'Destroying the environment is also a war crime', *The Canberra Times* (Canberra), 6 November 2008, 23

Amalie Frese, 'The Least Bad Option: A Pluralistic Theory of Discrimination', unpublished thesis, University of Copenhagen, 29 June 2012

T.D. Gill, 'The "Genocide" case: Reflections on the ICJ's Decision in Bosnia-Herzegovina v Serbia', *The Hague Justice Portal* (The Netherlands), 26 March 2007 <www.haguejusticeportal.net/index.php?id=7266>

Katy Glassborow, 'Court urged to consider environmental crimes', *Institute for War and Peace Reporting* (United Kingdom), 30 August 2007 <http://iwpr.net/?p=acr&s=f&o=33 8199&apc_state=henh#>

Ellen Goodman, 'A crime against humanity and nature', *The Chicago Tribune*, 3 February 1991 <http://articles.chicagotribune.com/1991-02-03/features/9101100755_1_civil-war-sea-turtles-enemy-territory>

Fiona Harvey and Andrew Bounds, 'Regulatory regime put to the test at borders', *Financial Times* (London), 7 February 2007, 2

Stephanie Holmes, 'Delicate territory: the landmine ban' *BBC News*, 2 December 2007 <http://news.bbc.co.uk/2/hi/in_depth/7121419.stm>

Thomas Homer-Dixon, 'Terror in the Weather Forecast', *The New York Times* (New York), 24 April 2007 <www.nytimes.com/2007/04/24/opinion/24homerdixon.html?_r=1&th&emc=th&oref=slogin>

Chris Hogg, 'Nagasaki bombing labeled a crime' (18 December 2006) *BBC News* <http://news.bbc.co.uk/2/hi/asia-pacific/6189489.stm>

International Association of Penal Law, Resolutions of the XVth International Congress on Penal Law (1995) 66 1:2 *International Review of Penal Law* 52

International Committee of the Red Cross, 'Anti-personnel Landmines: Friend or Foe? – A Study of the Military Use and Effectiveness of Anti-personnel Mines' (1996) <www.icrc.org/eng/assets/files/other/icrc_002_0654.pdf>

International Committee of the Red Cross, 'Commentary on the Additional Protocols of 8 June 1977 to the Geneva Conventions of 12 August 1949' <www.icrc.org/applic/ihl/ihl.nsf/Comment.xsp?viewComments=LookUpCOMART&documentId=78C0DA9A7B459ACEC12563CD0042F649&action=openDocument>

International Committee of the Red Cross, 'Custom as a Source of International Humanitarian Law' (14 September 2007) <www.icrc.org/web/eng/siteeng0.nsf/html/customary-law-publication-140907>

International Committee of the Red Cross, 'Customary IHL: Rule 156. Definition of War Crimes' <www.icrc.org/customary-ihl/eng/docs/v1_cha_chapter44_rule156>

International Committee of the Red Cross, 'Guidelines for Military Manuals and Instructions on the Protection of the Environment in Times of Armed Conflict' (1995) <www.icrc.org/eng/resources/documents/misc/57jn38.htm>

International Committee of the Red Cross, 'Protocol Additional to the Geneva Conventions of 12 August 1949, and relating to the Protection of Victims of International Armed Conflicts (Protocol I), 8 June 1977; Commentaries' <www.icrc.org/ihl.nsf/COM/470–750070?OpenDocument>

International Committee of the Red Cross, 'State Parties to the Following International Humanitarian Law and Other Related Treaties as at 15 November 2012' <www.icrc.org/IHL.nsf/%28SPF%29/party_main_treaties/$File/IHL_and_other_related_Treaties.pdf>

International Committee of the Red Cross, 'What treaties make up international humanitarian law?' (October 2002) <www.icrc.org/Web/eng/siteeng0.nsf/html/5KZGE6>

Institute for Environmental Security, Military Advisory Council, 'Climate Change and the Military – First Statement of the Military Advisory Council' (29 October 2009)

International Law Association, 'Final Report of the International Committee on the Formation of Customary (General) International Law; Principles Applicable to the Formation of General Customary International Law', in *Report of the Sixty-Ninth Conference* (2000), 712

Florian Jessberger, Report to Chairman of Degree Committee regarding dissertation submitted by Steven Robert Freeland to the University of Maastricht, 23 July 2014

Kingdom of Belgium, Foreign Affairs, Foreign Trade and Development Corporation, 'Cluster Munitions' (2010) <http://diplomatie.belgium.be/en/policy/policy_areas/

peace_and_security/disarmament_non_proliferation/human_security/cluster_munitions/>

Helmut Kohl, Statement by the German Chancellor, *Bulletin* (Bonn), 9 April 1991, 255

Bernard Kouchner, 'The savaging of humanitarian law', *The International Herald Tribune* (Paris), 29 January 2009, 8

Rory McCarthy, 'Fatal burns never seen before Israel's war', *The Guardian* (London), 21 January 2009, 24

Seth Mydans, 'Vietnam Sees War's Legacy in Its Young', *The New York Times* (New York), 16 May 1999 <www.nytimes.com/1999/05/16/world/vietnam-sees-war-s-legacy-in-its-young.html>

Stephanie Nebehay, 'Dirty Water Provokes Hepatitis Outbreak in Darfur', *Reliefweb*, 9 August 2004 <http://reliefweb.int/report/sudan/sudan-dirty-water-provokes-hepatitis-outbreak-darfur>

Richard Norton-Taylor, 'Britain to ban some cluster bombs', *The Sydney Morning Herald* (Sydney), 20 March 2007 <www.smh.com.au/news/world/britain-to-ban-some-cluster-bombs/2007/03/19/1174152969983.html>

Jonathan Pearlman, 'Dead baby and teenager ruled to be 'collateral casualties'', *The Sydney Morning Herald* (Sydney) 13 May 2008 <www.smh.com.au/news/ national/dead-children-ruled-collateral-casualties/2008/05/12/1210444338885.html>

Ed Pilkington, 'UK to raise climate talks as security council issue', *The Guardian* (London), 16 April 2007, 8

James Randerson, 'Study suggests cancer risk from depleted uranium', *The Guardian* (London), 8 May 2007, 9

Andrew C. Revkin, 'Bush's environmental legacy in play: Obama reshaping some policies and building on others: Climate Change', *The International Herald Tribune* (Paris), 7–8 February 2009, 5

Paul Rincon, 'Obama lays out climate action plan' (25 June 2013) *BBC News* <www.bbc.co.uk/news/science-environment-23032890>

Andrew Rosenthal, 'War in the Gulf: The President; Bush Calls Gulf Oil Spill a 'Sick' Act by Hussein', *The New York Times* (New York), 26 January 1991 <www.nytimes.com/1991/01/26/world/war-in-the-gulf-the-president-bush-calls-gulf-oil-spill-a-sick-act-by-hussein.html>

Royal Swedish Ministry for Foreign Affairs, 'Conventional Weapons, their Deployment and Effects from a Humanitarian Aspect: Recommendations for the Modernization of International Law' (1973)

David Scheffer, 'States Parties Approve New Crimes for International Criminal Court', 22 June 2010 <www.asil.org/insights100622.cfm>

Heather Sharp, 'Iran's 'devastated' Marsh Arabs', *BBC News*, 3 March, 2003 <http://news.bbc.co.uk/2/hi/middle_east/2807821.stm>

Joe Sills, Jerome C. Glenn, Elizabeth Florescu and Theodore J. Gordon, 'Environmental Crimes in Military Actions and the International Criminal Court (ICC) – United Nations Perspectives', April 2001 <http://millennium-project.org/millennium/es-icc.html>

David Smith, 'ICC Chief Prosecutor shelves Darfur war crime probe', *The Guardian* (UK), 14 December 2014, 15 December 2014 <www.theguardian.com/ world/2014/dec/14/icc-darfur-war-crimes-fatou-bensouda-sudan>

The Special Tribunal for Lebanon, 'Creation of the STL' <www.stl-tsl.org/en/about-the-stl/creation-of-the-stl>

University of Oregon, 'International Environmental Agreements (IEA) Database Project' <http://iea.uoregon.edu/page.php?query=summary&type=MEA>

John Vidal, 'Climate may force 1 billion from their homes', *The Sydney Morning Herald* (Sydney), 15 May 2007 <www.smh.com.au/news/environment/climate-may-force-1-billion-from-their-homes/2007/05/14/1178995077364.html>

D) OTHER UNATTRIBUTED DOCUMENTS

'A timeline of the EU' (12 March 2007) *BBC News* <http://news.bbc.co.uk/2/hi/europe/3583801.stm>

'Arms Treaty Blocked by Iran, North Korea and Syria' (29 March 2013) *BBC News* <www.bbc.co.uk/news/world-us-canada-21972700>

"Chemical Ali' Sentence Confirmed' (4 September 2007) *BBC News* <http://news.bbc.co.uk/2/hi/middle_east/6977842.stm>

'Ecocide: The Missing 5th Crime Against Peace – End Ecocide in Europe' (undated) <http://eradicatingecocide.cmail1.com/t/ViewEmail/r/7EDF8501F7FB82C7/9E223A7C45EC01860CC2E775D3CF5869>

'End Ecocide', (2013) <http://eradicatingecocide.cmail5.com/t/ViewEmail/r/9F38DE90606F959E/9E223A7C45EC01860CC2E775D3CF5869>

'Hariri court opens at The Hague' (1 March 2009) *BBC News* <http://news.bbc.co.uk/2/hi/middle_east/7917294.stm>

'Lebanon backs UN Hariri tribunal' (13 November 2006) *BBC News* <http://news.bbc.co.uk/2/hi/middle_east/6144162.stm>

'Muddy Waters' (18 June 2007) *The New York Times* (Editorial) <www.nytimes.com/2007/06/18/opinion/18mon2.html>

'North Korea 'readies rocket force' after US stealth flights' (29 March 2013) *BBC News* <www.bbc.co.uk/news/world-asia-21972936>

'Profile: 'Chemical Ali'' (25 January 2010) *BBC News* <http://news.bbc.co.uk/2/hi/middle_east/2855349.stm>

Restatement of the Law: Third Restatement of U.S. Foreign Relations Law, Volume 2 (1987), 165

'U.N chief: Darfur a victim of climate change' (16 June 2007) *NBC News* <www.nbcnews.com/id/19268452/#.UU8qVFf-jdc>

'U.S. Warns Sudan Death Toll Could Reach 300,000 To 1 Million' (3 June 2004) *UN Wire* <http://medilinkz.org/news/news2.asp?NewsID=6776>

'Warming and Global Security' (20 April 2007) *The New York Times* (Editorial) <www.nytimes.com/2007/04/20/opinion/20fri2.html>

E) PUBLIC SPEECHES

Daniel Bethlehem, 'International Humanitarian Law and the Challenge to Contemporary Conflicts', (speech delivered at seminar entitled 'Recent Developments in International Humanitarian Law', Wellington, New Zealand, 1 September 2007)

Professor Charles Vörösmarty, 'Drought as a Contributor to Civil War: Results from a Global Spatial Analysis', (speech delivered at seminar entitled 'Climate-Security Connections: An Empirical Approach to Risk Assessment', Washington DC, USA, 6 March 2007)

Scott Spence, 'The Biological and Toxin Weapons Convention and its Implementation through National Law', (speech delivered at the T.M.C. Asser Institute, as part of a seminar series entitled 'The Hague Initiative on Law and Armed Conflict', The Hague, The Netherlands, 25 November 2008)

F) TELEVISION BROADCASTS

BBC Television, 'Protests over Serbia Pollution', *BBC News Player*, 19 March 2007 <http://news.bbc.co.uk/player/nol/newsid_6460000/newsid_6465300/6465347.stm>

CURRICULUM VITAE

Steven Freeland was born in Sydney, Australia on 13 August 1957. He has a Bachelor of Commerce in Industrial Relations (with merit) (1980), a Bachelor of Laws (1981) and a Master of Laws in International Law (2001) from the University of NSW, Australia. He has been admitted to practice as a solicitor before the High Court of Australia and the Supreme Court of NSW (1981), and as a solicitor in England, Ireland and Wales (1985). He worked as a commercial lawyer (1980–1986) and an investment banker (1986–2000) prior to entering academic life.

He is a Professor of International Law at the University of Western Sydney, Australia. He is also a Visiting Professor at the University of Vienna, a Permanent Visiting Professor at the iCourts Centre of Excellence for International Courts, University of Copenhagen, Denmark, and a Member of Faculty of the London Institute of Space Policy and Law. He has also been a Visiting Professional within the Appeals Chamber at the International Criminal Court (ICC), and a Special Advisor to the Danish Foreign Ministry in matters related to the ICC.

He is Co-Editor of the book series *Annotated Leading Cases of International Criminal Tribunals*. He also sits on the Editorial Board of the *Australian Journal of Human Rights*, the *Australian International Law Journal*, the Canada-based *Annals of Air and Space Law*, the German-based *German Journal of Air and Space Law / Zeitschrift fur Luft- und Weltraumrecht* and the China-based *Space Law Review*, and on the Advisory Board of the India-based *Asian Journal of Air and Space Law* and the UK-based *Journal of Philosophy of International Law*, as well as a series of books entitled *Studies in Space Law*.

www.ingramcontent.com/pod-product-compliance
Ingram Content Group UK Ltd.
Pitfield, Milton Keynes, MK11 3LW, UK
UKHW051926100926
8556IPUK00022B/1389

9 781780 683140